Samuel H. Wright, A. M., Ph. D.

THE

ELEMENTS

OF

ANALYTICAL GEOMETRY;

COMPREHENDING THE

DOCTRINE OF THE CONIC SECTIONS,

AND THE

GENERAL THEORY OF CURVES AND SURFACES OF THE SECOND ORDER.

INTENDED FOR THE USE OF

MATHEMATICAL STUDENTS IN SCHOOLS AND UNIVERSITIES.

BY J. R. YOUNG,

Author of "An Elementary Treatise on Algebra," "Elements of Geometry," &c

REVISED AND CORRECTED BY

JOHN D. WILLIAMS,

AUTHOR OF "KEY TO HUTTON'S MATHEMATICS."

PHILADELPHIA:

BUBLISHED BY E. H. BUTLER & CO.

1850.

ADVERTISEMENT
TO
THE AMERICAN EDITION.

THE work now submitted to the notice of the American public, will, it is hoped, supply in part the want that has long been felt by the heads of instruction in this country, of a good elementary treatise, in our own language, upon that all important branch of mathematics—the application of analysis to the solution of Geometrical Problems. The Professors of several of our public institutions, convinced of the absolute necessity, to the student, of a thorough knowledge of this subject, prior to his entering upon the study of the Calculus, and its varied applications, have been induced to place in the hands of their pupils the works of the French writers in their native tongue. Among others, the *Essai de Geometrie Analytique* of Biot, the *Theorie des Courbes* of Boucharlat, and the *Application de l'Analyse à la Geometrie* of Bourdon, have been used to much advantage. Indeed it may be questioned if the use of the French authors as models be not almost absolutely necessary to the writer of a work on this subject; for nowhere else can we find that simple, and, at the same time, elegant and highly finished analysis, for which they are so justly distinguished in the scientific world.

Mr. Young, as will be seen by his preface, has drawn largely from these sources; and the eminent superiority of his elementary treatises on the mathematical sciences, is mainly to be attributed to the liberality of spirit with which—casting off the trammels imposed upon themselves by the countrymen of Newton—he has freely availed himself of every discovery and improvement in analysis, though such have been chiefly made on the French side of the channel.

In the present edition few alterations or additions could have been made which would improve the original, with the exception of a careful correction of the typographical errors—and whether or not the Editor has faithfully executed his task, the work itself will show.

JOHN D. WILLIAMS.

NEW YORK, August, 1833.

PREFACE.

THE application of algebra to the theory of curves and surfaces may be regarded as the fundamental branch of modern analytical science, and as the principal instrument, in conjunction with the differential and integral calculus, with which the continental mathematicians have worked such wonders in almost every department of the mathematics. The remarkable contrivance, first introduced by *Descartes*, of representing lines and surfaces by algebraical equations, enables us to embody in such an equation every property and peculiarity belonging to any curve or surface, when we know the law of its description, or any of its distinguishing characteristics; and then, to develope these several properties of the curve or surface, we have only to perform so many easy, and generally obvious, transformations on the equation which represents it. The superiority of this method over the geometrical, both in ease and fertility, immediately led to its general adoption among the French mathematicians; and the method of co-ordinates, which the Cartesian geometry involved, was afterwards applied to mechanics, and, indeed, to every other part of mathematical physics, each of which has been improved and extended by its introduction.*

English mathematicians have, however, been singularly slow in appreciating these decided advantages; so slow, indeed, that, till the year 1823, when Dr. Lardner published the first part of his Algebraic Geometry, the English language possessed not a single book on this subject. Besides this work, a treatise on analytical geometry has also emanated from the University of Cambridge, which, although a work of much originality and ability, the ingenuous author has since publicly acknowledged to be unsuited to the purposes of elementary instruction. Dr. Lardner's book will, no doubt, when completed, present a valuable body of analytical science, accessible, however, to those only who have a knowledge of the differential and integral calculus.

---

* Maclaurin, in his "Treatise on Fluxions," first suggested this happy idea, which threw a new light on the entire theory of mechanics. But, unfortunately, this simple principle has been neglected by later English authors, and much of what our mathematicians at present know and practise of this method, we owe chiefly to the re-importation of it through the medium of modern French works; and many, perhaps, who are admiring the facility which is thus thrown into mechanical investigations of the greatest difficulty, are unconscious that this thought had its origin in our own land.

Encyclopædia Metropolitana, art. Mechanics, p. 5.

The present little volume is then an attempt to fill up a chasm which seems still to exist in our mathematical courses of instruction, and to supply the connecting link between elementary geometry and trigonometry, and some of the most interesting applications of the transcendental analysis. For such an undertaking, the French language offers copious materials; and I have, accordingly, carefully examined and freely used the performances of Biot, Lacroix, Boucharlat, Bourdon, &c.; and I shall consider myself particularly fortunate, if it be found that I have in any degree imbibed the spirit of these elegant writers.

As regards arrangement, however, I have differed from most other authors, adopting that which appeared to me most likely to facilitate the progress of the student, without waiting to consider whether a more strictly methodical disposition of parts might not be devised. Conformably, too, to this determination, I have, in one or two cases, not hesitated to introduce a geometrical property to supply the place of analysis, where such introduction appeared of unquestionable advantage in shortening the process; instances of this occur at pages 162 and 163. The total rejection of all geometrical aid in most of French books is perhaps carried to an injudicious extent, and seems to be, in some cases, the result of caprice or affectation, for such aid is obviously allowable, and even adviseable, where simplicity may be attained by it. As to the arrangement here adopted, it may be briefly stated as follows: The volume consists of two principal parts, Analytical Geometry of Two Dimensions, and Analytical Geometry of Three Dimensions. The first part contains an introductory section on the algebraical solution of geometrical problems, and on the geometrical construction of algebraical equations; then follows, in three sections, an examination of the various properties of the lines of the second order, deduced from the most simple forms of their several equations; these three sections are, therefore, complete in themselves, comprehending, in the compass of one hundred and thirty-eight pages, a pretty copious treatise on the *Conic Sections.*

The fourth section enters more at large into the theory of these curves, by discussing very fully the most general forms of their equations, their positions in reference to any assumed axes, the determination of their varieties, &c. and the use of these researches is illustrated in Chapter iii. by their application to a variety of interesting problems on geometric loci. This first part terminates with a supplementary chapter containing some very useful theorems, such, for instance, are those at pages 211 and 214, the former of which is necessary in one very elegant mode of establishing the fundamental problem of physical astronomy, viz. that the planets move in elliptic orbits, having the sun in one of their foci; and the other problem is the foundation of the method of interpolations, so useful in the construction of tables, and in practical astronomy.

The second part is devoted to the consideration of lines and surfaces in space, the developement of their properties, and the general discussion of their equations. As to the first part, so here, a supplementary chapter is appended, containing many curious and interesting applications of the preceding theory. Most of the problems in this chapter have appeared before, some in the Annales Mathématiques, others in Leybourn's Repository, &c. but the solutions here given are for the most part new, and I think improved.

By way of index to the various topics embraced in the work, I have prefixed to the volume a very copious table of contents, which indeed precludes the necessity of extending further these prefatory remarks. I therefore conclude with the hope that the little volume now submitted to notice, though its pretensions be as humble as the form which it has assumed, may yet prove of some service to the mathematical student, in the earlier stages of his progress.

J. R. YOUNG.

CONTENTS.

PART I.

ANALYTICAL GEOMETRY OF TWO DIMENSIONS.

SECTION I.

Application of Algebra to Geometry.

SECTION II.

On the Point, Straight Line and Circle.

SECTION III.

On Lines of the Second Order.

SECTION IV.

General examination of indeterminate equations of the Second Degree.

PART II.

ANALYTICAL GEOMETRY OF THREE DIMENSIONS.

SECTION I.

On the Point, Straight Line, and Plane, in Space.

SECTION II.

On Surfaces of the Second Order.

SECTION III.

On Orthogonal Projection, Transformation of Coordinates, Discussion of the General Equation, &c

---

PROBLEM XXIV.

Within the sides of a given angle is inscribed a straight line of a given length; what is the locus of the point which divides this line in a given ratio? *Ans.* An ellipse.

PROBLEM XXV.

To determine the curve of which each ordinate is a mean between the corresponding ordinates of two given straight lines.

ANALYTICAL GEOMETRY.

SECTION I.

INTRODUCTION.

APPLICATION OF ALGEBRA TO GEOMETRY.

CHAPTER I.

Article (1.) ALGEBRA has been properly defined as that branch of mathematics in which calculations are performed by symbols. The signification given to these symbols is quite arbitrary, so that, in the practical application of this science to our inquiries about real quantity, it matters not whether the subject relate to time or space, number or motion. Whatever in nature can be submitted to calculation may always, and generally most commodiously, be treated algebraically; and hence Sir Isaac Newton and others have, with great propriety, called algebra *universal arithmetic*.

In the present introductory chapter we propose to show how algebra may be applied to the solution of a geometrical problem; in the next chapter will be explained how Geometry may be applied to the construction of an algebraical expression.

PROBLEM I.

(2.) Knowing the base and altitude of a plane triangle, to find the side of the square inscribed in it.

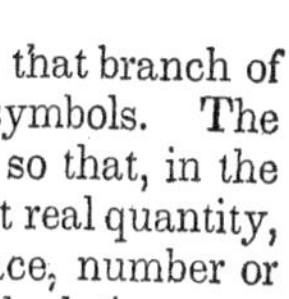

A
D E
H
B C
F I G

Let ABC be the triangle, and put the altitude $AI=a=78$, the base $BC=b=42$, and the side of the inscribed square $DG=x$.

Then, because in similar triangles the bases are as the altitudes, (*see Young's Geometry, p.* 94,) we have BC : DE :: AI : AH, or $b : x :: a : a-x$.

that is $ax=ab-bx$, or $ax+bx=ab$, that is $x=\dfrac{ab}{a+b}=27.3$

Hence the side of the inscribed square will be a fourth proportional to the three lines $a+b$, a and b.

PROBLEM II.

(3.) To divide a given straight line in extreme and mean proportion. (*Geom. p.* 113.)

F′ A F B

Let AB be the given straight line and call it a. Suppose F to be the point of division, and put AF$=x$, then x is to be determined, so that

$$a : x :: x : a - x \ldots\ldots (1)$$

hence we must have $x^2 = a^2 - ax \therefore x^2 + ax = a^2 \ldots (2)$,

$\therefore x = -\frac{a}{2} \pm \sqrt{a^2 + \frac{a^2}{4}} \ldots (3)$. Of these two values of x, one we perceive is negative, viz.

$$x = -\frac{a}{2} - \sqrt{a^2 + \frac{a^2}{4}}$$

and is in absolute magnitude greater than a, the whole line; this value, therefore, although it does fulfil the algebraical condition, (1,) cannot answer the geometrical conditions of the question, for the point of division, F, must necessarily fall between the extremities, A, B, of the proposed line, that is, AF, or x, must be less than AB, or a. The other value of x, viz. $x = -\frac{a}{2} + \sqrt{\left(a^2 + \frac{a^2}{4}\right)}$ is less than a, and therefore properly determines the point, F, required by the question. The reason why we have been furnished with two values of x instead of one, is that, having had to determine x, so that the condition (1) might exist, the algebraical process very properly led us to not only the one value sought by the question, but to every value that could fulfil that condition: and it afterwards remained for us to select that value as a solution to the question which involved no geometrical absurdity. However, although the negative value of x does not come within the geometrical restrictions of the question, yet it admits of a geometrical representation. In order to explain this, we may remark that a negative quantity in algebra may always be considered as resulting from the subtraction of a greater quantity from a less, thus the negative quantity $-q$ may be conceived to result from subtracting the greater quantity $(p+q)$ from the less, p, for it may always be supplied by the expression $p-(p+q,)$ whatever quantities the symbols p and q represent. Applying these remarks to the case in which the symbols denote lines, and taking the present problem as an example, we have AF the positive value of x equal to BA $-$ BF, and for the negative value, BF must exceed BA, that is, F must be on the other side of A, as at F′, hence making AF′ equal to the absolute value of the negative root of the equation (2); the two roots of that equation will be geometrically represented by AF and AF′. By taking the negative value of x, the condition (1) becomes

$$AB : AF' :: AF : BF'$$

so that the equation (3) is the complete solution to the question thus modified, viz.: Given two points A, B, to find on the line AB, or on its prolongation, a point such that its distance from the point A may be a mean proportional between its distance from B and the distance between A and B. The one point F answering these conditions is given by the positive value of x, and the other point F', on the opposite side of A, is given by the negative value; and in like manner, whenever it is required to determine the distance of a sought point from a given point, measured along a fixed straight line, and the solution furnishes both positive and negative values, if the positive values be taken in one direction from the fixed point, and the negative values in the opposite direction, every point so determined will solve the problem, and every possible solution will be obtained.

PROBLEM III.

(4.) From a given point without a circle, to draw a secant such that the intercepted chord may have a given length.

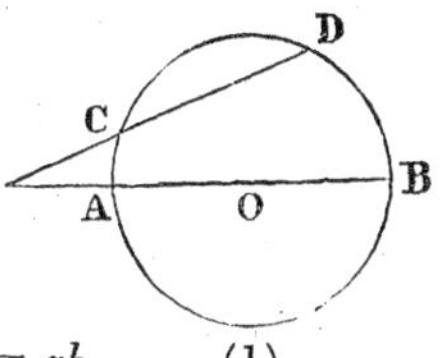

Let ACDB be the given circle, and P the given point. Draw PAB through the centre, and let PCD represent the required line. Put PB $= a$, PA $= b$, P CD $= c$, and PC $= x$, then (*Geom. p.* 106.) or *Euclid* III. 36 *Cor.*

$$PD \cdot PC = PB \cdot PA, \text{ that is } (x + c)\, x = ab \ \ldots\ldots (1)$$

and, solving this quadratic, we get $x = -\dfrac{c}{2} \pm \sqrt{\left(ab + \dfrac{c^2}{4}\right)}$

The positive value expresses the length of PC, the negative value gives no geometrical solution, although it fulfils the algebraical condition (1).

PROBLEM IV.

(5.) To divide a given straight line so that the rectangle of the two parts may be equivalent to a given square.

Let AB be the given line, which call a, put x for one of the required parts, and c for the side of the given square, then we have $(a - x)\, x = c^2$ or $x = \dfrac{a}{2} \pm \sqrt{\left(\dfrac{a^2}{4} - c^2\right)}$

Both these values being positive, the line may be divided in *two* points, F, F', as the problem requires. These points are obviously equidistant from the extremities of the line. If c exceeds $\dfrac{a}{2}$, the value of x is impossible.

PROBLEM V.

(6.) Given the perimeter of a right-angled triangle, and the radius of the inscribed circle, to determine the triangle.

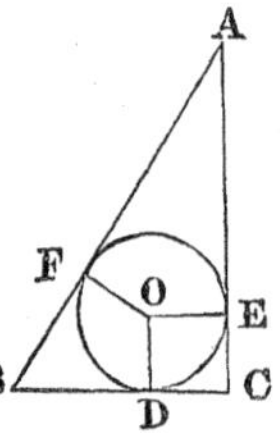

Let ABC be the triangle, and D, E, F, the points where the inscribed circle touches its sides, then (*Geom. p.* 106,) AF = AE, BF = BD, CD = EC = radius OD of the circle; hence, putting the perimeter $= p$, OE $= r$, AF $= x$ FB $= y$, we have $2x + 2y + 2r = p$, hence $y + r = \frac{1}{2}p - x \ldots\ldots (1.)$ Now (*Geom. p.* 22.) pr is twice the area of the triangle, but $(x + r)\ (y + r)$ is also equal to twice the area therefore $(x + r).\ (y + r) = pr \ldots\ldots (2)$ that is, by substituting for $y + r$ its value in equation (1.)

$$(x + r).\ (\tfrac{1}{2}p - x) = pr \text{ or } x^2 - (\tfrac{1}{2}p - r)\,x = -\tfrac{1}{2}pr$$

$$\therefore x = \tfrac{1}{2}\ (\tfrac{1}{2}p - r) \pm \surd\ \{\tfrac{1}{4}\ (\tfrac{1}{2}p - r)^2 - \tfrac{1}{2}pr\}$$

$$\therefore y = (\tfrac{1}{2}p - r) - x = \tfrac{1}{2}\,(\tfrac{1}{2}p - r) \mp \surd\ \{\tfrac{1}{4}\ (\tfrac{1}{2}p - r)^2 - \tfrac{1}{2}pr\}$$

Adding r to each of these expressions, we get

$$AC = \tfrac{1}{2}\ (\tfrac{1}{2}p + r) \pm \surd\ \{\tfrac{1}{4}\,(\tfrac{1}{2}p - r)^2 - \tfrac{1}{2}pr\}$$

$$BC = \tfrac{1}{2}\ (\tfrac{1}{2}p + r) \mp \surd\ \{\tfrac{1}{4}\,(\tfrac{1}{2}p - r)^2 - \tfrac{1}{2}pr\}$$

The double sign showing merely that if AC be made equal to any one of these values, BC will be equal to the other.

PROBLEM VI.

(7.) Given the chords of two arcs to find the chord of their sum.

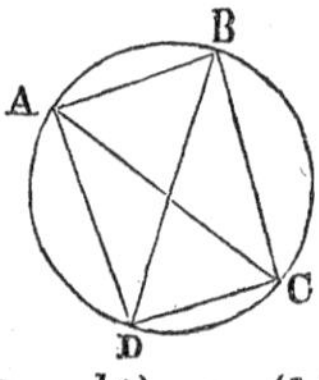

Let AB, BC, be the given chords, then it is required to determine the chord AC.

Draw the diameter BD and join AD, CD, then (*Geom. p.* 211,) AB·CD + AD·BC = AC·BD, that is, putting BC $= a$, AB, $= b$, AC $= c$, BD $= 2r$, and recollecting that AD $= \surd\ (BD^2 = AB^2,)$ and CD $= \surd\ (BD^2 - BC^2,)$ we have $b \surd\ (4r^2 - a^2) + a \surd\ (4r^2 - b^2) = 2cr$, or $\frac{b}{2r} \surd\ (4r^2 - a^2) + \frac{a}{2r} \surd (4r^2 - b^2) = c \ldots (1.)$ the expression sought. If the given chords are equal, then the expression for the chord of the sum is $\frac{a}{r} \surd\ (4r^2 - a^2)$

From equation (1) we may determine a when b and c are given; that is, when the chords of two arcs are known, we may find the expression for the chord of their difference.

PROBLEM VII.

(8.) Given the three sides of a triangle to determine the radius of the circumscribing circle.

Let us represent the three sides by a, b, c, and call the radius sought r, then we shall have to determine r from equation (1,) last problem.

By transposing, we have $\frac{b}{2r}\sqrt{(4r^2-a^2)} = c - \frac{a}{2r}\sqrt{(4r^2-b^2)}$

and by squaring and transposing $a^2 - \frac{ac}{r}\sqrt{(4r^2-b^2)} = b^2 - c^2$

$$\therefore r\,(a^2+c^2-b^2) = ac\sqrt{\{4r^2-b^2\}}$$

Squaring again $r^2\,(a^2+c^2-b^2)^2 = 4a^2c^2\,r^2 - a^2b^2c^2$

Hence $r = \sqrt{\frac{abc}{\{4a^2c^2-(a^2+c^2-b^2)^2\}}}$ the expression sought.

PROBLEM VIII.

(9.) Having given the three sides of a triangle to determine the expression for its surface.

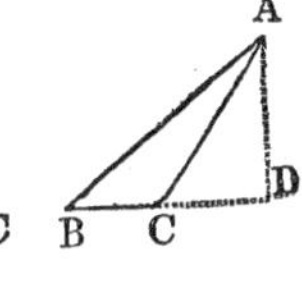

Put $BC = a$, $AC = b$, $AB = c$, perp. $AD = y$, $BD = x$, then $DC = a - x$, or $-a + x$, according as the perpendicular falls within or without the triangle.

by geometry we have $y^2 + x^2 = c^2 \ldots (1.)$

$$y^2 + (a-x)^2 = b^2 \ldots (2.)$$

The second equation is not altered by substituting $(x-a$ for $(a-x.)$

Subtracting equation (2) from equation (1), we have

$$-a^2 + 2ax = c^2 - b^2 \therefore x = \frac{a^2+c^2-b^2}{2a}$$

Putting this value of x in equation (1) we get

$$y^2 = c^2 - \left(\frac{a^2+c^2-b^2}{2a}\right)^2 \therefore y = \frac{1}{2a}\sqrt{4a^2c^2-(a^2+c^2-b^2)^2}$$

Now, calling the surface of the triangle S, we have $S = \frac{BC \cdot AD}{2} = \frac{ay}{2}$

therefore, substituting for y the value just found, we have

$S = \frac{1}{4}\sqrt{\{4a^2c^2-(a^2+c^2-b^2)^2\}}$ (3.) for the expression required.

(10.) Since the quantity under the radical is the difference of two squares, we may substitute for it the product of the sum, and difference of their roots. This sum and difference is $(2ac + a^2 + c^2 - b^2$ and $(2ac - a^2 - c^2 + b^2)$ which is the same as $(a+c)^2 - b^2$ and $b^2 - (a-c)^2$ and since each of these is also the difference of two squares, they may, in like manner, be replaced by the products.

$(a + c + b)\ a + c - b)$ and $(b + a - c)\ (b - a + c)$

Hence the expression (3) is the same as

$$S = \frac{1}{4}\sqrt{(a+b+c)\,(a+c-b)\,(b+a-c)\,(b+c-a)}$$

or putting, for shortness, $2p$ for the perimeter,

$$S = \sqrt{\{p(p-a)(p-b)(p-c).\}}$$

(11.) *Cor.* The expression for the radius of a circle circumscribing the triangle has been found (*Prob.* 7) to be

$$r = \frac{abc}{\sqrt{\{4a^2c^2 - (a^2+c^2-b^2)^2\}}}$$

therefore, putting for the denominator of this fraction its value 4S, as given by equation (3,) we have $r = \dfrac{abc}{4S}$.

PROBLEM IX.

The three sides of a triangle being given, to find the segments formed by letting fall a perpendicular from a vertical angle upon the base, the perpendicular itself, the area of the triangle, and radii of inscribed and circumscribed circles?

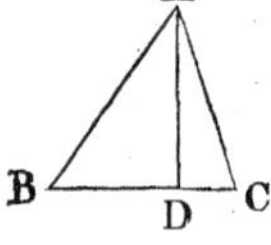

Let ABC be the given triangle, call BC, a; AC, b; AB, c. If AD is a perpendicular from the vertical angle on the base, we have by Prop. 13 Book 2, Euclid,

$\overline{AC}^2 = \overline{AB}^2 + \overline{BC}^2 - 2BC \times BD$; therefore $BD = \dfrac{a^2+c^2-b^2}{2a}$ = one of the segments formed by the perpendicular. The other segment $DC = BC - BD = \dfrac{a^2+b^2-c^2}{2a}$. The above value of BD gives $\overline{AB}^2 - \overline{BD}^2$ or $AD^2 = c^2 - \left(\dfrac{a^2+c^2-b^2}{2a}\right)^2$

$$= \frac{4a^2c^2-(a^2+c^2-b^2)^2}{4a^2}; \quad \therefore AD = \frac{\sqrt{\{4a^2c^2-(a^2+c^2-b^2)^2\}}}{2a}$$

Let S represent the area of the triangle, and we shall have $S = \frac{1}{2}BC \times AD$, Consequently $S = \frac{1}{4}\sqrt{\{4a^2c^2 - (a^2+c^2-b^2)^2\}} = \frac{1}{4}\sqrt{(2a^2b^2+2a^2c^2+2b^2c^2-a^4-b^4-c^4)}$. But this formula may be exhibited in a shape better adapted for logarithmic computation; to this end we may observe that the quantity $4a^2c^2-(a^2+c^2-b^2)^2$ is the product of the two factors $2ac+(a^2+c^2-b^2)$ and $2ac-(a^2+c^2-b^2)$; the first $(a+c)^2-b^2=(a+c+b)(a+c-b)$; the second $=b^2-(a-c)^2$ $=(b+a-c)(b-a+c)$; therefore we shall have $S=\frac{1}{4}\sqrt{\{(a+b+c)(a+b-c)(a+c-b)(b+c-a).\}}$ Now if we make $\dfrac{a+b+c}{2}=p$, we find $a+b+c=2p$, $a+b-c=2p-2c$, $a+c-b=2p-2b$, $b+c-a=2p-2a$ and by substituting these values in the above formula for S, and reducing, we finally obtain $S=\sqrt{(p.\overline{p-a}.\overline{p-b}.\overline{p-c})}$.

From this we see that to find the area of a triangle whose three sides are given, we must find the half-sum of the three sides, subtract from the half-sum successively each of the three sides, which will give three remainders, multiply continually these three remainders and the half-sum of the three sides, and lastly extract the square root of the product: this root will be the area of the triangle.

Let now z represent the radius of the circumscribed circle, and u the radius of the inscribed circle, then by Prop. C, Euclid, Book vi. $BA \times AC = z \times AD$, consequently

$$2z = \frac{BA \times AC}{AD} = \frac{\frac{1}{2} BC \times BA \times AC}{\frac{1}{2} AD \times BC} = \frac{\frac{1}{2} abc}{S},$$

and therefore *z, or the radius of the cirumscribed circle* $= \frac{1}{4} \frac{abc}{S}$

Or by writing for S its value $z = \frac{\frac{1}{4} abc}{\sqrt{\{p(p-a)(p-b)(p-c)\}}}$

Lastly it appears from Prop. 14, Euclid, Book iv. that if from the 3 angles of any triangle we draw 3 straight lines to the centre of the inscribed circle, we shall divide the triangle into three right-angled triangles, the area of which will be respectively represented by

$\frac{au}{z}, \frac{bu}{z}, \frac{cu}{z}$, and therefore $S = \frac{a+b+c}{z} u = pu$, and consequently u,

or the radius of the inscribed circled $= \frac{s}{p}$. Or by substituting for S its value $u = \sqrt{\left(\frac{(p-c)(p-b)(p-a)}{p}\right)}$

PROBLEM X.

Having given two contiguous sides a, b, of a parallelogram, and one of its diagonals, m, to find the other diagonal. Or the adjacent sides, a, b, and the diagonal, d, of a parallelogram, to find the other diagonal, x.

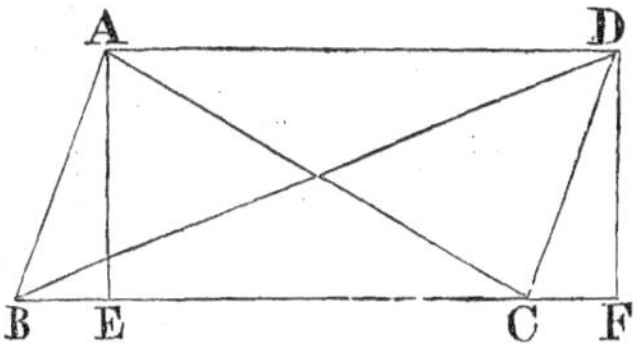

Let ABCD be a parallelogram whose sides AB, $AD = a$, b, respectively.

Suppose $BD = m$, and $AC = x$. Draw AE, DF perp. to BC. Because ABCD is a parallelogram, $AD = BC$, but $AD = EF$,

A D B E C F

$\therefore BC = EF$ and $BE = CF$.

Now $AC^2 = AB^2 + BC^2 - 2CB \cdot BE$ by Prop. 13. B. 2. Euclid. And $BD^2 = BC^2 + CD^2 + 2CB \cdot CF$ by Prop. 12.

$\therefore$ by addition $AC^2 + BD^2 = 2AB^2 + 2BC^2$

i. e. $x^2 + m^2 = 2a^2 + 2b^2 \therefore x = \{2a^2 + 2b^2 - m^2\}^{\frac{1}{2}} = AC.$

PROBLEM XI.

Given the altitude (a), the base (b), and (s) the sum of the sides of a plane triangle, to find the sides.

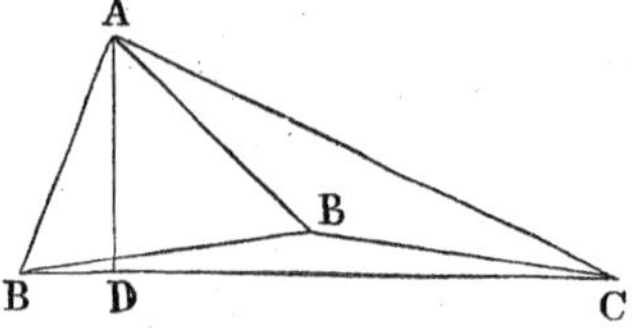

Let ABC be a triangle whose base BC$=b$, and altitude a. Let DB $= x$, then AB$=\sqrt{a^2+x^2}$; also DC$=b-x$ $\therefore$ AC$=\sqrt{a^2+(b-x)^2}$

Now AB+AC$=s$

$\therefore \{a^2+x^2\}^{\frac{1}{2}}+\{a^2+(b-x)^2\}^{\frac{1}{2}}=s.$

$\therefore \{a^2+(b-x)^2\}^{\frac{1}{2}}=s-\{a^2+x^2\}^{\frac{1}{2}}$. Square both sides,

$$\text{And } a^2+(b-x)^2=s^2-2s.\{a^2+x^2\}^{\frac{1}{2}}+a^2+x^2$$

$$\text{Or } b^2-2bx=s^2-2s\{a^2+x^2\}^{\frac{1}{2}}$$

$$\text{Transpose and } b^2-2bx-s^2=-2s\{a^2+x^2\}^{\frac{1}{2}}$$

Square both sides,

$$\text{And } (b^2-s^2)^2-4bx(b^2-s^2)+4b^2x^2=4s^2a^2+4s^2x^2$$

Arrange the quantities and divide by $4(s^2-b^2)$

$$\text{Then } x^2-bx=\frac{s^2-b^2}{4}-\frac{a^2s^2}{s^2-b^2}$$

$$\text{And } x=\frac{b}{2}\pm\frac{s}{2}\sqrt{\left\{1-\frac{4a^2}{s^2-b^2}\right\}}=\text{BD.}$$

Hence AB which $=\sqrt{\{a^2+x^2\}}$ is found, and AC which $=\sqrt{\{a^2+(b-x)^2\}}$ can be determined. If $a=4$, $b=8$, and $s=12$

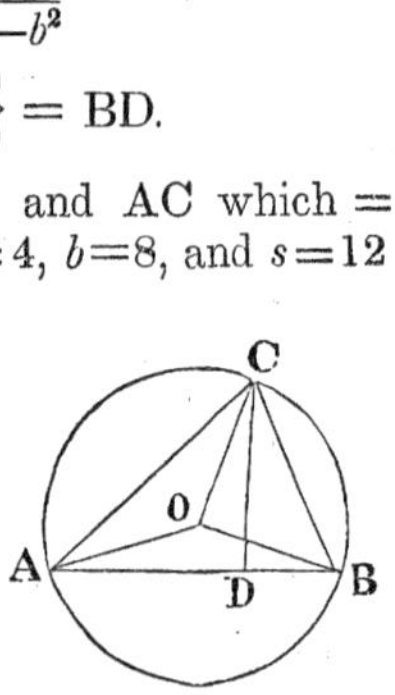

Otherwise, Let S$=6=\frac{1}{2}$ sum of the sides and let $x=\frac{1}{2}$ difference. Then AC $=$ S $+x$ CB $=$ S $-x$. Put AB $=b=8$ and CD $=4=p$. Then AD $=\sqrt{\{(S+x)^2-p^2\}}$ And DB $=\sqrt{\{(S-x)^2-p^2\}}$ By the quest. $\sqrt{\{(S+x)^2-p^2+\}}\sqrt{\{(S-x)^2-p^2\}}=b$. Or $\sqrt{\{(S+x)^2-p^2\}}=b-\sqrt{\{(S-x)^2-p^2\}}$ by squaring $(S+x)^2-p^2=b^2-2b\sqrt{\{(S-x)^2-p^2\}}+(S-x)^2-p^2$. Or $4Sx=b^2-2b\sqrt{\{(S-x)^2-p^2\}}$ that is $2b\sqrt{\{(S-x)^2-p^2\}}=b^2-4Sx$; squaring again we have $4b^2(S-x)^2-4b^2p=b^4-8b^2Sx+16S^2x^2$; Or $4b^2S^2+4b^2x^2-4b^2p^2=b^4+16S^2x^2$

$$\therefore x=\sqrt{\left(\frac{b^2(b^2-4S^2+4p^2)}{4\cdot(b^2-4S^2)}\right)}=\frac{b}{2}\sqrt{\left(1+\frac{4p^2}{b^2-4S^2}\right)}$$

$=4\sqrt{(1-\frac{4}{5})}=4\sqrt{\frac{1}{5}}=\frac{4}{5}\sqrt{5}$, and consequently AC $=6+\frac{4}{5}\sqrt{5}$ and BC $=6-\frac{4}{5}\sqrt{5}$.

PROBLEM XII.

Suppose the town A to be (a) miles from B, and B (b) miles from C, and C (c) miles from A, to find where a house, O, must be erected equally distant from A, B, and C.

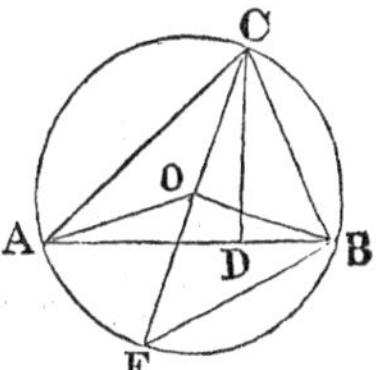

Construction.—Make a triangle ABC whose sides AB, BC, and AC, shall be equal to 30, 25, and 20, by (Euc. I. 22.) and about this triangle describe a circle by Euc. iv. 5. then OC, OA, or OB are an equal distance from each other.

In the triangle ABC, all the sides are given to find the angle BCD, as AB 30 : BC + AC 45 :: BC — AC 5 : BD — AD = 7.5. Whence $\frac{1}{2}$ (7.5 + 30) = 18.75 = BD. Again as 25 : 18.75 : : radius : sin. BCD = 48° 35′ 25″. Produce CO to the circumference in F and join BF, then in the right angled triangle BCF are given the angle BCF = ACD = 48° 35′ 25″, and AC = 20; As rad. : sec. 48° 35′ 25′, :: AC 20 : CF = 30.2371; then $\frac{1}{2}$CF = 15.1185 = CO distance required; or thus: Let AB $= a = 30$ BO $= b = 25$, CA $= c = 20$; and put BD $= x$ and AD $= y$; then $x + y = a$, and $x^2 - y^2 = c^2 - b^2$, by division, $x - y = \dfrac{c^2 - b^2}{a}$ by addition $2x = \dfrac{a^2 + c^2 - b^2}{a}$ or $x = \dfrac{a^2 + c^2 - b^2}{2a}$ = BD, and by Euc. I. 47. $CD = \sqrt{\left(c^2 - \dfrac{(a^2 + c^2 - b^2)^2}{4a^2}\right)} = \sqrt{\dfrac{4375}{16}} = \dfrac{25}{4}\sqrt{7}$ by Euc. vi. C. Diam. × CD = AC × CB; whence Diam. $\dfrac{500}{\frac{25}{4}\sqrt{7}} = \dfrac{80}{\sqrt{7}} = \dfrac{80\sqrt{7}}{7} = 30.2371578$; whence AO or CO or BO = 15 1185789 distance sought.

Draw AD perpendicular to the base, and let AE $= x$. then $a : b + c :: b - c :$ BD — DA ∴ BD — DA $= \dfrac{b^2 - c^2}{a}$? BD + DA $= a$

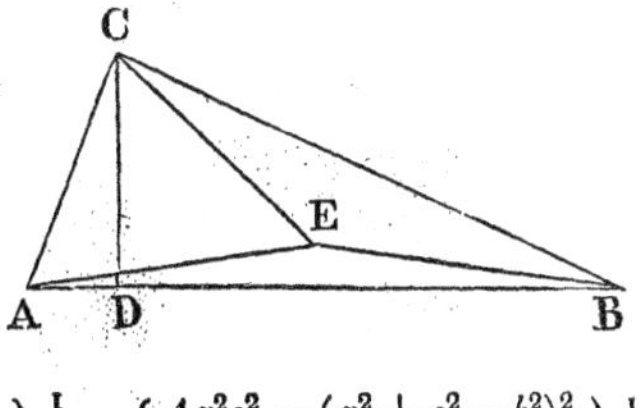

$$\therefore DA = \frac{a}{2} \quad \frac{b^2 - c^2}{2a} = \frac{a^2 - b^2 + c^2}{2a}$$

$$\text{Now } CD = \left\{ c^2 - \left(\frac{a^2 - b^2 + c^2}{2a}\right)^2 \right\}^{\frac{1}{2}} = \left\{ \frac{4a^2c^2 - (a^2 + c^2 - b^2)^2}{4a^2} \right\}^{\frac{1}{2}}$$

$$\text{But } CD = \frac{bc}{2x} \text{ by Prob. 13.} \ \therefore \frac{bc}{2x} = \frac{\{4a^2c^2 - (a^2 - b^2 + c^2)^2\}}{2a}$$

$$\therefore x = \frac{abc}{\sqrt{\{4a^2c^2 - (a^2 - b^2 + c^2)^2\}}} = \text{the distance of the house E}$$

from each of the angles A, B, C. If $a = 30$ miles, $b = 25$, $c = 20$

$$x = \frac{30 \times 25 \times 20}{\sqrt{(1440000 - 455626)}} = \frac{40}{\sqrt{7}} = 15.12 \text{ miles.}$$

The above is nothing more than having the three sides of a triangle given to find the radius of the circumscribing circle. See prob. 7.

PROBLEM XIII.

If a, b, c be the three sides of the plane triangle; R, r the radii of circumscribed and inscribed circles: show that $Rr = \dfrac{abc}{2(a+b+c)}$

Let AB $= a$, BC $= b$, AC $= c$, R $=$ NA in figure 2, and $r =$ NF in figure 1. Then, since NF, NE, NG are perpendicular to AB, AC, CB respectively, the area of

$\triangle$ ANC $= \frac{1}{2}rc$

The area of $\triangle$ ANB $= \frac{1}{2}ra$

The area of $\triangle$ BNC $= \frac{1}{2}rb$

Area of $\triangle$ ABC$=(a+b+c).\dfrac{r}{2}$

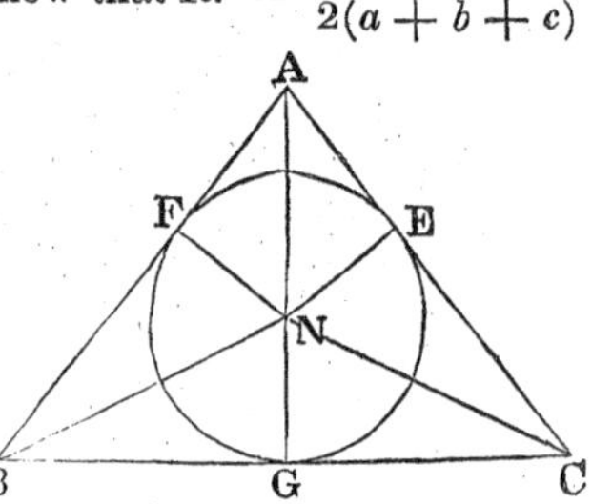

Now, if AD be drawn perpendicular to BC in fig. 2; since the angle BNF $= \frac{1}{2}$ angle BNA $=$ angle BCA, and the right angle BFN $=$ the right angle ADC, $\therefore$ the triangles BNF, ADC are similar; hence AC : AD :: BN : BF, Or c : AD :: R : $\dfrac{a}{2}$. $\therefore$ AD$=\dfrac{ca}{2R}$ Now the area of triangle ABC* $= \frac{1}{2}\{$AD $\times$ BC$\}$

$$\therefore \{a+b+c\}.\frac{r}{2} = \frac{abc}{4R}. \text{ And } Rr = \frac{abc}{2(a+b+c)}.$$

PROBLEM XIV.

The diagonal of a rectangle ABCD, and the perimeter, or sum of all its four sides, being given, to find the sides.

Let the diagonal AC $= d$, half the perimeter AB $+$ BC $= a$, and the base BC $= x$; then will the altitude AB $= a - x$.

And by Euc. I. 47. $AB^2 + BC^2 = AC^2$, we shall have $a^2 - 2ax + x^2 + x^2 = d^2$, or $x^2 - ax = \frac{1}{2}\{d^2 - a^2.\}$

Which last equation, being resolved, gives $x = \frac{1}{2}a \pm \frac{1}{2}\sqrt{(2d^2 - a^2)}$.

Where a must be taken greater than d and less than $d\sqrt{2}$,

*The reader must recollect that the triangle presented by ABC in each figure, is supposed to be of the same magnitude.

PROBLEM XV.

Given the three sides of a triangle, to determine its area.

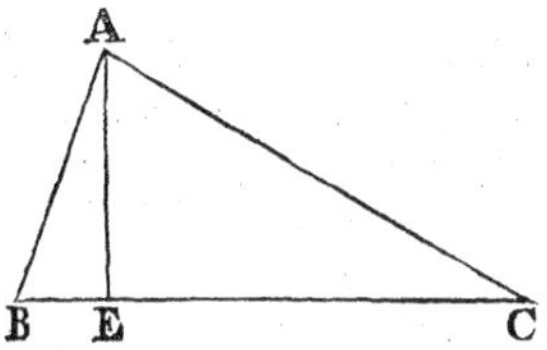

Let ABC be any triangle whose sides AB, AC, BC $= a, b, c$ respectively. Draw AE perpendicular to BC.

Then $b^2 = a^2 + c^2 - 2c.$ BE. (Prop. 13. b. 2. Euclid.)

$$\therefore \text{BE} = \frac{a^2 + c^2 - b^2}{2c} \qquad \text{Also AE} = \left\{ a^2 - \left(\frac{a^2 + c^2 - b^2}{2c} \right)^2 \right\}^{\frac{1}{2}}$$

$$= \left\{ a - \frac{a^2 + c^2 - b^2}{2c} \right\}^{\frac{1}{2}} \left\{ a + \frac{a^2 + c^2 - b^2}{2c} \right\}^{\frac{1}{2}}$$

$$= \sqrt{\left\{ \frac{b^2 - (a-c)^2}{2c} \right\}} \sqrt{\left\{ \frac{(a+c)^2 - b^2}{2c} \right\}}$$

$$= \sqrt{\left\{ \frac{(b+a+c)\,(b-a+c)}{2c} \right\}} \sqrt{\left\{ \frac{(a+c+b)\,(a+c-b)}{2c} \right\}}$$

$$= \frac{1}{2c} \sqrt{\left\{ (b+a-c)\cdot(b-a+c)\cdot(a+c+b)\cdot(a+c-b) \right\}}$$

$$= \frac{2}{c} \sqrt{\left\{ \left(\frac{a+b+c}{2}\right). \left(\frac{a+b+c}{2} - a\right). \left(\frac{a+b+c}{2} - b\right) \left(\frac{a+b+c}{2} - c\right) \right\}}$$

$$\therefore \frac{\text{AE}.c}{2} = \sqrt{\left\{ \left(\frac{a+b+c}{2}\right). \left(\frac{a+b+c}{2} - a\right). \left(\frac{a+b+c}{2} - b\right). \left(\frac{a+b+c}{2} - c\right) \right\}}$$

If $s = a+b+c$, then since the area $= \dfrac{\text{AE}.\,c}{2}$

We find the area $= \sqrt{\left\{ \frac{s}{2} \left(\frac{s}{2} - a\right) \left(\frac{s}{2} - b\right) \left(\frac{s}{2} - c\right) \right\}}$

Cor. by help of this problem we can determine the radius of a circle inscribed in a triangle in terms of the sides. If the figure were constructed it would readily appear that the

area of the triangle $= (a+b+c)\dfrac{r}{2}$, see Prob. 13.

$$\therefore (a+b+c)\,\frac{r}{2} = \frac{rs}{2} = \sqrt{\left\{ \frac{s}{2} \left(\frac{s}{2} - a\right) \left(\frac{s}{2} - b\right) \left(\frac{s}{2} - c\right) \right\}}$$

$$\therefore r = \frac{2}{s} \sqrt{\left\{ \frac{s}{2} \cdot \left(\frac{s}{2} - a\right) \left(\frac{s}{2} - b\right) \left(\frac{s}{2} - c\right) \right\}}$$

PROBLEM XVI.

To find the side of a regular octagon inscribed in a circle whose radius is known.

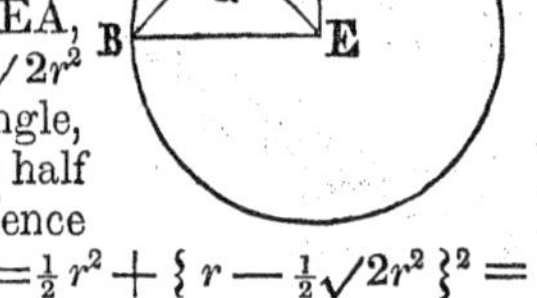

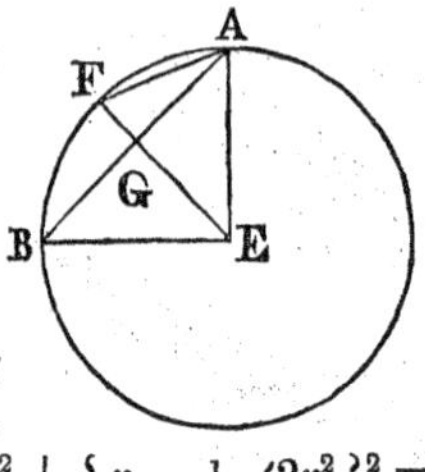

Let AB be the side of a square inscribed in the circle AFB, whose centre is E. Draw EG perpendicular to AB, then AG = GB, and the arc AF = arc FB. Join AF which is the side of a regular octagon. Let r=radius EA, and y=AF. Then $AB^2=2r^2$ $\therefore$ $AG=\frac{1}{2}\sqrt{2r^2}$

Now, since the $\angle$AGE is a right angle, and AEG half a right angle, $\therefore$ GAE is half a right angle, and $AG=GE=\frac{1}{2}\sqrt{2r^2}$ Hence $FG=r-\frac{1}{2}\sqrt{2r^2}$, But $y^2=AG^2+GF^2=\frac{1}{2}r^2+\{r-\frac{1}{2}\sqrt{2r^2}\}^2=r^2.\{2-\sqrt{2}\}$ $\therefore y=r.\sqrt{\{2-\sqrt{2}\}}$ the value required.

PROBLEM XVII.

To find the side of a regular decagon inscribed in a circle.

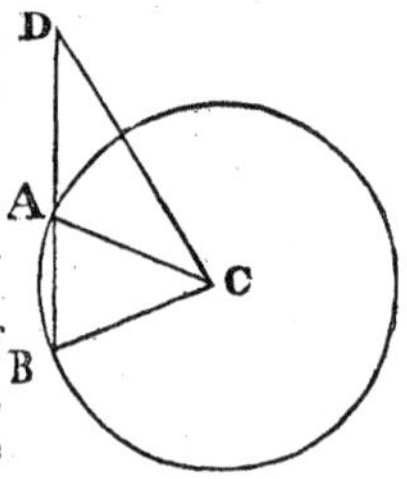

Suppose AB to be the side of a regular decagon. Join CA, CB and produce BA to D, making AD=AC. Join DC. Then since the $\angle$ ACB=$\frac{1}{10}$th of 4 right angles or $\frac{1}{5}$ of two right angles $\therefore$ each of the angles CAB, CBA is equal to $\frac{2}{5}$ of two right angles. Also the $\angle$CAD = $\frac{3}{5}$ of two right angles; now AD=AC, $\therefore$ each of the angles ADC, ACD is equal to $\frac{1}{5}$ of two right angles, and $\therefore$ the triangles BDC, BAC are equiangular. Let BC=r, AB=y. Hence BD:BC::BC:BA *i. e.* $r+y:r::r:y$ $\therefore y^2+ry=r^2$ And $y=\frac{1}{2}r\{\sqrt{(5)}-1,\}$ as required.

PROBLEM XVIII.

Having given the side of a regular decagon inscribed in a circle whose radius is known, to find the side of a regular pentagon inscribed in the same circle.

Let AB (see figure to Prob. XVI.) be the side of a regular pentagon, and AF the side of a regular decagon inscribed in the circle AFB. Suppose AF=a, AE=r, AB=y. Now $GE=\sqrt{(r^2-\frac{1}{4}y^2)}$ $\therefore FG=r-\sqrt{(r^2-\frac{1}{4}y^2)}$ $\therefore a^2=\frac{1}{4}y^2+r^2-2r\sqrt{(r^2-\frac{1}{4}y^2)}+r^2-\frac{1}{4}y^2=2r^2-2r\sqrt{(r^2-\frac{1}{4}y^2)}$ $\therefore 2r\sqrt{(r^2-\frac{1}{4}y^2)}=2r^2-a^2$, $4r^4-r^2y^2=4r^4-4a^2r^2+a^4$ $\therefore$ $y^2=4a^2-\{a^4\div r^2\}$ Now $a=\frac{1}{2}r(\sqrt{5}-1)$ by the last problem; $\therefore$ by substitution, $y^2=r^2+\frac{1}{4}r^2(6-2\sqrt{5})-r^2+a^2$ $\therefore y=\sqrt{(r^2+a^2.)}$

Hence the square of the side of a regular pentagon inscribed in a circle, is equal to the square of the side of a regular decagon, together with the square of the radius.

PROBLEM XIX.

To find the side of an equilateral triangle inscribed in a circle, whose radius is (a), and that of another circumscribed about the same circle.

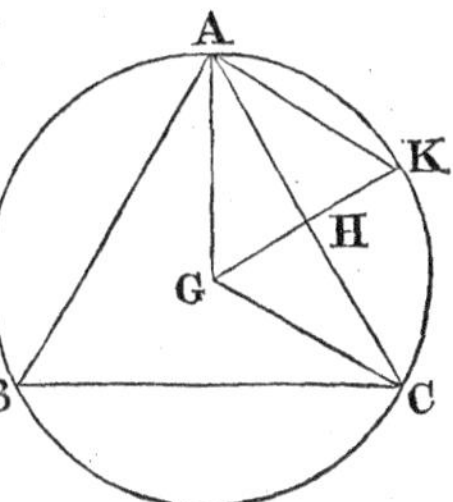

Suppose ABC an equilateral triangle inscribed in the circle ABC. Find the centre G, and from G draw GH perpendicular to AC, then AH = HC, Prop. 3. B. III Euclid. Produce GH to meet the circumference in K; and join AK. Then because AG, GH are equal to CG, GH, and AH = CH; $\therefore$ the $\angle$AGH = $\angle$CGH. Now the $\angle$AGC is one third of four right angles, $\therefore$ AGK = one third of two right angles, and each of the angles GAK, GKA is one-third of two right angles; hence the triangle AGK is equilateral; and since AH is perpendicular to GK, $\therefore$ GH = HK; *i. e.* GH = $\frac{1}{2}$ GA. Now by Prop. 47. I. Euclid,

$$GA^2 = GH^2 + HA^2 = \frac{AG^2}{4} + HA^2. \text{ Hence } HA = AG\sqrt{\tfrac{3}{2}}$$

$\therefore$ 2HA or AC = AG. $\sqrt{3} = a\sqrt{3}$ = a side of the inscribed triangle. By describing an equilateral triangle about the circle, if one of its sides touch the circle in the point K, it may be easily shown that AC : a side of the circumscribed triangle :: GH : GK :: 1 : 2, $\therefore$ a side of the equilateral triangle described about the circle$=2a\sqrt{3}$

PROBLEM XX.

From the given point C, to draw the straight line CF, which, together with two other right lines AE, AF given in position shall constitute a triangle AEF of a given magnitude.

Through C draw CD parallel to EA, meeting FA produced in D. Draw CB, EG perpendicular to DF. Now, because C is given in position with respect to EA, AF, $\therefore$ CB may be considered as given; as also AD. Let CB = b, AD = a, AF = x, and A = the area of the triangle EAF. Then by similar triangles DF : AF :: DC : AE :: BC : EG

$$\text{Or } a + x : x :: b : \frac{bx}{a+x} = EG \quad \therefore \frac{bx^2}{a+x} = 2A, \text{ and } bx^2 = 2Aa$$

$$+ 2Ax, \therefore x^2 - \frac{2A}{b}.x = \frac{2Aa}{b} \quad \text{Or } x^2 - \frac{2A}{b}.x + \frac{A^2}{b^2} = \frac{A^2 + 2Aab}{b^2}$$

$$\therefore x - \frac{A}{b} = \pm \frac{\sqrt{\{A^2 + 2Aab\}}}{b} \quad \therefore x = \frac{A \pm \sqrt{\{A^2 + 2Aab\}}}{b} \text{ which}$$

determines the point F; join CF, and the area of the triangle AEF = A.

PROBLEM XXI.

In a plain triangle, having given the perpendicular (p) and the radii (r, R) of its inscribed and circumscribed circles, to determine the triangle.

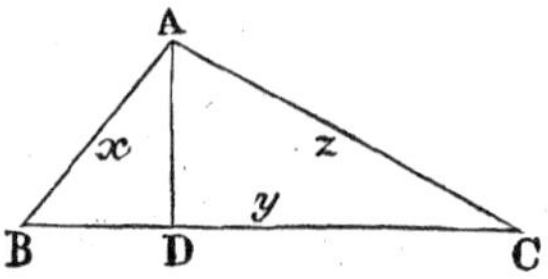

Let ABC be a triangle whose sides AB, BC, AC are respectively equal to x, y, z. Also let $p =$ AD, the perpendicular from the vertex. Then by Problem XIII. page 28, we have

A, B, D, C, x, y, z

$$x+y+z=\frac{py}{r}, \text{ also } Rr=\frac{xyz}{2(x+y+z)}, \therefore xz=2pR$$

Now, $x^2=y^2+z^2-2y.\text{CD} \therefore \text{CD}=\frac{y^2+z^2-x^2}{2y}$, And $p=\left\{z^2-\right.$

$$\left.\left(\frac{y^2+z^2-x^2}{2y}\right)^2\right\}^{\frac{1}{2}}=\left\{z+\frac{y^2+z^2-x^2}{2y}\right\}^{\frac{1}{2}}.\left\{z-\frac{y^2+z^2-x^2}{2y}\right\}^{\frac{1}{2}}$$

$$=\frac{1}{2y}.\{(z+y+x)\cdot(z+y-x)\cdot(x-y+z)\cdot(x+y-z)\}^{\frac{1}{2}}$$

$$=\frac{2}{y}.\left\{\frac{py}{2r}\left(\frac{py}{2r}-x\right).\left(\frac{py}{2r}-y\right).\left(\frac{py}{2r}-z\right)\right\}^{\frac{1}{2}}$$

$$\therefore\frac{py}{2}=\frac{1}{r}\left(\frac{py}{2r}-x\right).\left(\frac{py}{2r}-y\right).\left(\frac{py}{2r}-z\right)$$

$$\therefore\frac{p}{2}=\frac{1}{8r^4}(py-2rx)\cdot(py-2rz)\cdot(p-2r)$$

$$\therefore\frac{4pr^4}{p-2r}=p^2y^2-2pry.(x+z)+4r^2xz$$

But $x+z=\frac{py}{r}-y=\frac{p-r}{r}.y$, and $xz=2p\text{R}$

$$\therefore\frac{4pr^4}{p-2r}=p^2y^2-\{2p^2-2pr\}y^2+8p\text{R}r^2$$

$$\text{Or }\frac{4r^4}{p-2r}-8\text{R}r^2=\{2r-p\}.y^2$$

$$\because\frac{4r^4-8pr^2\text{R}+16r^3\text{R}}{p-2r}=\{2r-p\}.y^2$$

$$\text{Hence } y^2=\left(\frac{2r}{p-2r}\right)^2.\{2p\text{R}-4r\text{R}-r^2\}$$

$$\therefore y=\frac{2r}{p-2r}\{2p\text{R}-4r\text{R}-r^2\}^{\frac{1}{2}}=\text{the base}$$

The other sides are easily found from the equations $x+z=\frac{p-r}{r}.y$, and $xz=2p\text{R}$.

PROBLEM XXII.

Given the area of an equilateral triangle, CEF, whose base, EF, falls on the diameter, and its vertex C in the middle of a semicircular arc: required the diameter of the circle.

Let ACB be a semicircle whose diameter is AB and centre D. Draw DC perpendicular to AB and let ECF be an equilateral triangle whose area $= a$. Suppose DC $= x$.

Now $a = \frac{1}{2}x \times$ EF $\therefore$ EF $= 2a \div x$ Or EC $= 2a \div x$; but EC $= \sqrt{\{ED^2 + DC^2\}}$ $\therefore 2a \div x = \sqrt{\{(a^2 \div x^2) + x^2\}}$ Or $4a^2 \div x^2 = \{(a^2 \div x^2) + x^2\}$ and $x^4 = 3a^2$ $\therefore x = (3a^2)^{\frac{1}{4}}$. $\therefore 2x$ the diameter $= 2\ (3a^2)^{\frac{1}{4}}$

If the area be represented by 100, the diameter of the circle $= 20\sqrt{\{\sqrt{3}\}} = 20(3)^{\frac{1}{4}}$

PROBLEM XXIII.

Through a given point P, in a given circle ACBD, to draw a chord CD, of a given length.

Draw the diameter APB; and put CD $= a$, AP $= b$, PB $= c$, and CP $= x$; then will PD $= a - x$. But, by the property of the circle (Euc. III. 35.) CP $\times$ PD $=$ AP $\times$ PB; whence $x\ (a - x) = bc$, or $x^2 - ax = -bc$. which equation, being resolved in the usual way, gives $x = \frac{1}{2}a \pm \sqrt{(\frac{1}{4}a^2 - bc)}$; Where x has two values, both of which are positive.

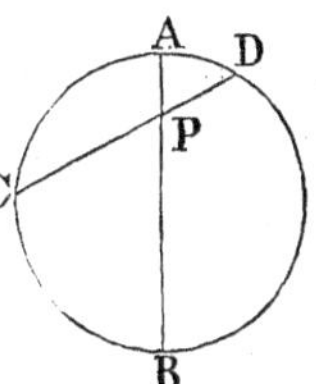

PROBLEM XXIV.

The base BC, of any plane triangle ABC, the sum of the sides AB, AC, and the line AD, drawn from the vertex to the middle of the base, being given, to determine the triangle.

Put BD or DC $= a$, AD $= b$, AB + AC $= s$, and AB $= x$; then will AC $= s - x$. (Geometry B II., Prop. 13,) $AB^2 + AC^2 = 2BD^2 + 2AD^2$; whence $x^2 + (s - x)^2 = 2a^2 + 2b^2$, or $x^2 - sx = a^2 + b^2 - \frac{1}{2}s^2$.

Which last equation, being resolved as in the former instances, gives $x = \frac{1}{2}s \pm \sqrt{(a^2 + b^2 - \frac{1}{4}s^2)}$, for the values of the two sides AB and AC of the triangle; taking the sign $+$ for one of them, and $-$ for the other, and observing that $a^2 + b^2$ must be greater than $\frac{1}{4}s^2$.

PROBLEM XXV.

The two sides AB, AC, and the line AD, bisecting the vertical angle of any plane triangle ABC, being given, to find the base BC.

5

Put AB $= a$, AC $= b$, AD $= c$, and BC $= x$; then, by Euc. VI. 3, (See last Problem,) we shall have AB (a) : AC (b) : : BD : DC. And consequently, by the composition of ratios (Euc. V. 18,) $a + b : a :: x :$ BD $= \{ax \div (a+b)\}$, and $a + b : b :: x :$ DC $= \{bx \div (a + b.)\}$, But, by Euc. VI. 13, DC $\times$ BD $+$ AD^2 $=$ AB $\times$ AC; therefore, also, $\{abx^2 \div (a+b)^2\} + c^2 = ab$, or $abx^2 = (a+b)^2 \times (ab - c^2)$. From which last equation we have $x = (a+b)\sqrt{\{(ab-c^2) \div ab\}} =$ BC.

PROBLEM XXVI.

Determine a triangle; having given the base, the line bisecting the vertical angle, and the diameter of the circumscribing circle.

Let ABC be the triangle, and BD the line bisecting the vertical angle.

Draw the diameter FG at right angles to the base. If BD be produced it will meet the circumference in G, because equal angles stand on equal arcs. Put a for BD, b for AE or EC, c for FG, and x for EG. Then (Geom. 90) $x\ (c - x) = b^2$, hence $x = \frac{1}{2}\,c \pm \sqrt{(\frac{1}{4}\,c^2 - b^2)}$; for this value of x put e; and join BF, also let y represent DG.

The angle GBF being (Geom. 52) a right angle, the triangles GDE, GFB, are [Geom. 96] similar; $\therefore y : e :: c : a+y$, that is, $ay + y^2 = ec$, hence $y = \pm \sqrt{\{ec + \frac{1}{4}a^2\}} - \frac{1}{2}a$. Put f for this value of y; and $z =$ DC. Then [Geom. 108] $af = 2bz - z^2$, that is, $z = b \pm \sqrt{\{b^2 - af.\}}$ Put this value of $z = g$, and from B draw BH at right angles to FG. As GD . GB : : DE : BH, that is,

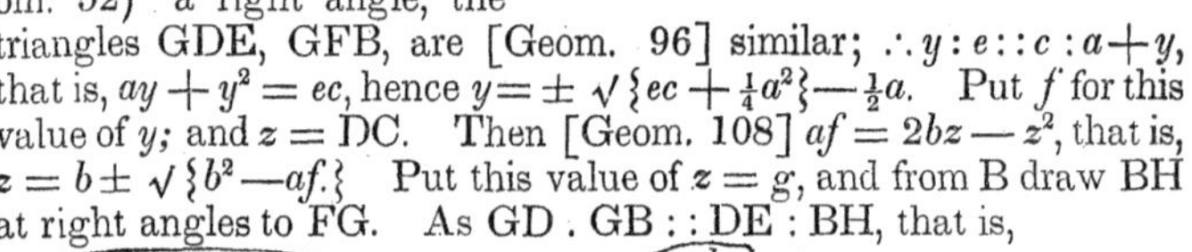

As $f : (a+f) :: (b-g) : (b - g + \frac{ab + ag}{f}) =$ BH the distance of the perpendicular from the middle of the base, $=$ EP. Join BP GE : EH (=BP) : : GD : DB, that is, As e : BP : : f : a, or BP $= ae \div f =$ the perpendicular of the triangle ABC. But

$$AB = \sqrt{\{AP^2 + PB^2\}} = \sqrt{\left(\left\{b + (b - g + \frac{ab+ag}{f})\right\}^2 + p^2\right)}$$

[where p represents PB]; And BC $= \sqrt{\{PC^2 + BP^2\}} =$

$$\sqrt{\left(\left\{b - (b - g + \frac{ab+ag}{f})\right\}^2 + p^2.\right)}$$ as required.

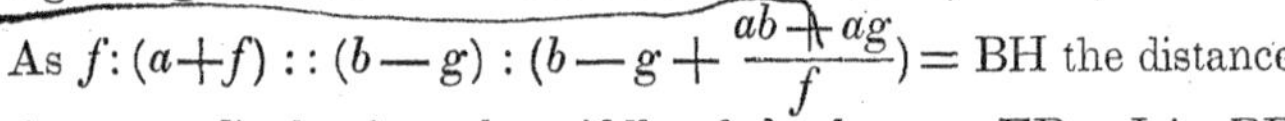

PROBLEM XXVII.

In a right angled triangle having given the side of its inscribed square (12) and the radius of its inscribed circle (7) to determine the triangle.

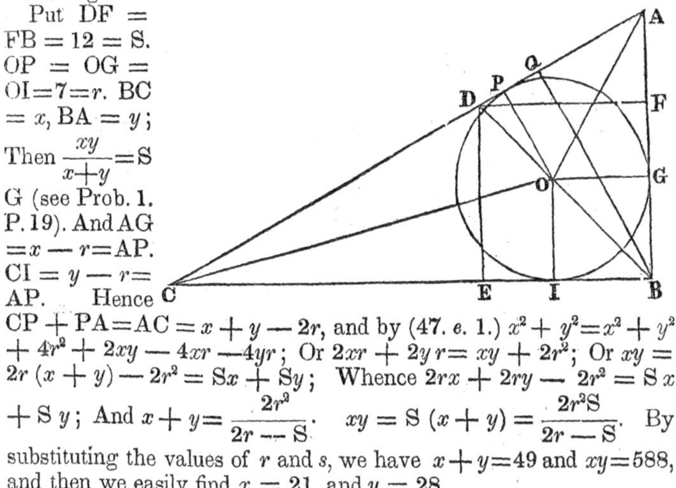

Put DF $=$ FB $= 12 =$ S. OP $=$ OG $=$ OI$=7=r$. BC $= x$, BA $= y$; Then $\frac{xy}{x+y}=$S G (see Prob. 1. P. 19). And AG $=x-r=$AP. CI $= y-r=$ AP. Hence CP $+$ PA$=$AC $=x+y-2r$, and by (47. *e.* 1.) $x^2+y^2=x^2+y^2+4r^2+2xy-4xr-4yr$; Or $2xr+2yr=xy+2r^2$; Or $xy = 2r\,(x+y)-2r^2=$ S$x+$ Sy; Whence $2rx+2ry-2r^2=$ Sx $+$ Sy; And $x+y=\frac{2r^2}{2r-\mathrm{S}}$. $xy=$ S $(x+y)=\frac{2r^2\mathrm{S}}{2r-\mathrm{S}}$. By substituting the values of r and s, we have $x+y=49$ and $xy=588$, and then we easily find $x=21$. and $y=28$.

Let ABC be the proposed triangle, BFDE the inscribed square, OG and OP radii of the inscribed circle at the right angles to AB and AC, and BD a diagonal of the inscribed square; also let BQ be perpendicular to AC from the right angle. Put a for the side of the given square, b for the radius of the given circle, and x for the segment AQ of the base AC by the perpend. BQ,

Then FG$=a-b$ since GB$=$OG; and $a-b : a :: b : \{ab \div (a-b).\}$ $=$ BQ (*because* GF : BF :: OD : BD :: OP : BQ). Therefore (since BD $=\sqrt{2a^2}$), DQ $=\sqrt{\{2a^2-\frac{a^2b^2}{(a-b)^2}.\}}$ Let this value of DQ to be recognized in c, and put d for $\frac{ab}{a-b}$. It is as $x : d :: d :$

$\frac{d^2}{x}=$CQ [*Geom. p.*]. And $x+c : \frac{d^2}{x}-c :: x : d.$*

Hence $dx+dc=d^2-cx$. That is $x=\frac{d^2-dc}{d+c}=$ AQ, *therefore the triangle is determined.*

* For the triangles AFD, DEC are similar; also the triangle AQB is similar to AFD, and consequently to DEC. Therefore AD : DC :: AF : DE ($=$ DF) :: AQ : QB.

PROBLEM XXVIII.

To determine the radii of three equal circles, described in a given circle, to touch each other and also the circumference of the given circle.

Let ABC be the given circle, whereof O is the centre. Inscribe in it the equilateral triangle PQR (Eucl. iv. 15); join OQ; OR, OP; and produce OP till PS equal the half of PR, or of PQ. Draw SR, and parallel to SR through the point P draw PM meeting OR, in M; and through M draw ML parallel to RP; through L, LN parallel to PQ; and join MN.

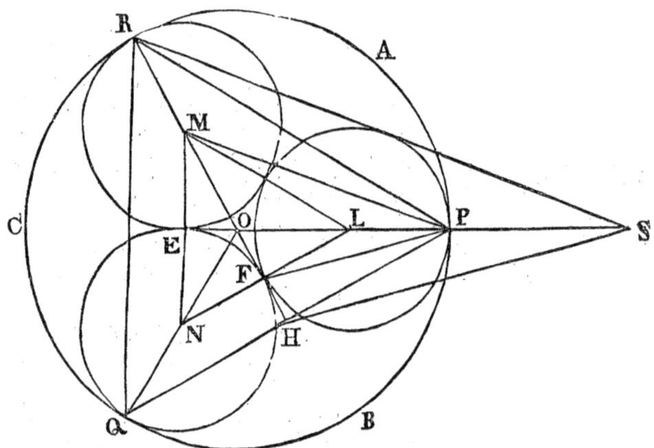

L, M, N are the centres of three circles that shall touch one another, and the circumference of the given circle ABC.

For bisect PQ in H, and join SH; and parallel to SH draw PF. Because POQ is an isoceles triangle, and that, LN is parallel to the base, PL is equal to QN: And because SH, PF are parallel, (PH, LF being also parallel,) and that, SP, PH are equal, PL, LF are equal. In the same manner it may be proved that QN is equal to NF, to NE, to EM, to MR, &c. Moreover it is evident that, NF, FL are in the same straight line.

Putting, therefore, a = radius of the given circle, and x = radius of one of the inscribed circles; It is, (*because* NF $= \frac{1}{2}$ MN, *and, by similar triangles* OE $= \frac{1}{2}$ OM), $(a - x)^2 - \left(\frac{a - x}{2}\right)^2 = x^2$.

Whence $x = -3a \pm \sqrt{12a^2} = 2a\sqrt{3} - 3a$. *Which was required.*

PROBLEM XXIX.

Given the base and difference of the sides to determine the triangle, when the rectangle of the longest side and difference of the segments of the base is equal to the square of the shortest side.

Let $b =$ the base, $x =$ the shorter side, $d =$ the difference of the sides; then (by Geom. *p.* 36.) as $b : 2x + d :: d : \frac{d}{b}(a + 2x) =$ the difference of the segments of the base, and by the question, this $\times$ by the longer side, or $\frac{d}{b}(d + 2x).(d + x) = x^2$; and reduced we have $x^2 - \frac{3d^2x}{b - 2d} = \frac{d^3}{b - 2d}$, or, $x = \sqrt{\left\{\frac{d^3}{b - 2d} + \frac{36d^4}{(b - 2d)^2}\right\}} \pm \frac{6d^2}{b - 2d}$

PROBLEM XXX.

When a parish was inclosed, the allotment of one of the proprietors consisted of two pieces of ground; one of which was in the form of a right-angled triangle; the other was a rectangle, one of the sides of which was equal to the hypothenuse of the triangle, the other to half the greater side: but, wishing to have his land in one piece, he exchanged his allotments for a square piece of ground of equal area, one side of which equalled the greater of the sides of the triangle which contained the right angle. By this exchange, he found that he had saved 10 poles of railing. What are the respective areas of the triangle and rectangle; and what is the length of each of their sides;

Let $2x =$ the greater side of the triangle, and $y =$ the less; that is $\sqrt{\{4x^2 + y^2\}} =$ the hypothenuse; and also the greater side of the rectangle, and $x =$ the less side of the rectangle; $\therefore xy =$ the area of the triangle, and $x\sqrt{(4x^2 + y^2)} =$ the area of the rectangle; $\therefore 4x^2 = xy + x\sqrt{(4x^2 + y^2)}$, or $4x - y = \sqrt{(4x^2 + y^2)}$; also $8x + 10 = 2x + y + \sqrt{(4x^2 + y^2)} + 2x + 2\sqrt{(4x^2 + y^2)}$, or $4x + 10 = y + 3\sqrt{(4x^2 + y^2)}$; in which equation substituting the value of $\sqrt{(4x^2 + y^2)}$ found above; $\therefore 4x + 10 = y + 3(4x - y) = 12x - 2y$, that is by transposition, $2y = 8x - 10$, and $y = 4x - 5$; $\because$ from the first equation, $5 = \sqrt{\{4x^2 + (4x - 5)^2\}}$, and $25 = 4x^2 + 16x^2 - 40x + 25$; by transposition $40x = 20x^2$; that is $2 = x$ and $y = 4x - 5 = 3$; the sides of the triangle are 3, 4, and 5; the sides of the rectangle are 2 and 5; and the area of the triangle and rectangle are 6, and 10 respectively.

PROBLEM XXXI.

In an oblique-angled plane triangle; there is given the difference of the sides which includes the angle of 71° 10,′ equal to 11, and the line that bisects the said angle is equal to 24; from whence is required a theorem that will determine the base and sides of the said triangle. The figure can be supplied by the reader.

Let ABC be the triangle, and make CH = CA the shorter side, draw AH, and it will cut the bisecting line CD at right angles in P; make PE = PD and EH = DH = AD, then will the triangles CEH and CDB be similar. Put CD = 24 = b, HB = 11 = d, the triangle BCD = ACD = 35° 35′, whose cosine call q, and let x = CH = CA. Then as rad. $(=1) : x :: q : qx =$ PC; Hence $2qx - b =$ EC; then by similar triangles $2qx - b : b :: x : x + d =$ CB; therefore $bx = 2qx^2 - bx + 2dqx - bd$, this equation being solved will give $x = \sqrt{\left(\frac{d^2q^2 + b^2}{4q^2}\right)} - \frac{dq + b}{2q} = 25.00218$; and hence we get AB $= 36.60737$, and BC $= 36.00218$.

PROBLEM XXXII.

Given the base, the perpendicular, and the ratio, of the two sides of a triangle; to find the sides.

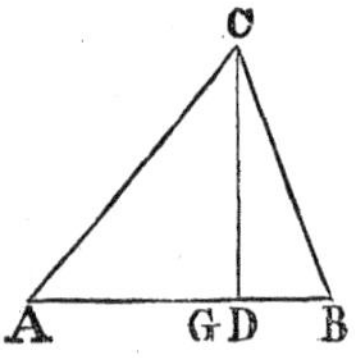

Call the base AB $= a$; the perpendicular CD, $= b$; and AD $= x$; and let the given ratio of AC to CB be that of m to n. Then BD $= a - x$; AC^2 = CD^2 + AD^2 $= b^2 + x^2$, and by (Euclid I. 47). BC^2 = CD^2 + BD^2 $= b^2 + a^2 - 2ax + x^2$ and therefore $m^2 : n^2 :: b^2 + x^2 : b^2 + a^2 - 2ax + x^2$. Multiplying extremes and means, we obtain $m^2b^2 + m^2a^2 - 2m^2ax + m^2x^2 = n^2b^2 + n^2x^2$.

That is, $x^2 - \frac{2am^2}{m^2 - n^2} x = -b^2 - \frac{a^2m^2}{m^2 - n^2}$, which equation resolved,

gives $x = \frac{am^2}{m^2 - n^2} \pm \sqrt{\left\{\frac{a^2m^4}{(m^2 - n^2)^2} - \frac{a^2m^2}{m^2 - n^2} - b^2.\right\}}$

$r = \sqrt{\left(b^2 + \left(\frac{am^2}{m^2 - n^2} \pm \sqrt{\left\{\frac{a^2m^4}{(m^2 - n^2)^2} - \frac{a^2m^2}{m^2 - n^2} - b^2\right\}}\right)^2.\right)}$ and

$s = \frac{n}{m}\sqrt{\left(b^2 + \left(\frac{am^2}{m^2 - m^2} \pm \sqrt{\left\{\frac{a^2m^4}{(m^2 - n^2)^2} - \frac{a^2m^2}{m^2 - n^2} - b^2\right\}}\right)^2.\right)}$

Let a represent the base, x one of the segments of the base by the perpendicular, r the side of the triangle adjacent to x, s the other side of the triangle, b the perpendicular; and $m : n$ the ratio of $r : s$. Then $a - x$ is the other segment of the base. Also $x^2 + b^2 = r^2$, and $x^2 + b^2 + a^2 - 2ax = s^2$. But $m^2 : n^2 :: (x^2 + b^2) : (x^2 + b^2 + a^2 - 2ax)$ Therefore $(m^2 - n^2)x^2 - 2am^2x = (n^2 - m^2)\, b^2 - a^2m^2$ as before.

PROBLEM XXXIII.

To find the area of a plane triangle, when two of its sides and the included angle are given.

Let ABC be the triangle of which the area is required; BC, AC, the given sides and C the given angle; from A draw AD perpendicular to BC, or BC produced; then by trigonometry *p.* 681.

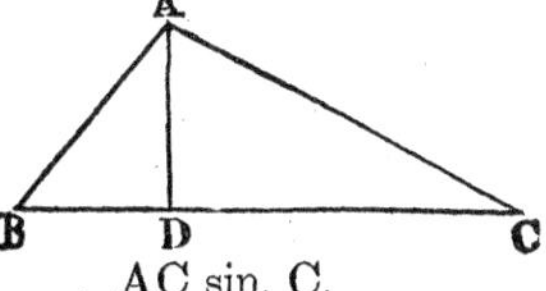

as R : sin. C : : AC : AD, and therefore $AD + \frac{AC \text{ sin. } C.}{R}$;

and $\frac{1}{2} AD \times BC = \frac{AC.BC. \text{ sin. } C}{2R}$ = the area of the triangle.

PROBLEM XXXIV.

In a right-angled plane triangular field, the legs are $3x^x$ and x^{3x}; and the line that bisects the right angle $= x^{3x}$ chains: what is the content in acres?

Put $y = x^x$; then $3y = 3x^x = BC$, $y^2 = x^{2x} = CD$, and $y^3 = x^{3x} = AC$. Let fall the perpendicular DE and DF; then it is evident (per question) that the ∠ DCA = the ∠ DCB, and the sides $FC = CE = ED = DF = \sqrt{\left(\frac{y^4}{2}\right)} = \frac{y^2}{\sqrt{2}}$. Put the $\sqrt{2} = m$; then by similar triangles, as AF : DF :: AC : BC; that is as $y^3 - \frac{y^2}{m} : \frac{y^2}{m} :: y^3 . 3y$; therefore $3y^4 - \frac{3y^3}{m} = \frac{y^5}{m}$, or $3y - \frac{3}{m} = \frac{y^2}{m}$; whence by completing the square, &c. we find $y = 3.346065$, $3y = 10.038195 = BC$, $y^2 = 11.196152 = DC$ and $y^3 = 37.463053 = BC$ and the content $= \frac{1}{2} \{BC \times AC =\} (\frac{1}{2} 4y^4 = 2y^4 = 188.030715$ chains.

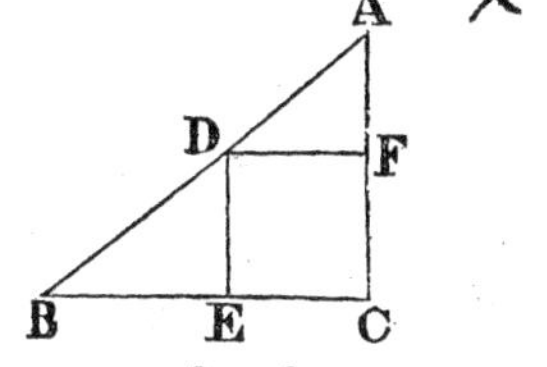

PROBLEM XXXV.

The area of a right-angled triangle, whose sides are in arithmetical progression, being given equal to 216; to determine the triangle.

Call the least side $x - y$, and let the common difference be y, then the three sides will be $x - y$, x, and $x + y$. Now by the nature of a right-angled triangle we have $(x - y)^2 + x^2 = (x + y)^2$, and by question the area $= \frac{1}{2} x (x - y) = a^2 = 216$. From the former equation we get $2x^2 - 2xy + y^2 = x^2 + 2xy + y^2$ and by reduction $4y = x$, substituting this value for x in the second equation we obtain $2y \times 3y = a^2$, from which $y = \sqrt{\frac{1}{6}}\, a^2$, and $x = 4 \sqrt{\frac{1}{6}} a^2$, consequently the 3 sides are $3 \sqrt{(\frac{1}{6} a^2)} = 18$; $4 \sqrt{(\frac{1}{6} a^2)} = 24$; and $5 \sqrt{(\frac{1}{6} a^2)} = 30$.

PROBLEM XXXVI.

To produce a given straight line (a) so that the rectangle under the given line and the whole line produced, may be equal to the square of the part produced.

Let $x =$ the part produced, then per question $(x + a)\ a = x^2$ or $x^2 - ax = a^2$ or by completing the square $x^2 - ax + \frac{1}{4}a^2 = \frac{5}{4}a^2$ and by extracting the root $x - \frac{1}{2}a = \frac{1}{2}\sqrt{(5)}$, or $x = \frac{1}{2}a\ (\sqrt{5} + 1) =$ the part required.

PROBLEM XXXVII.

Find the side of an equilateral and equiangular dodecagon in a circle whose radius $= r$.

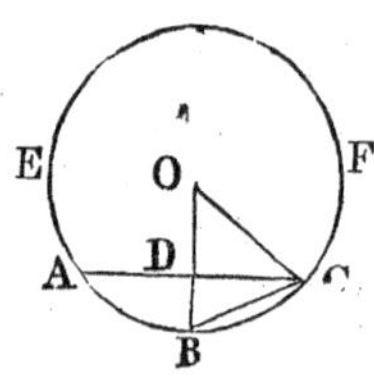

Let EBF denote the given circle, and apply the radius from A to C then AC = the side of a regular hexagon inscribed in the circle (Legendre's Geometry Art. 271.) bisect the arc AC in B, and join BC, and BC is the side of the regular dodecagon sought; Join BO and it bisects AC perpendicularly, (Legendre Art. 106.) Join OC, then since $DC = \frac{1}{2}OC$ and $OC^2 = OD^2 + DC^2$ (Legendre 186.) $\therefore OC^2 = \frac{1}{4}OC^2 + OD^2$ or $OD = \frac{1}{2}(OC\sqrt{3})$ and $BD = OC - OD = \frac{1}{2}OC\ (2 - \sqrt{3})$ $\therefore$ (since $OC = r$,) $DC = \frac{1}{2}r$, $DB = \frac{1}{2}r\ (2 - \sqrt{3})$ and $BC^2 = DC^2 + BD^2 = r^2 (2 - \sqrt{3}.)$ $\therefore BC = r\sqrt{(2 - \sqrt{3})} =$ the line sought.

PROBLEM XXXVIII.

There is a triangular piece of ground whose area = 525 square yards, and two of the sides measure 30 and 42 yards respectively find the remaining side.

Let $p =$ the perpendicular to the side 42 from its opposite angle, then $\frac{1}{2}\{p \times 42\} =$ the area $= 525$ (Legendre art. 176.) $\therefore p = 25$ yards, and $30^2 - 25^2 = 5^2 \times 11$ or $5\sqrt{11} =$ the segment of the side 42 adjacent to the angle formed by the given sides, and $42 - (5\sqrt{11})$ = the remaining segment of the side 42, this segment and p are the legs of a right angled triangle, of which the side sought is the hypothenuse, hence if $x =$ the side sought, there results the equation, $x^2 = (42 - 5\sqrt{11})^2 + p^2 = (42 - 5\sqrt{11})^2 + 25^2 = 2^2(666 - 105\sqrt{11})$ and $x = 2\sqrt{(666 - 105\sqrt{11})} = 35.65$ yards nearly.

NOTE.—It is evident that there is another triangle which answers this question, whose remaining side being denoted by y will be found by the equation $y = 2\sqrt{\{666 + 105\sqrt{11}\}}$, in this case the perpendicular (p) falls without the triangle. See Prob. VIII. p. 23.

PROBLEM XXXIX.

The four sides of a field, whose diagonals are equal to each other, are 25, 35, 31, and 19 poles respectively: what is the area.

Since the sums of the squares of the opposite sides of the trapezium ABCO are equal (by the question), the diagonals will cut each other at right angles in D; so that putting BC $= 25 = a$, BA $= 35 = b$, OA $= 31 = c$, and CO $= 19 = d$, and AC $=$ BO $= x$, we shall have $x : b + a :: b - a :$ AD $-$ CD $= (b^2 - a^2) \div x$; and $x : b + c :: b - c :$ BD $-$ OD $= (b^2 - c^2) \div x$; whence AD $= \frac{x}{2} + \frac{b^2 - a^2}{2x}$, and BD $= \frac{x}{2} + \frac{b^2 - c^2}{2x}$; and; $\therefore (\frac{x}{2} + \frac{b^2 - a^2}{2x})^2 + (\frac{x}{2} + \frac{b^2 - c^2}{2x})^2 = b^2$; from which equation $2x^4 - \{a^2 + c^2\} \times 2x^2 + \{b^2 - a^2\}^2 + \{b^2 - c^2\} = 0$, this solved gives $x = \frac{1}{2}(a^2 + c^2) \pm \sqrt{\{b^2d^2 - \frac{1}{4} \times (c^2 - a^2)^2\}} = 37.9$, and the area $= \frac{1}{2}x^2 = \frac{1}{2}(37.9)^2$ as required.

PROBLEM XL.

In a right-angled triangle, there are given the ratio of the sides as 3 to 4, and the difference between the area of its inscribed circle and inscribed square $= 20.2825$. Required the area and sides of the triangle.

Let $20.2825 = a$, $3x =$ the perpendicular, $4x =$ the base, and $5x =$ the hypothenuse; also $2x =$ circle's diameter; Put $y =$ the side of the inscribed square; therefore $4x - y =$ the base minus the side of the inscribed square, and by similar triangle we have as $4x : 3x :: 4x - y : y$; $\therefore 4xy = 12x^2 - 3xy$, or $7xy = 12x^2$. therefore $y = \frac{12}{7}x$; and by the question $\frac{(2x)^2}{49} = a$; and $x = 10$, and the sides are 30, 40 and 50, respectively.

PROBLEM XLI.

There is a triangular field, whose content is known to be $= 15$ acres, 2 roods, and 16 perches; the perimeter 78 chains; and one of the angles 126° 52′ 12″. It is required to find the sides of the field separately, by a general theorem, that may be of use to the practical surveyor.

Put $a =$ area of the triangle $= 156$ square chains, $r =$ the sum of all the sides $= 78$ chains, and s and q for the sine and cosine of half the sum of the unknown angles respectively; also let $\frac{s}{q} = t$; then is the cosine of half the difference of the said unknown angles $=$

$\frac{4aqt + qz^2}{z^2 - 4at} = .9911223 = 7° 38' 25''$, which being added to and subtracted from the half sum, gives 34° 12′ 19″, and 18° 55′ 29″ for the two unknown angles. Now the longest side AC $= \frac{z^2 - 4at}{2z} =$ $= 37$ chains; consequently AB $= 26$, and BC$=15$. Or if it be put equal to the cotangent of half the given ∠ ABC, then will $\frac{z}{2} - \frac{2at}{z}$ $=$ HC $= 37$ chains as before. This question may likewise be solved by finding the radius of the inscribed circle, which is equal to the area divided by the perimeter.

PROBLEM XLII.

There is a triangular garden, the length of whose sides are 200, 198, and 178 yards. Now there is a dial so placed in the garden, that, if walks be made from each of the three angles to the dial, they will exactly divide the said garden into three equal parts: from whence is required the length of each walk.

Let ABD be the proposed angular garden, in which having the sides given I find (by plane trig. *p.*   ), the angle BAD$=53° 7' 48''$ (whose sine call s and cosine c), the perpendicular Cd = 52.8, and $Ch = 53\frac{1}{3}$. Now $b = Cd = 52.8$, $Ch = 53\frac{1}{3}$, x and y = sine and cosine of ∠ CAd; then $sy - cx$ = sine ∠ CAh, and as $x : b :: 1 : \frac{b}{x}$ =AC; also as $1 : \frac{b}{x} :: sy - cx : \frac{bsy - bcx}{x} = Ch = d$; therefore $bcx + dx = bsy$; that is $\frac{x}{y} = \frac{bs}{ba + d}$ $=.496875$, the natural tangent of the angle CAD $= 26° 25' 18''$, fere; from hence is found CD $=$ 107.5838; AC $= 118.6587$, and BC $= 106.3425$, the length of each walk as required.

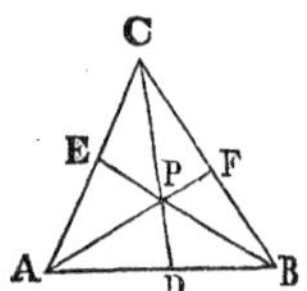

PROBLEM XLIII.

In any right-angled triangle, the area ($= 294$), and the difference between the hypothenuse and perpendicular ($= 14$), being given; to find all the other parts of a triangle by a simple equation.

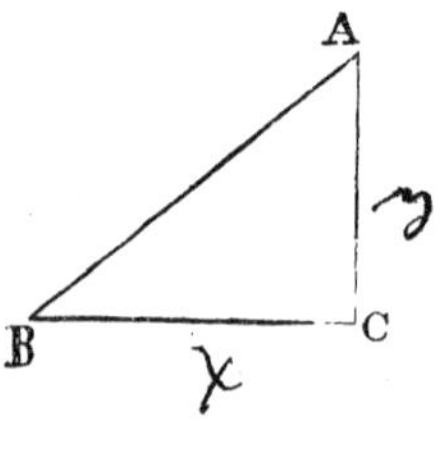

Given AB — AC $= 14 = a$, and BC × AC $= 588 = b$, put x and y for the sine and cosine of the ∠ at B; then (By trigon. *p.*   ) I have $1 - x : a :: x : \frac{ax}{1-x}$ = BC, and $1 - x :$ $a :: y : \frac{ay}{1-x}$ = AC: $\therefore \frac{a^2xy}{(1-x)^2} = b$; consequently $\frac{(1-x)^2}{xy} = \frac{1}{3}$, the tangent of half the angle BAC; $\therefore$ the side AC = 21, BC = 28, and BA = 35.

PROBLEM XLIV.

In a plane triangle ABC, there is given the sides AC and BC equal to 24 and 30 poles respectively; and supposing a circle inscribed in the same, so as to touch all its sides, a line drawn from the angle C, to the centre thereof is found to measure 12 poles. Find, the base of the triangle by a simple equation.

Let AC $= q = 24$, BC $= d = 30$, CO $= m = 12$, AC + BC $= n = 54$ and AB $= y$; then $n : y :: q : \text{AD} = \frac{qy}{n}$, and $:: d : \text{DB} = \frac{dy}{n}$, $q : m :: q + \frac{qy}{n} : \text{CD} = m + \frac{my}{n}$; Hence we have $dq - \frac{dqy^2}{n^2} = (m + \frac{my}{n})^2$, which reduced gives

$$y = \frac{(dq - m^2)n}{dq + m^2} = 36, \text{ the base as required.}$$

PROBLEM XLV.

Having the perpendiculars let fall from each angle of a triangle to the opposite sides given, to find the opposite sides and area.

Here is given the perpendicular BE$=100=m$, AF$=98=n$, CD$=95=a$; let $x=$AB, then ax =twice the triangle's area, therefore AC$=\frac{ax}{m}$; and BC$=\frac{ax}{n}$; per (trig. $p.$) x : $\left(\frac{a}{n}+\frac{a}{m}\right) \times x :: \left(\frac{a}{n}-\frac{a}{m}\right) \times x : \left(\frac{a^2}{n^2}-\frac{a^2}{m^2}\right) \times x =$ DB $-$ AD; hence DB $= \left(\frac{1}{2}+\frac{a^2}{2n^2}-\frac{a^2}{2m^2}\right) \times x$, but $\frac{ax}{n} : 1 ::$ DB : cosine $\angle$ B $= \frac{n}{2a}+\frac{a}{2n}-\frac{an}{2m^2} = .5349834$, the cosine of $57^\circ\ 39'\ 27'' = \angle$ B; hence AB $= 115.9951$, BC $= 112.4443$, AC $=110.1953$, and the area $= 5509.765$.

Otherwise algebraically.

Put BE $= b = 21$, AF $= c = 20$, CD $= d = 19$, AB $= x$, AD $= z$, then by similar triangle (Euclid I. 47.) equating the different processes, completing the square, &c. we get

$$x = \sqrt{\left(\frac{4b^4c^2d}{2b^4c^2d^2 + 2b^2c^2d^4 + 2b^2c^4d^2 - 2b^4d^4 - c^4d^4}\right)} = 24.34.$$

PROBLEM XLVI.

From the vertex A of a triangle, draw a straight line meeting the base produced in D, so that the rectangle BD.DC $= AD^2$.

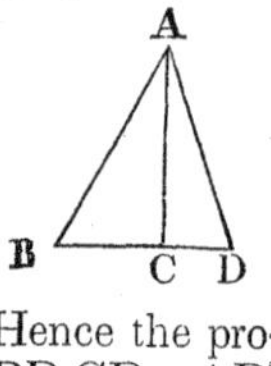

Let ABC be the given triangle, whose base is BC, and let AB be greater than AC, then make the angle CAD = ABC and produce BC to intersect AD in D, then is AD the line which was to be drawn. For since the angle D is common to the triangle ADB, ADC, and because the angle B=CAD the remaining angles of those triangles are equal, (Legendre 74.) Hence the proportion CD : AD : : AD : BD (Legendre 202.) Or BD.CD=AD^2 as required. CD is easily calculated, for put a = AB, c = AC, b = CB, then because the triangle ABD, ADC are similar there results the proportion $AB^2 : AC^2 : : BD^2 : AD^2 = BD.CD$, or $AB^2 : AC^2 : :$ DB : CD and by division of proportion $AB^2 - AC^2 : AC^2 : :$ BC : CD, or in symbols we have $a^2 - c^2 : c^2 : : b : CD$, $\therefore CD = \frac{bc^2}{a^2 - c^2}$ *Ans.*

Note, that $BD = BC + CD = b + \frac{bc^2}{a^2 - c^2} = \frac{ba^2}{a^2 - c^2} =$ the whole side produced.

THEOREM XLVII.

In any quadrilateral figure whose diagonals intersect at right angles, if S.D be respectively the sum and difference of two opposite sides, and S, d the sum and difference of the other sides, then $S + d : s + D : : s - D : S - d$. Required the proof.

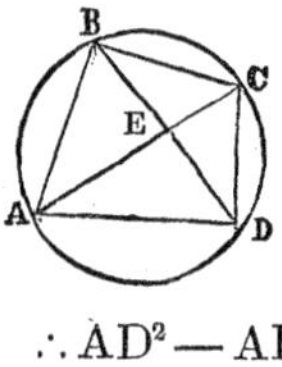

Let ABCD denote any quadrilateral whose diagonals AC, BD intersect at right angles in F and suppose that AD is greater than AB then $AD^2 = AE^2 + DE^2$ and $AB^2 = AE^2 + BE^2$ and by subtraction $AD^2 - AB^2 = DE^2 - BE^2$, in the same manner $DC^2 - CB^2 = DE^2 - BE^2$, $\therefore AD^2 - AB^2 = DC^2 - BC^2$. Put AD + BC=S, AD − BC=D, then AD=(S+D) ÷ 2 and BC=(S − D) ÷ 2; also put DC+AB=s, DC − AB=d, then as before DC =($s + d$) ÷ 2, AB = ($s - d$) ÷ 2, by substituting these values in the equation $AD^2 - AB^2 = DC^2 - CB^2$ it becomes $(S + D)^2 - (s - d)^2 = (s + d)^2 - (S - D)^2$, or by reduction $S^2 + D^2 = s^2 + d^2$ or $S^2 - d^2 = s^2 - D^2$, or $(S + d).(S - d) = (s + D).(s - D)$ which gives $S + d : s + D : : s - D : S - d$ as required.

PROBLEM XLVIII.

Having given the perimeter (p) of a rhombus, and the sum (s) of its two diagonals, to find the diagonals

Since a rhombus is a parallelogram whose sides are all equal, we have $\frac{1}{4}p$= to one of the sides, put x=one of the diagonals then $s - x$

=the other; hence (Legendre 195.) $(s-x)^2+x^2=4\left(\frac{1}{4}p\right)^2$, or by reduction $2x^2-2sx=\frac{1}{4}p^2-s^2$ and $x^2-sx=\frac{1}{8}(p^2-4s^2)$ $\because$ by quadratics $x-\frac{s}{2}=\frac{\pm\sqrt{(p^2-2s^2)}}{2\sqrt{2}}$ and $x=\frac{1}{2\sqrt{2}}\left(s\sqrt{2}\pm(p^2-2s^2)^{\frac{1}{2}}\right)$ hence $s-x=\frac{1}{2\sqrt{2}}\left(s\sqrt{2}\mp\sqrt{(p^2-2s^2)}\right)$ as required.

PROBLEM XLIX.

Given the area (a^2) of a right angled triangle, whose sides are in geometrical progression to find the sides.

Let x, y be the two legs $y > x$, then $\sqrt{\{y^2+x^2\}}$ = the hypothenuse and (per question.) $\sqrt{\{y^2+x^2\}}:y::y:x$ or $x\sqrt{\{y^2+x^2\}}=y^2$ and $x^2y^2+x^4=y^4$ also $xy=2a^2$, that is $x^2y^2=4a^4$, and $y^4=\frac{16a^8}{x^4}$, hence $4a^4+x^4=\frac{16a^8}{x^4}$ or $x^8+4a^4x^4=16a^8$ and by quadratics $x^4+2a^4=\sqrt{\{20a^8\}}=2a(5)^{\frac{1}{4}}$ or $x=a\sqrt{(2\sqrt{5}-2)^{\frac{1}{4}}}$ then $y=\frac{2a^2}{x}=\frac{2a}{(2\sqrt{5}-2)^{\frac{1}{4}}}=a(2\sqrt{5}+2)^{\frac{1}{4}}$ as required.

Otherwise, call the perpendicular x, then the base will be $\frac{2a^2}{x}$, and the hypothenuse $\sqrt{\left(x^2+\frac{4a^4}{x^2}\right)}$. And by the question $\frac{2a^2}{x}:x::x:\sqrt{\left(x^2+\frac{4a^4}{x^2}\right)}$. Hence $x^2=\frac{2a^2}{x}\sqrt{\left(x^2+\frac{4a^4}{x^2}\right)}$, and squaring we have the final equation, $x^4=\frac{4a^4}{x^2}\left(x^2+\frac{4a^4}{x^2}\right)$. or $x^8-4a^4x^4=16a^8$; completing the square and extracting the square root we have, $x^4=a^4\{2+\sqrt{20}\}$, and $x=a(2+\sqrt{20})^{\frac{1}{4}}$ as before.

PROBLEM L.

Having given the base (b) of a plane triangle, its area (a^2) and the ratio of the two sides as $m:n$. Required the values of these sides.

Draw a perpendicular from the vertical angle to the base then $\frac{2a^2}{b}$ = the perpendicular, let x = the lesser segment of the base, then $\sqrt{\left\{\frac{4a^4}{b^2}+x^2\right\}}$ = the lesser of the remaining sides, and $\sqrt{\left\{\frac{4a^4}{b^2}+(b-x)^2\right\}}$ =the greater; hence the proportion $m^2:n^2::\frac{4a^4}{b^2}+x^2:$

$\frac{4a^4}{b^2}+(b-x)^2$ which gives $x^2-\frac{2bm^2x}{m^2-n^2}=\frac{4a^4(n^2-m^2)-b^4m^2}{b^2(m^2-n^2)}$, this quadratic solved gives x, and thence the sides become known.

PROBLEM LI.

Having given the base (b), the area (a^2), and (c) the difference of the two sides, to find the sides and the perpendicular altitude of the triangle.

By using the same notation, &c., as in the last question the perpendicular $=\frac{2a^2}{b}$, the lesser side $=\sqrt{\left\{\frac{4a^4}{b^2}+x^2\right\}}$ and the greater $=\sqrt{\left\{\frac{4a^4}{b^2}+(b-x)^2\right\}}$ $\therefore$ (per question.) $\sqrt{\left\{\frac{4a^4}{b^2}+x^2\right\}}+c$ $=\sqrt{\left\{\frac{4a^4}{b^2}+(b-x)^2\right\}}$ or $2c\sqrt{\left\{\frac{4a^4}{b^2}+x^2\right\}}+c^2=b^2-2bx$, or $2c\sqrt{\left\{\frac{4a^4}{b^2}+x^2\right\}}=b^2-c^2-2bx$ or $\frac{16a^4c^2}{b^2}+4c^2x^2=b^4+c^4+$ $4b^2x^2-4b^3x+4c^2bx-2b^2c^2$ or $4(b^2-c^2)x^2-4b(b^2-c^2)x=$ $\frac{16a^4c^2}{b^2}-(b^2-c^2)^2=\frac{16a^4c^2-b^2(b^2-c^2)^2}{b^2}$ and $4x^2-4bx=$ $\frac{16a^4c^2-b^2(b^2-c^2)^2}{b^2(b^2-c^2)}$, hence $x=\frac{1}{2}b-\frac{c}{2b}\sqrt{\left\{\frac{16a^4+b^2(b^2-c^2)}{b^2-c^2}\right\}}$ then $\sqrt{\left\{\frac{4a^4}{b^2}+x^2\right\}}=\sqrt{\left(\frac{4a^4}{b^2-c^2}+\frac{1}{4}b^2\right)}-\frac{1}{2}c=$ the lesser side and $\sqrt{\left\{\frac{4a^4}{b^2}+(b-x)^2\right\}}=\sqrt{\left(\frac{4a^4}{b^2-c^2}+\frac{1}{4}b^2\right)}+\frac{1}{2}c=$ the greater side as required.

PROBLEM LII.

Having the lengths of three perpendiculars, EF, EG, EH, drawn from a certain point E, within an equilateral triangle ABC, to its three sides, to determine the sides.

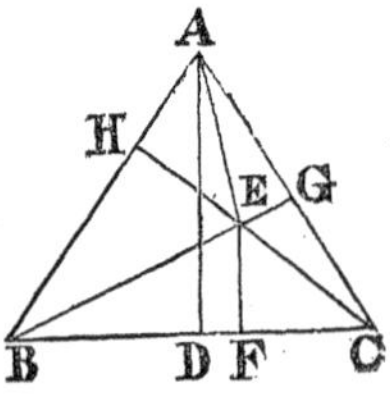

Draw the perpendicular AD, and having joined EA, EB, and EC, put EF $=a$, EG $=b$, EH $=c$, and BD (which is $\frac{1}{2}$BC) $=x$. Then, since AB, BC, or CA, are each $=2x$, we shall have, by Euc. I, 47, AD $=\sqrt{(\text{AB}^2-\text{BD}^2)}$ $=\sqrt{(4x^2-x^2)}=\sqrt{3x^2}=x\sqrt{3}$.

And because the area of any plane triangle is equal to half the rectangle of its base and perpendicular, it follows that

$$\triangle\,\text{ABC}=\tfrac{1}{2}\text{BC}\times\text{AD}=x\times x\sqrt{3}=x^2\sqrt{3}$$
$$\triangle\,\text{BEC}=\tfrac{1}{2}\text{BC}\times\text{EF}=x\times a\quad=ax,$$

$\triangle$ AEC $= \frac{1}{2}$AC $\times$ EG $= x \times b \quad = bx$,
$\triangle$ AEB $= \frac{1}{2}$AB $\times$ EH $= x \times c \quad = cx$.

But the last three triangles BFC, AEC, AEB, are together equal to the whole triangle ABC; whence $x^2 \sqrt{3} = ax + bx + cx$.

And consequently, if each side of this equation be divided by x, we shall have $x \sqrt{3} = a + b + c$, or $x = (a + b + c) \div \sqrt{3}$.

Which is, therefore half the length of either of the three equal sides of the triangle.

Cor. Since, from what is above shown, AD is $= x \sqrt{3}$, it follows, that the sum of all the perpendiculars, drawn from any point in an equilateral triangle to each of its sides, is equal to the whole perpendicular of the triangle.

PROBLEM LIII.

Having given the lengths of two lines AE, DC, drawn from the acute angles of a right-angled triangle ABC, to the middle of the opposite sides, it is required to determine the triangle.

Put CD $= a$, AE $= b$, BD or $\frac{1}{2}$ AB $= x$, and BE or $\frac{1}{2}$BC $= y$; then, since (Euc. I. 47,) BD^2 + BC^2 = CD^2 and BE^2 + AB^2 = AE^2, we shall have $x^2 + 4y^2 = a^2$, and $y^2 + 4x^2 = b^2$.

Whence, taking the second of these equations from four times the first, there will arise $15y^2 = 4a^2 - b^2$ or $y = \sqrt{(4a^2 - b^2) \div 15}$.

And, in like manner, taking the first of the same equations from four times the second, there will arise $15x^2 = 4b^2 - a^2$, or $x = \sqrt{(4b^2 - a^2) \div 15}$.

Which values of x and y are half the lengths of the base and perpendicular of the triangle, observing that b must be less than $2a$ and greater than $\frac{1}{2}a$.

PROBLEM LIV.

Having given the ratio of the two sides of a plane triangle ABC, and the segments of the base, made by a perpendicular falling from the vertical angle, to determine the triangle.

Put BD $= a$, DA $= b$, BC $= x$, AC $= y$, and the ratio of the sides as m to n.

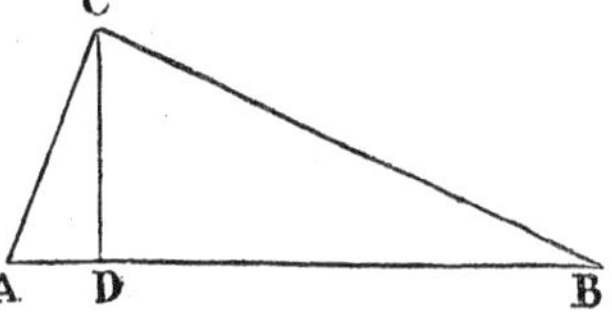

Then, since by the question, BC : AC :: $m : n$, and by (Geom. p. 36) CB^2 — AC^2 = BD^2 — DA^2, we shall have $x : y :: m : n$, and $x^2 - y^2 = a^2 - b^2$.

But, by the first of these expressions, $nx = my$, or $y = nx \div m$; whence, if this be substituted for y in the second, there will arise, $x^2 - (n^2 \div m^2)\, x^2 = a^2 - b^2$, or $(m^2 - n^2)\, x^2 = m^2 (a^2 - b^2)$.

And consequently, by division and extracting the square root, we shall have $x = m\sqrt{\dfrac{a^2 - b^2}{m^2 - n^2}}$ and $y = n\sqrt{\dfrac{a^2 - b^2}{m^2 - n^2}}$; which are the values of the two sides BC, AC, of the triangle, as was required.

PROBLEM LV.

Given the hypothenuse of a right-angled triangle ABC, and the sides of its inscribed square DE, to find the other two sides of the triangle.

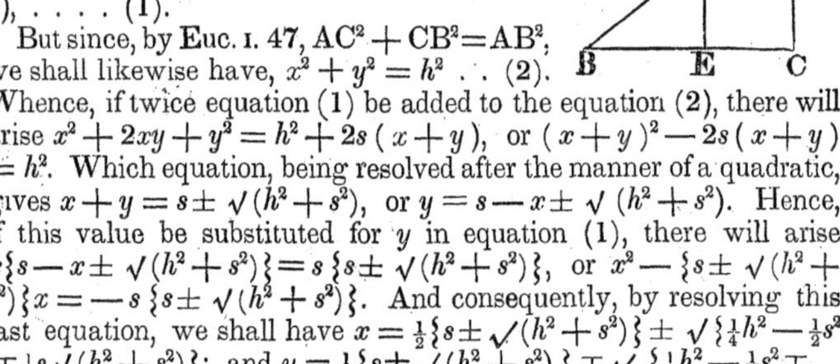

Put AB $= h$, DF or DE $= s$, AC $= x$, and CB $= y$; then, by similar triangles, we shall have AC (x) : CB (y) : : AF $(x - s)$: FD (s). And consequently, by multiplying the means and extremes, $xy - sy = sx$, or $xy = s\,(x + y)$, ... (1).

But since, by Euc. I. 47, $AC^2 + CB^2 = AB^2$, we shall likewise have, $x^2 + y^2 = h^2$ . . (2). Whence, if twice equation (1) be added to the equation (2), there will arise $x^2 + 2xy + y^2 = h^2 + 2s\,(x + y)$, or $(x + y)^2 - 2s\,(x + y) = h^2$. Which equation, being resolved after the manner of a quadratic, gives $x + y = s \pm \sqrt{(h^2 + s^2)}$, or $y = s - x \pm \sqrt{(h^2 + s^2)}$. Hence, if this value be substituted for y in equation (1), there will arise $x\{s - x \pm \sqrt{(h^2 + s^2)}\} = s\,\{s \pm \sqrt{(h^2 + s^2)}\}$, or $x^2 - \{s \pm \sqrt{(h^2 + s^2)}\}x = -s\,\{s \pm \sqrt{(h^2 + s^2)}\}$. And consequently, by resolving this last equation, we shall have $x = \frac{1}{2}\{s \pm \sqrt{(h^2 + s^2)}\} \pm \sqrt{\{\frac{1}{4}h^2 - \frac{1}{2}s^2 \mp \frac{1}{2}s\sqrt{(h^2 + s^2)}\}}$; and $y = \frac{1}{2}\{s \pm \sqrt{(h^2 + s^2)}\} \mp \sqrt{\{\frac{1}{4}h^2 - \frac{1}{2}s^2 \mp \frac{1}{2}s\sqrt{(h^2 + s^2)}\}}$; which are the values of the perpendicular AC and base BC, as was required.

PROBLEM LVI.

Find the radius of that circle in which the side of a regular pentagon is 1.

Let $r =$ the radius, then $\sqrt{(r^2 - \frac{1}{4})} =$ the perpendicular from the centre of the circumscribed circle to the side of the pentagon.

Hence $\sqrt{\{\frac{1}{4} + (r - \sqrt{(r^2 - \frac{1}{4})})^2\}} =$ the side of the inscribed decagon, but (Legendre 273.) $r : \sqrt{\{\frac{1}{4} + (r - \sqrt{(r^2 - \frac{1}{4})})^2\}} :: \sqrt{\{\frac{1}{4} + (r - \sqrt{(r^2 - \frac{1}{4})})^2\}} : r - \sqrt{\{\frac{1}{4} + (r - \sqrt{(r^2 - \frac{1}{4})})^2\}}$ put $\sqrt{\{\frac{1}{4} + (r - \sqrt{(r^2 - \frac{1}{4})})^2\}} = x$, then $r : x :: x : r - x$ or $x^2 = r^2 - rx$ and $x^2 + rx = r^2$ $\therefore x = \frac{1}{2}r(\sqrt{5} - 1) = \sqrt{\{\frac{1}{4} + (r - \sqrt{(r^2 - \frac{1}{4})})^2\}}$ or $\frac{1}{2}r^2(3 - \sqrt{5}) = \frac{1}{4} + (r - \sqrt{(r^2 - \frac{1}{4})})^2 = \frac{1}{4} + r^2 - 2r\sqrt{\{r^2 - \frac{1}{4}\}} + r^2 - \frac{1}{4}$ and by reduction $r\,(1 + \sqrt{5}) = 4\sqrt{(r^2 - \frac{1}{4})}$ or $r^2(3 + \sqrt{5}) = 8r^2 - 2$ and $r^2 = \dfrac{2}{5 - \sqrt{5}} = \dfrac{5 + \sqrt{5}}{10}$ or $r = \sqrt{\left(\dfrac{5 + \sqrt{5}}{10}\right)}$ as required.

PROBLEM LVII.

Given the lengths of two chords, cutting each other at right angles in a circle, and the distance of the point of their intersection from the centre, to determine the diameter of the circle.

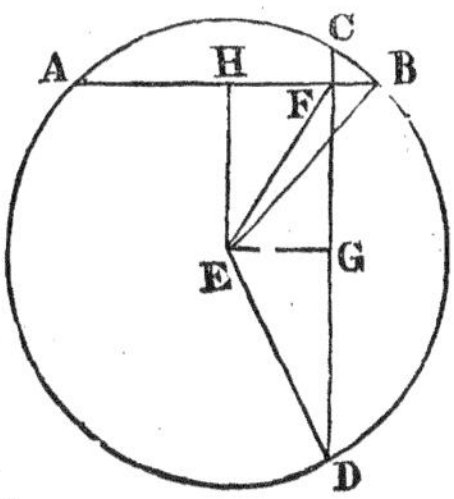

Let AB, CD be two chords within the circle ABD, cutting each other at right angles in F. Take the centre E, and draw EG, EH respectively perpendicular to CD, AB. Then CD, AB are bisected in the points G, H. Join EB, ED, EF. Let AB $= a$, CD $= b$, EF $= m$, and $x =$ EB or ED. Then $EB^2 = EH^2 + HB^2$, or $x^2 = EH^2 + \frac{a^2}{4}$

Also $ED^2 = EG^2 + GD^2$, or $x^2 = HF^2 + \frac{b^2}{4}$ $\therefore 2x^2 = EH^2 + HF^2 + \frac{1}{4}(a^2 + b^2) = m^2 + \frac{1}{4}(a^2 + b^2)$ $\therefore x = \left\{ \frac{m^2}{2} + \frac{1}{8}(a^2 + b^2) \right\}^{\frac{1}{2}}$

.. $2x$, or the diameter $= \sqrt{\{2m^2 + \frac{1}{2}(a^2 + b^2)\}}$.

PROBLEM LVIII.

Given the ratio of the two sides, together with both the segments of the base, made by a perpendicular from the vertical angle; to determine the sides of the triangle.

Put a and b for the two segments of the base, x for the side adjacent to a, and y for the other side of the triangle; and $m : n$, the ratio x has to y. Then, (since $m : n :: x : y$) it follows that $y = \frac{n}{m} x$. But $x^2 - a^2 = y^2 - b^2 = \frac{n^2x^2}{m^2} - b^2$. That is, $m^2x^2 - n^2x^2 = m^2a^2 - m^2b^2$. Hence $x = \sqrt{\left\{\frac{m^2a^2 - m^2b^2}{m^2 - n^2}\right\}}$, And $y = \frac{n}{m}\sqrt{\left\{\frac{m^2a^2 - m^2b^2}{m^2 - n^2}\right\}}$ which were to be determined.

PROBLEM LIX.

Given the difference of the segments of the base made by a line bisecting the vertical angle, the difference of the sides, and the difference of their squares, to construct the triangle.

Let a, b, and c, be the three differences as per quest., then $c \div b =$ the sum of the sides, and $\{b^2 + c\} \div 2b$, and $\{c - b^2\} \div 2b$ are the two sides; $b : a ::$ the greater side: the greater segment $= \frac{b^2 + c}{2b^2} \times a$, and $\frac{c - b^2}{2b^2} \times a =$ the less, hence the base $= \frac{ca}{2b^2}$, and thus we have theorems for all the sides.

PROBLEM LX.

Given the area, or measure of the space, of a rectangle, inscribed in a given triangle; to determine the sides of the rectangle

If a be put for the perpendicular, and b for the base of the triangle; c for the area of the rectangle; and x for the length of the rectangle parallel to the base of the triangle, it will be as $b : a :: x : ax \div b$ the difference of the altitudes of the rectangle and triangle

Therefore the altitude of the rectangle is $a - \frac{a}{b}x$.

But $c = x(a - \frac{a}{b}x) = ax - \frac{a}{b}x^2$, $\therefore x = \frac{1}{2}b \pm \sqrt{\left\{\frac{b^2}{4} - \frac{b}{a}c\right\}}$ the length of the rectangle; consequently $\frac{c}{\frac{1}{2}b \pm \sqrt{\left\{\frac{b^2}{4} - \frac{b}{a}c\right\}}}$ = the breadth; which were to be determined.

PROBLEM LXI.

The base of a right-angled triangle being given $= 24$, and the difference between the perpendicular and hypothenuse $= 12$; required the other sides.

Put $24 = a$, $12 = d$, and the perpendicular $= x$; then the hypothenuse $= x + d$, and by 47 Euclid, Book I. we have $a^2 + x^2 = (x + d)^2$, that is $a^2 + x^2 = x^2 + 2dx + d^2$ and $x = \frac{a^2 - d^2}{2d}$, and $x + d = \frac{a^2 + d^2}{2d}$, Substituting for a and d their values we find the perpendicular $= 18$, and hypothenuse $= 30$.

PROBLEM LXII.

The area of a certain isosceles triangular field is eight acres and a half; find the sides and angles, when the area of the inscribed circle is equal half the field.

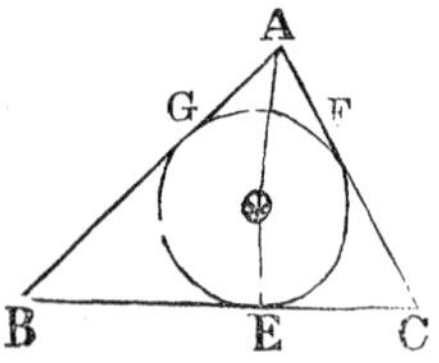

A

G F

B E C

Put $1360 = 2a$, = the area of the triangle in perches; $29.424 = b$, = the dia. of the inscribed circle; draw GF parallel to BC, and put $x =$ BC; then $\frac{4a}{x} =$ AE; $\therefore \frac{4a - bx}{x}$ = AO; and by similar triangles, as $\frac{4a}{x} : x :: \frac{4a - bx}{x} : \frac{4ax - bx^2}{4a}$ = GF. $\therefore \frac{4ax^2 - bx^3}{4a} = b^2$; which brought out of fractions, and reduced, gives $x^3 - 92.44x^2 = -80033.28$.

By converging sines x is found $= 38.5296 =$ BC, and $70.566 =$ AE. Then (by trig. *p.*) $\angle A = 15° 16'$, and $\angle B = \angle C = 74° 44'$.

PROBLEM LXIII.

To determine a right angled triangle; having given the hypothenuse, and the radius of the inscribed circle.

Construct a right angled triangle, and inscribe in it a circle; also draw right lines from the centre to the points of contact, and to the two acute angles, as in the figure ABCDEFG.

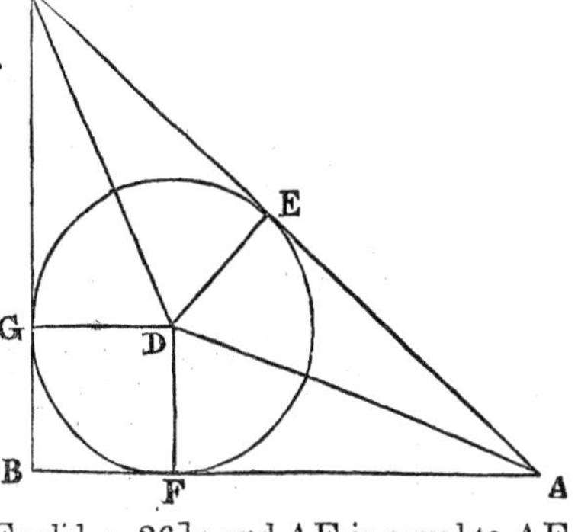

Put x for the base (AB), y for the perpendicular (BC) and let a represent the given radius (DG, DE, or DF), b the hypothenuse (AC.) Then, because GB and DF are equal, $y - a$ is the expression for CG; and $x - a =$ AF. Also CE and CG are equal [Euclid, I. 26]; and AF is equal to AE. But AE + CE = AC; that is, $(y - a) + (x - a) = b$, Or $x + y = 2a + b$. Now $x^2 + y^2 = b^2$; comparing, therefore, the double of this equation, with the square of the preceding, $x^2 - 2xy + y^2 = b^2 - 4ab - 4a^2$. Hence $x - y = \sqrt{\{b^2 - 4ab - 4a^2,\}}$ consequently $x = a + \frac{1}{2}b \pm \frac{1}{2}\sqrt{(b^2 - 4ab - 4a^2)}$, and $y = a + \frac{1}{2}b \mp \frac{1}{2}\sqrt{(b^2 - 4ab - 4a^2)}$.

PROBLEM LXIV.

To find the side of an equilateral triangle, inscribed in a circle, whose diameter is d; and that of another circumscribed about the same circle.

Let AFGC be the given circle, and ABC the required inscribed equilateral triangle.

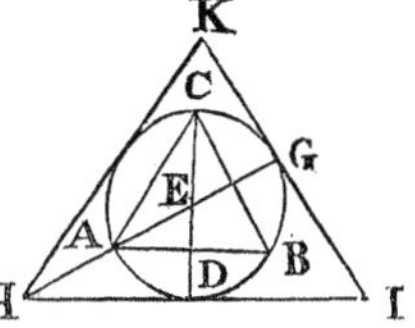

Join A and the centre E; also join CE, and produce it to D. Then by (Euc. IV. 2.) the angle D is a right angle, and the triangles ADE and ADC are similar. But AD $= \frac{1}{2}$AC, therefore also DE $= \frac{1}{2}$ AE.

Let then AE = radius $= \frac{1}{2}d$; and consequently ED $= \frac{1}{2}$AE $= \frac{1}{4}d$; also put AB $= x$, or = AD $= \frac{1}{2}x$. Then by (Euc. I. 47.) $\frac{1}{4}x^2 = \frac{1}{4}d^2 - \frac{1}{16}d^2 = \frac{3}{16}d^2$; Whence $x = \sqrt{\frac{3}{4}d^2} = \frac{1}{2}d\sqrt{3}$, the side of the inscribed triangle. Again produce CD to F, and AE both ways to G and H; and draw HI, IK, perpendicular to EF and EG.

Then it is obvious from the proposition above referred to, and (Euc.

IV. 3.) that $EF = \frac{1}{2}HE$, and $HF = \frac{1}{2}HI$. Let now HI (the side of the circumscribed triangle) $= y$ and we shall have $HE^2 = EF^2 + HF^2$; and since $EF = \frac{1}{2}d$, $HE = d$, and the above becomes $d^2 = \frac{1}{4}d^2 + \frac{1}{4}y^2$, or $y^2 = 3d^2$, or $y = d\sqrt{3}$, the side of the circumscribing triangle. (See Prob. xix. p. 31.)

PROBLEM LXV.

To describe a circle through two given points A, B, that shall touch a right line CD given in position.

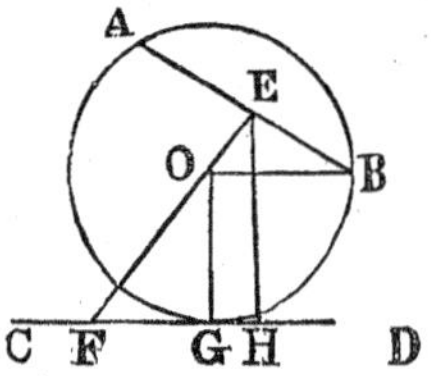

Join AB; and through O, the assumed centre of the required circle, draw FE perpendicular to AB; which will bisect it in E (Euc. III. 3). Also join OB; and draw EH, OG, perpendicular to CD; the latter of which will fall on the point of contact G (Euc. III. 18). Hence, since, A, E, B, H, F, are given points, put $EB = a$, $EF = b$, $EH = c$, and $EO = x$; which will give $OF = b - x$. Then, because the triangle OEB is right angled at E, we shall have $OB^2 = EO^2 + EB^2$; or $OB = \sqrt{(x^2 + a^2)}$. But, by similar triangles, $FE : EH :: FO : OG$ or OB, or $b : c :: b - x : OB$; whence, also, $OB = \frac{c}{b}(b - x)$. And consequently, if these two values of OB be put equal to each other, there will arise, $\sqrt{(x^2 + a^2)} = \frac{c}{b}(b - x)$.

Or, by squaring each side of this equation, and simplifying the result, $(b^2 - c^2)x^2 + 2bc^2x = b^2(c^2 - a^2)$.

Which last equation, when resolved in the usual manner, gives,

$x = -\frac{bc^2}{b^2 - c^2} + b\sqrt{\left\{\frac{c^4}{(b^2 - c^2)^2} + \frac{c^2 - a^2}{b^2 - c^2}\right\}}$, for the distance of the centre O from the chord AB; where b must evidently be greater than c, and c greater than a.

PROBLEM LXVI.

The three lines AO, BO, CO, drawn from the angular points of a plane triangle ABC, to the centre of its inscribed circle, being given, to find the radius of the circle, and the sides of the triangle.

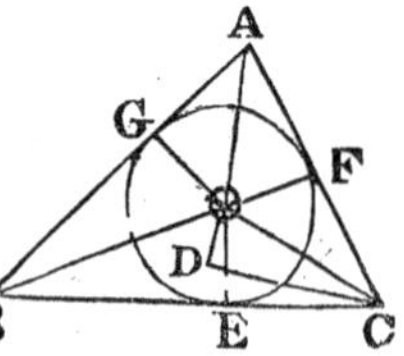

Let O be the centre of the circle, and on AO produced, let fall the perpendicular CD; and draw OE, OF, OG, to the points of contact E, F, G. Then, because the three angles of the triangle ABC are, together, equal to two right angles, (Euc. I. 32.) the sum of their halves $OAC + OCA + OBE$ will be

equal to one right angle. But the sum of the two former of these, OAC + OCA, is equal to the external angle DOC; whence the sum of DOC + OBE, as also of DOC + OCD, is equal to a right angle; and consequently, OBE = OCD. Let therefore, AO = a, BO = b, CO = c, and the radius OE, OF, or OG = x. Then, since the triangles BOE, COD, are similar; BO : OE : : CO : OD, or $b : x :: c :$ OD; which gives OD $= \frac{cx}{b}$, and CD $= \sqrt{(c^2 - \frac{c^2x^2}{b^2})}$, or $\frac{c}{b}\sqrt{(b^2 - x^2)}$. Also, because the triangle AOC, is obtuse angled at O, we shall have (Euc. II. 12.) AC^2 = AO^2 + CO^2 + 2AO × OD; or AC $= \sqrt{(a^2 + c^2 + \frac{2acx}{b})}$, or $\sqrt{(\frac{b(a^2+c^2)+2acx}{b})}$. But the triangles ACD, AOF, being similar, AC : CD : : AO : OF, or $\sqrt{(\frac{b(a^2+c^2)+2acx}{b})} : \frac{c}{b}\sqrt{(b^2 - x^2)} :: a : x$. Whence, multiplying the means and extremes, and squaring the result, there will arise $bx^2 \{b(a^2 + c^2) + 2acx\} = a^2c^2(b^2 - x^2)$. Or, by collecting the terms together, and dividing by the coefficient of the highest power of x, we have the equation, $x^3 + \left(\frac{ab}{2c} + \frac{ac}{2b} + \frac{bc}{2a}\right)x^2 = \frac{abc}{2}$

From which last equation x may be determined, and thence the side of the triangle.

PROBLEM LXVII.

From one of the extremities A, of the diameter of a given semicircle ADB, to draw a right line AE, so that the part DE, intercepted by the circumference and a perpendicular drawn from the other extremity, shall be of a given length.

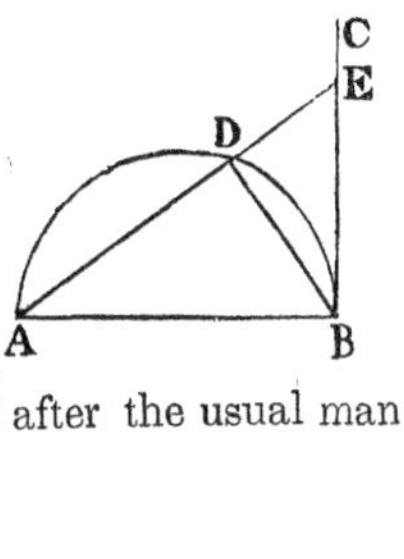

Let the diameter AB = d, DE = a, and AE = x; and join BD. Then, because the angle ADB is a right angle, (Euc. III. 31.) the triangles ABE, ABD, are similar. And consequently, by comparing their like sides, we shall have AE : AB : : AB : AD, or $x : d :: d : x - a$. Whence, multiplying the means and extremes of these proportionals, there will arise $x^2 - ax = d^2$. Which equation, being resolved after the usual manner, gives $x = \frac{1}{2}a + \sqrt{(\frac{1}{4}a^2 + d^2)}$.

PROBLEM LXVIII.

A gentleman has a triangular piece of land, whose sides are in the proportion of 3, 4, and 5, and the area of it is equal to the cube of one-fifth part of the base; required the sides and area in numbers.

The given proportion denotes a right-angled triangle; let therefore $3x$ be the base, then $4x$ and $5x$ will be the hypothenuse and perpendicular, and, by the question, the cube of three-fifths of the base $= \{\frac{3}{5}x\}^3 = \frac{1}{2}\{3x \times 4x\}$, whence $x \times 3 = 83\frac{1}{3} =$ the base, $111\frac{1}{9} =$ the perp. and $139\frac{8}{9} =$ the hypoth. the area being $4629\frac{17}{27}$. If $4x$ be called the base, then the three sides will be, $35\frac{15}{32}$, $46\frac{7}{8}$, and $58\frac{19}{32}$, area $823\frac{499}{512}$; but 18, 24, and 30, area 216, if the hypoth. be base. And had the triangle been oblique, an equation might have been formed and the sides found, with the same ease.

PROBLEM LIX.

To find the area and the sides of a rectangle of which the perimeter and the diagonal are given.

Let ABCD be the proposed rectangle call the perimeter p, and the diagonal BD, d.

D C

A B

Let AB $= x$, AD $= y$, and we shall have $x + y = \frac{1}{2}p$. and $x^2 + y^2 = a^2$. Squaring the first equation, we obtain $x^2 + 2xy + y^2 = \frac{1}{4}p^2$, and therefore $(x^2 + 2xy + y^2) - (x^2 + y^2) = 2xy = \frac{1}{4}p^2 - a^2$. Consequently xy, or the area $= \frac{1}{8}pp - \frac{1}{2}a^2$, and $x = \text{AB} = \frac{1}{4}p + \surd(\frac{1}{2}a^2 - \frac{1}{16}p^2)$, and $y = \text{AD} = \frac{1}{4}p - \surd(\frac{1}{2}a^2 - \frac{1}{16}p^2)$.

PROBLEM LX.

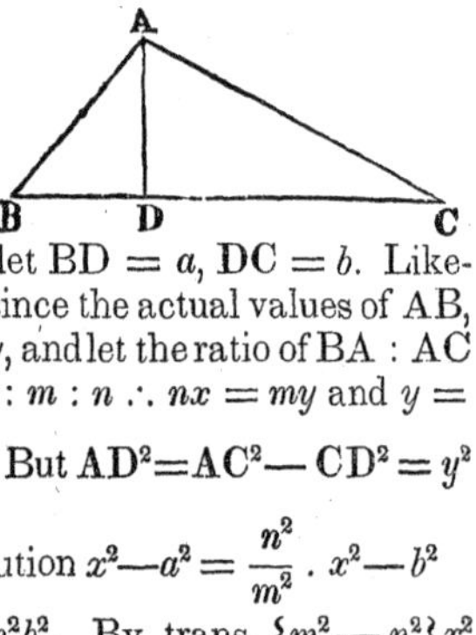

Having given the segments of the base, made by a perpendicular falling from the vertical angle, and likewise the ratio of the two sides, to determine the triangle; *i. e.* to determine the actual value of the sides of the triangle.

A

B D C

In this problem, BD, DC are given; $\therefore$ let BD $= a$, DC $= b$. Likewise the *ratio* of BA : AC is given: but since the actual values of AB, AC are unknown, let BA$=x$, and AC$=y$, and let the ratio of BA : AC be that of $m : n$. Now BA : AC :: $x : y :: m : n \therefore nx = my$ and $y = \frac{n}{m}x$. Also $\text{AD}^2 = \text{AB}^2 - \text{BD}^2 = x^2 - a^2$, But $\text{AD}^2 = \text{AC}^2 - \text{CD}^2 = y^2 - b^2$, $\therefore x^2 - a^2 = y^2 - b^2$, And by substitution $x^2 - a^2 = \frac{n^2}{m^2} \cdot x^2 - b^2$ $\times$ by m^2, and $m^2x^2 - m^2a^2 = n^2x^2 - m^2b^2$, By trans. $\{m^2 - n^2\}x^2 = m^2\{a^2 - b^2\} \therefore x = m\sqrt{\left\{\frac{a^2 - b^2}{m^2 - n^2}\right\}} =$ the value of AB.

By a similar process $y = n.\sqrt{\left\{\frac{a^2 - b^2}{m^2 - n^2}\right\}} =$ the value of AC.

PROBLEM LXXI.

Given all the three sides of the triangle, to find the radius of the inscribed circle.

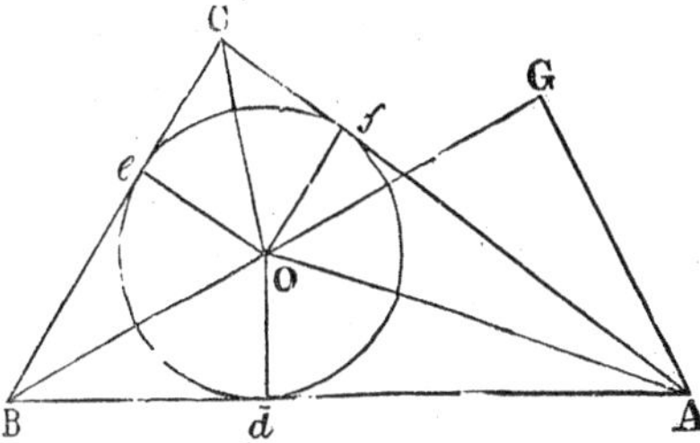

Conceiving the figure constructed (in the accompanying diagram) draw lines from the centre to each of the angles, and to the points of contact. Produce either of the lines joining the angular points and centre (as BO) indefinitely through the opposite side, and on it produced let fall a perpendicular (as AG) from that angular point from which the perpendicular falls without the triangle.

Put a for Ad, b for dB, and c for Cf. Also x for Od = Oe = Of.

Then $\sqrt{\{b^2+x^2\}}$ [BO] : x [Od] : : $a+b$ [AB] : $\frac{ax+bx}{\sqrt{\{b^2+x^2\}}}$ [AG].

But BO : Bd : : AB : BG. that is, $\frac{ab+b^2}{\sqrt{\{b^2+x^2\}}} - \sqrt{\{b^2+x^2\}} = \frac{ab-x^2}{\sqrt{\{b^2+x^2\}}}$. And AG : OG : : C$f$: Of, whence $ax+bx : ab-x^2 : : c : x$. Therefore $x = \sqrt{\left(\frac{abc}{a+b+c}\right)}$. Which was required.

PROBLEM LXXII.

To determine a right-angled triangle; having given the hypothenuse, and the difference of two lines drawn from the two acute angles to the centre of the inscribed circle.

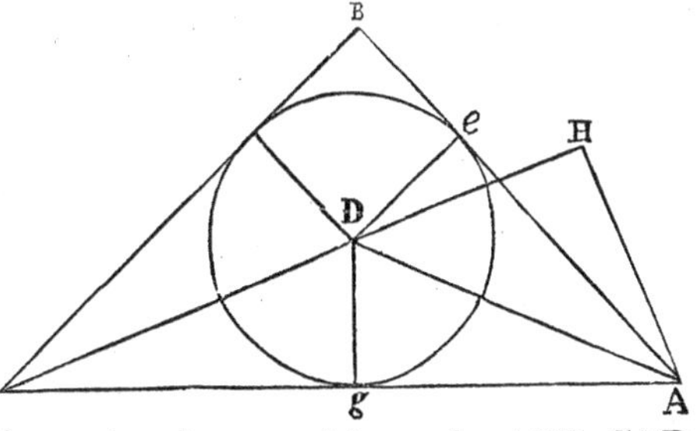

In the annexed fig. let CD be the greater, and AD the less of the two lines of which the difference is given, and let DH be a production of CD, and AH perpendicular to CH, AH is equal to HD, because the angle ADH is equal to the sum of the angles ACD, CAD, together equal to half a right angle, and the angle at H a right-angle.

If a be put for AC the hypothenuse, x for CD, y for AD, b for the difference of x and y, r for DH, and s for AH. Then (Geom. *p.* 34) $r = s = y \div \sqrt{2}$. But $x^2 + y^2 + 2rx = a^2 = x^2 + y^2 + xy\sqrt{2}$.

Now substituting $x - b$ for y, and c for the $\sqrt{2}$, it will be $x^2 - 2bx + b^2 + x^2 + cx^2 - cbx = a^2$, that is, $(2+c)\,x^2 - (2+c)\,bx + b^2 = a^2$. Whence $x = \frac{1}{2}b \pm \sqrt{\left\{\frac{a^2 - b^2}{2+c} + \frac{1}{4}b^2\right\}}$, and

$y = \sqrt{\left\{\frac{a^2-b^2}{2+c} + \frac{1}{4}b^2\right\}} - \frac{1}{2}b$. Consequently the radius of the inscribed circle is known, *and the triangle determined.* For put $m = \frac{1}{2}b \pm \sqrt{\left\{\frac{a^2-b^2}{2+c} + \frac{1}{4}b^2\right\}} = \text{CD}$. $n = \sqrt{\left\{\frac{a^2-b^2}{2+c} + \frac{1}{4}b^2\right\}} - \frac{1}{2}b = \text{AD}$. And let DE, Df, Dg, be three radii at right angles to the sides of the triangle; likewise put w for Ag, and z for Cg. $z^2 - w^2 = m^2 - n^2$. Also $z + w = a$; and by division $z - w = \frac{m^2 - n^2}{a}$, that is, $z = \frac{a^2 + m^2 - n^2}{2a}$; $w = \frac{a^2 - m^2 + n^2}{2a}$,

And $\text{Dg} = \sqrt{\left\{m^2 - \frac{(a^2 + m^2 - n^2)^2}{4a^2}\right\}}$ = *the radius of the inscribed circle.*

PROBLEM LXXIII.

Given the perpendicular, base, and sum of the sides of an obtuse angled plane triangle ABC, to determine the two sides of the triangle.

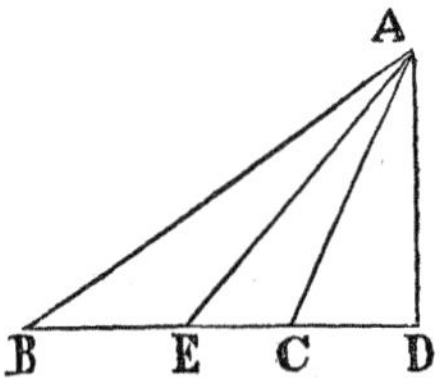

Let the perpendicular AD $= p$, the base BC $= b$, the sum of AB and AC $= s$, and their difference $= x$. Then, since half the difference of any two quantities added to half their sum, gives the greater, and, when subtracted, the less, we shall have, $\text{AB} = \frac{1}{2}(s + x)$, and $\text{AC} = \frac{1}{2}(s - x)$.

But, by (Euc. I. 47,) $\text{CD}^2 = \text{AC}^2 - \text{AD}^2$, or $\text{CD} = \sqrt{\{\frac{1}{4}(s-x)^2 - p^2\}}$; $\therefore$ by Geom. *p.* 37) $\text{AB}^2 = \text{BC}^2 + \text{AC}^2 + 2\text{BC} \times \text{CD}$; whence $\frac{1}{4}(s+x)^2 = b^2 + \frac{1}{4}(s-x)^2 + 2b\sqrt{\{\frac{1}{4}(s-x)^2 - p^2\}}$, or $sx - b^2 = 2b\sqrt{\{\frac{1}{4}(s-x)^2 - p^2\}}$ And if each of the sides of this last equation be squared, there will arise, by transposition and simplifying the result, $(s^2 - b^2)x^2 = b^2(s^2 - b^2) - 4b^2p^2$, or $x = b\sqrt{(1 - \frac{4p^2}{s^2 - b^2})}$. Whence, by addition and subtraction, we shall have, $\text{AB} = \frac{1}{2}s + \frac{1}{2}b\sqrt{(1 - \frac{4p^2}{s^2 - b^2})}$, and $\text{AC} = \frac{1}{2}s - \frac{1}{2}b\sqrt{(1 - \frac{4p^2}{s^2 - b^2})}$ as required.

THEOREM LXXIV.

If two chords in a circle intersect each other at right angles, the sum of the squares described upon the four segments, is equal to the square described upon the diameter.

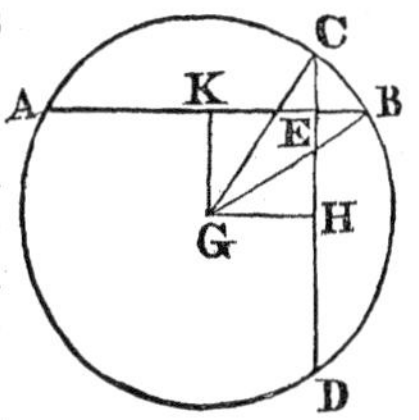

Let ADC be a circle; and let the two chords AB, CD cut each other at right angles in E. Find the centre G; join GC, GB; and draw GK, GH perpendicular to AB, CD. Then AK = KB and CH = HD. (Euc. III. 3.) Now because AB is divided equally in K and unequally in E, $\therefore AE^2 + EB^2 = 2AK^2 + 2KE^2$ (Euc. II. 9.) Also $DE^2 + EC^2 = 2DH^2 + 2HE^2$. $\therefore$ by addition $AE^2 + EB^2 +$

$$DE^2 + EC^2 = 2\,AK^2 + 2KE^2 + 2DH^2 + 2HE^2$$
$$= 2(BK^2 + KG^2) + 2(GH^2 + CH^2)$$
$$= 2\,BG^2 + 2GC^2$$
$$= 4\,GC^2 = (\text{diameter})^2.$$

PROBLEM LXXV.

Upon a given straight line as an hypothenuse, describe a right-angled triangle which shall have its three sides in continued proportion.

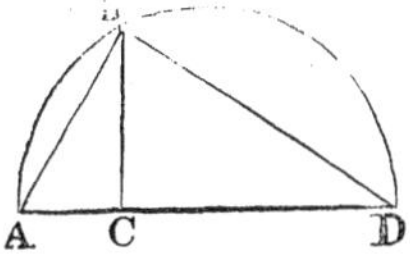

Let AD be the given straight line; upon it describe a semicircle ABD. Let AD $= a$, AC $= x$. Then AD : BD : : BD : DC. (Euc. VI. 8.) But by the question, AD : BD : : BD : AB, $\therefore$ AB=DC. Hence $\sqrt{\{ax = a - x\}}$ $\therefore ax = a^2 - 2ax + x^2$, or $x^2 - 3ax = -a^2$, Complete the square, and $x^2 - 3ax + \frac{9}{4}a^2 = \frac{5}{4}a^2$, Or $x - \frac{3}{2}a = \pm\frac{1}{2}(a\sqrt{5})$ $\therefore x = \frac{1}{2}\{3a \pm a\sqrt{5}\} = \frac{1}{2}a.\{3 \pm \sqrt{5}\}$ which determines the point C. Draw BC perpendicular to AD, join BA, AD, and the triangle ABD will have its sides in geometric progression

PROBLEM LXXVI.

Having given the perimeter of a right-angled triangle ABC, and the perpendicular CD, falling from the right angle on the hypothenuse, to determine the triangle.

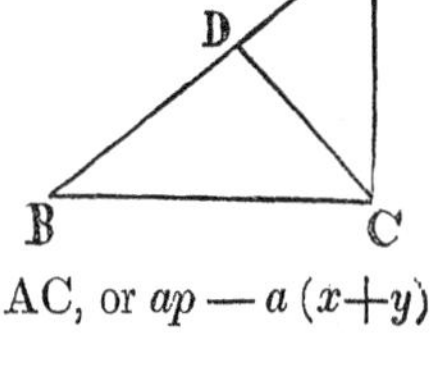

Put $p =$ perimeter, CD $= a$, AC $= x$, and BC $= y$; then AB $= p - (x + y)$. But, by right-angled triangles (Euc. I. 47,) $AC^2 + BC = AB^2$; whence $x^2 + y^2 = p^2 - 2p\,(x + y) + x^2 + 2xy + y^2$. Or, by transposing the terms and dividing by 2, $p(x+y) - \frac{1}{2}p^2 = xy$. . (1).

And since, by similar triangles, AB : BC : : AC : CD, we shall also have, by multiplying the means and extremes, AB $\times$ CD = BC $\times$ AC, or $ap - a\,(x+y)$

$= xy$. . . . (2). Whence, by comparing equation (1) with equation (2), there will arise $(a+p)\times(x+y) = ap+\frac{1}{2}p^2$.

Whence, $x+y=\frac{p(a+\frac{1}{2}p)}{a+p}$, or $y=\frac{p(a+\frac{1}{2}p)}{a+p}-x$.

And if these values be now substituted for $x+y$ and y in equation (2), the result, when simplified and reduced, will give $(a+p)x^2 - p(a+\frac{1}{2}p)\,x = -\frac{1}{2}ap^2$. From which last equation, and the value of y above found, we shall have,

$$x \text{ or AC} = \frac{p(a+\frac{1}{2}p)}{2(a+p)} \pm \frac{p}{2(a+p)}\sqrt{\{(a-\tfrac{1}{2}p)^2-2a^2\}}$$

$$\text{and } y \text{ or BC} = \frac{p(a+\frac{1}{2}p)}{2(a+p)} \mp \frac{p}{2(a+p)}\sqrt{\{(a-\tfrac{1}{2}p)^2-2a^2)\}}.$$

And if the sum of these two sides be taken from p, the result will give $AB = p-(x+y) = \frac{p^2}{2(a+p)}$. Which expressions are, therefore, respectively equal to the values of the three sides of the triangle.

PROBLEM LXXVII.

To find the side of a regular pentagon, inscribed in a circle, whose diameter is d.

It appears, from (Euc. IV. 10,) that the side of an equilateral and equiangular decagon inscribed in a circle, is found by dividing the radius of the circle into extreme and mean ratio, the greater part of which is the side of the decagon.

Hence calling the radius $OB = r$, and the greater part $OD = x$, we must have $r(r-x) = x^2$, or $x^2+rx=r^2$; whence $x = -\frac{1}{2}r+\frac{1}{2}\sqrt{5r^2}$, or $x = \frac{1}{2}r(-1+\sqrt{5})$, That is, BC or AB in the above figure $= \frac{1}{2}r(-1+\sqrt{5})$ Produce BO to E, and join EC; then by (Euc. I. 47,) $EC^2 = EB^2 - BC^2 = r^2(\frac{5}{2}+\frac{1}{2}\sqrt{5})$, or $EC = r\sqrt{(\frac{5}{2}+\frac{1}{2}\sqrt{5})}$, Again, as EB : EC : : EC : $ED = r(\frac{5}{4}+\frac{1}{4}\sqrt{5})$, and EB : BC : : BC : $BD = r(\frac{3}{4}-\frac{1}{4}\sqrt{5})$, whence $DC = \sqrt{(ED\times DB)} = \sqrt{\{r^2(\frac{5}{4}+\frac{1}{4}\sqrt{5})(\frac{3}{4}-\frac{1}{4}\sqrt{5})\}} = \frac{1}{4}r\sqrt{(10-2\sqrt{5},)}$ or $AC = \frac{1}{4}d\sqrt{\{10-2\sqrt{5}\}}$. the side of the pentagon required.

PROBLEM LXXVIII.

To find the side of a square, inscribed in a given semicircle, whose diameter is d.

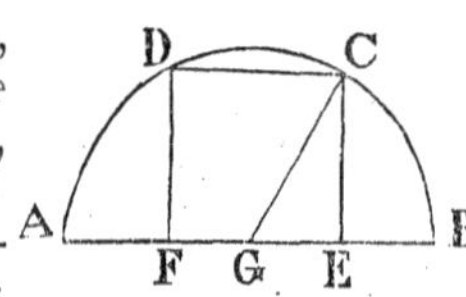

Let ABCD be the given semicircle: AB, its diameter: G, its centre; and CDFE, the required square. Then, since $DF = CE$, we have $FG = GE$. Let therefore $AB = d$ or $CG = \frac{1}{2}d$; also $CE = x$, and consequently $GE = \frac{1}{2}x$; then by (Euc. I. 47,) $CG^2 =$

$GE^2 + CE^2$, or $x^2 + \frac{1}{4}x^2 = \frac{1}{4}d^2$. Whence $5x^2 = d^2$, or $x = d\sqrt{\frac{1}{5}} = \frac{1}{5}d\sqrt{5}$.

Otherwise, let ADCB be a semicircle whose diameter $AB = a$. Suppose CDFE to be a square inscribed in the semicircle. Find G, the centre of the circle, and join GC. Let $GE = x$. Then $AE.BE = CE^2$ by (Euc. III. 35.) Or $(AG + GE).(GB - GE) = CE^2$ *i. e.* $\{\frac{1}{2}a + x\}.\{\frac{1}{2}a - x\} = CE^2$. Or $\frac{1}{4}a^2 - x^2 = CE^2$.

Now since FECD is a square, $CD = DF = FE = FG + GE = 2x$, $\because \frac{1}{4}a^2 - x^2 = 4x^2$, $x^2 + 4x^2 = \frac{1}{4}a^2$, or $5x^2 = \frac{1}{4}a^2$, $x = a\sqrt{\{\frac{1}{20}\}}$ $\therefore 2x = 2a\sqrt{\{\frac{1}{20}\}} = a\sqrt{\{\frac{4}{20}\}} = a\sqrt{\{\frac{1}{5}\}}$ = a side of the square.

PROBLEM LXXIX

The lengths of three lines drawn from the three angles of a plane triangle to the middle of the opposite sides, being 18, 24, and 30, respectively: it is required to find the sides.

Let ABC be the required triangle, and AF, BE, and CD, the three given lines bisecting the three sides CB, AC, and AB. Make $AF = a$, $BE = b$, $CD = c$, also $CB = x$, $AC = y$, and $AB = z$.

Now it is a well known property of triangles, that "double the square of a line drawn from any angle of a triangle to the opposite side, together with double the square of half that side, is equal to the sum of the squares of the other two sides;" that is $2a^2 + \frac{1}{2}x^2 = y^2 + z^2$, $2b^2 + \frac{1}{2}y^2 = x^2 + z^2$ and $2c^2 + \frac{1}{2}z^2 = x^2 + y^2$, Or $y^2 + z^2 - \frac{1}{2}x^2 = 2a^2$, $x^2 - \frac{1}{2}y^2 + z^2 = 2b^2$, and $x^2 + y^2 - \frac{1}{2}z^2 = 2c^2$. From whence, by taking the former of these equations from twice the sum of the two latter, there comes out $4x^2 + \frac{1}{2}x^2 = 2(2b^2 + 2c^2 - a^2)$; $\therefore x = \frac{2}{3}\sqrt{(2b^2 + 2c^2 - a^2)}$ In like manner $y = \frac{2}{3}\sqrt{(2a^2 + 2c^2 - b^2)}$, and $z = \frac{2}{3}\sqrt{(2a^2 + 2b^2 - c^2)}$; Where, by substituting the given values of a, b, and c, viz. $a = 18$, $b = 24$, $c = 30$, we have $x = 34.176$, $y = 28.844$, and $z = 20$, which are the sides required.

PROBLEM LXXX.

Given the base (194) of a plane triangle, the line that bisects the vertical angle (66), and the diameter (200) of the circumscribing circle, to find the other two sides.

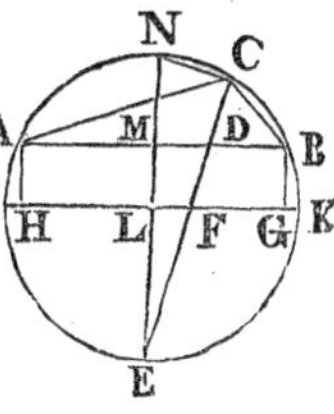

Let ABC be the proposed triangle, AB its base $= 194 = b$; IK the diameter of the circumscribing circle $= 200 = d$, drawn parallel to AB; and DC the bisecting line $= 66 = a$. Then we shall have $HI = GK = \frac{1}{2}(IK - AB) = 3$. $\therefore AH = GB = \sqrt{(IG \times GK,)} = \sqrt{197} \times 3 = \sqrt{591} = c$. Let now CD be produced to meet the circle in E; Then, because CD bisects the angle ACB, it will bisect the arc

AEB, and therefore the perpendicular ELM will pass through the centre L; $\therefore$ EM $= 100 + \sqrt{519} = e$ is also known, as is also MN $= 100 - \sqrt{519} = f$. Now let DE $= x$; then, since the two triangles ENC and MDE are similar, we have ME : DE : : CE : NE, or $e : x : : a + x : d$; whence $x^2 + ax = de$, or $x = -\frac{1}{2}a + \sqrt{(\frac{1}{4}a^2 + de)}$, which thus becomes known; and consequently the rectangle CD $\times$ DE, or $a \times \{-\frac{1}{2}a + \sqrt{(\frac{1}{4}a^2 + de)}\} = r$ is also known.

But this latter rectangle CD $\times$ DE $=$ AD $\times$ DB; therefore AD $\times$ DB, and AD $+$ DB, are both known. Assume now, AD $= y$, and BD $= z$, then we have $y + z = b$, $yz = r$, whence is readily found, $y = \frac{1}{2}b + \frac{1}{2}\sqrt{(b^2 - 4r^2)} = m$, and $z = \frac{1}{2}b - \frac{1}{2}\sqrt{(b^2 - 4r^2)} = n$, Again, calling AC $= v$, and BC $= u$, and we have $v : u : : m : n$, or $v = um \div n$; also $vu = a^2 + mn$. Substituting $um \div n$ for v in the second equation, gives $mu^2 \div n = a^2 + mn$.

Whence $u = \sqrt{(\frac{a^2n + mn^2}{m})}$, $\therefore v = \frac{um}{n} = \sqrt{(\frac{a^2m + nm^2}{n})}$, the sides required. The numerical values of which may be found by substituting those of a, b, and d, in the original equations.

PROBLEM LXXXI.

It is required to draw a right line BFE from one of the angles B of a given square BD, so that the part FE, intercepted by DE and DC, shall be of a given length.

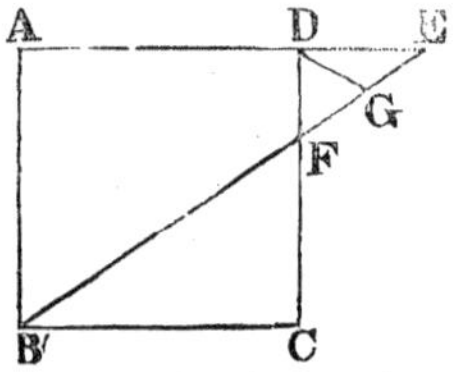

Bisect FE in G, and put AB or BC $= a$, FG or GE $= b$, and BG $= x$; then will BE $= x + b$ and BF $= x - b$. But since, by right-angled triangles, AE$^2 =$ BE$^2 -$ AB^2, we shall have AE $= \sqrt{\{(x + b)^2 - a^2\}}$. And because the triangles BCF, EAB, are similar, BF : BC : : BE : AE, or $a(x + b) = (x - b)\sqrt{\{(x + b)^2 - a^2\}}$. By squaring each side of this equation, and arranging the terms in order, there will arise $x^4 - 2(a^2 + b^2)x^2 = b^2(2a^2 - b^2)$. Which equation, being resolved after the manner of a quadratic, will give $x = \sqrt{\{a^2 + b^2 \pm a\sqrt{(a^2 + 4b^2)}\}}$ And consequently, by adding b to, or subtracting it from, this last expression, we shall have BE $= \sqrt{\{a^2 + b^2 \pm a\sqrt{(a^2 + 4b^2)}\}} + b$, or BF $= \sqrt{\{a^2 + b^2 \pm a\sqrt{(a^2 + 4b^2)}\}} - b$.

Which values, by determining the point E, or F, will satisfy the problem. Where it may be observed, that the point G lies in the circumference of a circle, described from the centre D, with the radius FG, or half the given line.

PROBLEM LXXXII.

In a plane triangle, there is given a base (50), the area (796), and the difference of the sides (10) to find the sides and the perpendicular.

Let ABC be the proposed triangle, of which the base AB is given, $= 50 = 2b$,

Then since the area is also given $= 796$, the perpendicular $= \frac{796}{25} = p$ is also known. Make now BE = half base $= b$, and CD $= p$, and ED $= x$; then BD $= b + x$, and AD $= b - x$: BC $= \sqrt{\{p^2 + (b+x)^2\}}$, and AC $= \sqrt{\{p^2 + (b-x)^2\}}$. Whence, calling the given difference $= 10 = d$, we have $\sqrt{\{p^2 + (b+x)^2\}} - d = \sqrt{\{p^2 + (b-x)^2\}}$, which equation, squared, gives $p^2 + (b+x)^2 - 2d\sqrt{\{p^2 + (b+x)^2\}} + d^2 = p^2 + (b-x)^2$; and this, by reduction, becomes $4bx + d^2 = 2d\sqrt{\{p^2 + (b+x)^2\}}$. And this, again squared, produces $16b^2x^2 + 8bd^2x + d^4 = 4d^2 \times (p^2 + b^2 + 2bx + x^2)$; Or $16b^2x^2 + d^4 = 4d^2 \times (b^2 + p^2) + 4d^2x^2$.

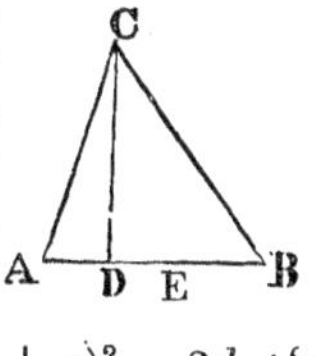

Whence $x = \sqrt{\left(\frac{4d^2 \times (b^2 + p^2) - d^4}{16b^2 - 4d^2}\right)}$. Where, by substituting the numeral values for b, p, and d, the answers for the sides will be found.

PROBLEM LXXXIII.

In a right-angled triangle given the hypothenuse 30, and the difference between the base and perpendicular $= 6$; required those two sides.

Let the hypothenuse $= 30 = a$, and the given difference $= b$; and call the base x, then the perpendicular $= x - b$. Now by Euclid, I. 47.,) $x^2 + (x-b)^2 = a^2$, whence $x^2 - bx = \frac{a^2 - b^2}{2}$, completing the square and extracting the root $x = \sqrt{(\frac{1}{2}a^2 - \frac{1}{2}b^2)} + \frac{1}{2}b$. Substituting for a and b their values, $x = 24$, = the base, and the perpendicular $x - b = 24 - 6 = 18$.

PROBLEM LXXXIV.

Given the three sides AB, BC, CD, of a trapezium ABCD, inscribed in a semicircle, to find the diameter, or remaining side AD.

Let AB $= a$, BC $= b$, CD $= c$, and AD $= x$; then, by (Euc. VI. Prop. D). AC $\times$ BD $=$ AD $\times$ BC $+$ AB $\times$ CD $= bx + ac$. But ABD, ACD, being right angles, (Euc. III. 31) we shall have AC $= \sqrt{(\text{AD}^2 - \text{DC}^2)}$, or $\sqrt{(x^2 - c^2)}$, and BD $= \sqrt{(\text{AD}^2 - \text{AB}^2)}$, or $\sqrt{(x^2 - a^2)}$. Whence, by substituting these two values in the former expressions, there will arise $\sqrt{(x^2 - c^2)} \times \sqrt{(x^2 - a^2)} = bx + ac$. Or, by squaring each side, and reducing the result, $x^3 - (a^2 + b^2 + c^2)x = 2abc$. From which last equation the value of x may be found, as in the last problem. (p. 52.)

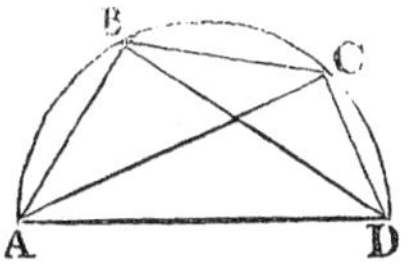

PROBLEM LXXXV.

Given the perpendicular (24), the line bisecting the base (40), and the line bisecting the vertical angle (25), to determine the triangle.

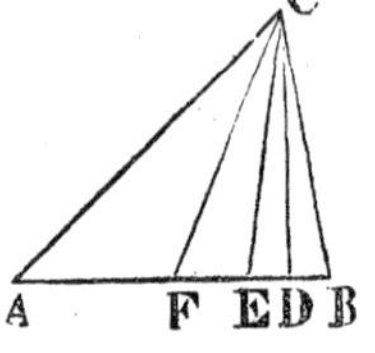

Let ABC be the proposed triangle, and make the perpendicular $CD = 24 = p$, CE the line bisecting the angle $ACB = 25 = b$, and CF, the line bisecting the base, $= 40 = c$. Then (Euc. I. 47.) $ED = \sqrt{(CE^2 - CD^2)} = 7 = m$, Also $FD = \sqrt{(FC^2 - CD^2)} = 32 = n$; And in order to simplify, let $EF = q$. Also let half the base $AF = FB = x$; then $AE = x + q$, $EB = x - q$; $AD = x + n$, $DB = x - n$;

Hence $AC = \sqrt{\{(x+n)^2 + p^2\}}$, $BC = \sqrt{\{(x-n)^2 + p^2\}}$. And from (Euc. VI. 3.), we have AC : BC :: AE : EB, or $\sqrt{\{(x+n)^2 + p^2\}} : \sqrt{\{(x-n)^2 + p^2\}} :: x + q : x - q$; Whence $\{(x+n)^2 + p^2\} \times (n-q)^2 = \{(x-n)^2\} + p^2 \times (x+q)^2$, Which by multiplying, cancelling, &c. becomes $nx(x^2 + q^2) = qx(x^2 + n^2 + p^2)$.

Where $x^2 = \dfrac{qn^2 + qp^2 - nq^2}{n-q}$, or $2x = 2\sqrt{\dfrac{qn^2 - nq^2 + qp^2}{n-q}}$, the base of the triangle; which, by substituting the proper numeral values of q, n, and p, gives $\frac{200}{7}\sqrt{14}$; from which and the given lines the other two sides are readily obtained.

PROBLEM LXXXVI.

Given the hypothenuse (10) of a right angled triangle, and the difference of two lines drawn from its extremities to the centre of the inscribed circle (2), to determine the base and perpendicular.

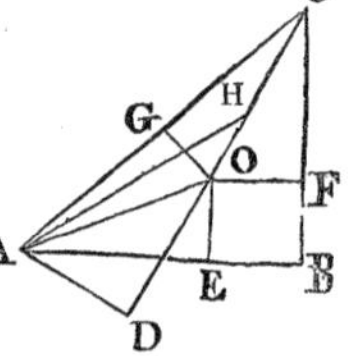

Let ABC be the proposed right angled triangle, and O the centre of its inscribed circle; and let $CO - AO = 2 = d$, and $AC = 10 = h$.

Produce CO to D, and let fall upon it the perpendicular AD; which put $= x$. Then, since CO and AO, bisect the two angles, C and A, and these two angles together are equal to a right angle, it follows that the two angles OAC and OCA = half a right angle. But the outward $\angle$ of any triangle, being equal to the two inward opposite $\angle$s, $\angle AOB = \angle OAC + \angle OCA$. Whence also AOD = half a right angle, and since D is a right angle, DAO is also = half a right angle.

Therefore $DO = AD = x$, and $AO = \sqrt{2x^2} = x\sqrt{2}$; and consequently $CO = x\sqrt{2} + d$, and $CD = x + x\sqrt{2} + d = (1 + \sqrt{2})x + d$. Now $AD^2 + DC^2 = AC^2$, or $x^2 + \{(1+\sqrt{2})x + d\}^2 = h^2$, or $\{1 + (1+\sqrt{2})^2\} x^2 + 2d(1+\sqrt{2})x = h^2 - d^2$, or $(4 + 2\sqrt{2})x^2 + 2d(1+\sqrt{2})x = h^2 - d^2$, or

$x^2+2d\frac{1+\sqrt{2}}{4+2\sqrt{2}}x=\frac{h^2-d^2}{4+2\sqrt{2}}$. Then by quadratics we have $x=-\frac{1+\sqrt{2}}{4+2\sqrt{2}}d\pm\sqrt{\{(\frac{1+\sqrt{2}}{4+2\sqrt{2}})^2d^2+\frac{h^2-d^2}{4+2\sqrt{2}}\}}$.

By reducing these surds to rational denominations, we have $x=-\frac{1}{4}d\sqrt{2}\pm\sqrt{\{\frac{1}{8}d^2+\frac{1}{4}(2-\sqrt{2})(h^2-d^2)\}}=-\frac{1}{4}d\sqrt{2}\pm\frac{1}{2}\sqrt{\{\frac{1}{2}d^2+(2-\sqrt{2})(h+d)(h-d)\}}=3.10850255=\text{AD}=\text{DO}$. Hence $\text{OA}=x\sqrt{2}=4.39608645=m$. $\text{OC}=6.39608645=n$. That is $\text{AO}=m$, and $\text{OC}=n$; now to find the segments AG and GC, (by Geometry, p, 360.) we have As $h:n+m::n-m:\text{CG}-\text{GA}=\frac{(n+m)\times(n-m)}{h}$, but $\text{CG}+\text{GA}=h$. Hence $\text{CG}=\frac{1}{4}h+\frac{(n+m)\times(n-m)}{2h}$ and $\text{GA}=\frac{1}{2}n-\frac{(n+m)\times(n-m)}{2h}$; now $\text{OG}=\sqrt{\{\text{OC}^2-\text{CG}^2\}}=\sqrt{\left\{n^2-\left(\frac{1}{2}h+\frac{(n+m)\times(n-m)}{4h^2}\right)^2\right\}}$ $=1.98822509$, $\text{CG}=\frac{1}{2}h+\frac{(n+m)\times(n-m)}{2h}=6.07921729$, & $\text{GA}=3.92078271$; consequently, $\text{AB}=\text{AG}+\text{OG}=5.9090078$, $\text{BC}=\text{CG}+\text{OG}=8.06744238$.

NOTE. The answers given in Bonnycastle's Algebra appear to be wrong, (from which this is taken.)

Proof. The sum of the squares of those numbers above is 99.9999997 which should be just 100, viz. the square of the hypothenuse 10, the error being $\frac{3}{10000000}$;

Otherwise,

Let x and y denote the base and perpéndicular, then as is well known $\frac{1}{2}(x+y-a)=$ the radius of the inscribed circle, (Simpson's Algebra, Lemma, Page 345, Published by Carey & Sons, Philadelphia,) or (Young's Geometry, just published by Carey & Lea,) put $\frac{1}{2}(x+y-a)=r$ then $\sqrt{\{(x-r)^2+r^2\}}=$ the line drawn from the extremity of the base to the centre of the inscribed circle; and $\sqrt{\{(y-r)^2+r^2\}}=$ the line from the centre to the extremity of the perpendicular. Hence supposing $x\,7\,y$ (per ques.) $\sqrt{\{x^2-2rx+2r^2\}}=b+\sqrt{\{y^2-2ry+2r^2\}}$, or $x^2-2rx=b^2+2b\sqrt{\{y^2-2ry+2r^2\}}+y^2-2ry$, or $x^2-y^2+2r(y-x)-b^2=2b\sqrt{\{y^2-2ry+2r^2\}}$, or restoring the value of r, $a(x-y)-b^2=2b\sqrt{\frac{1}{2}\{(a-x)^2+y^2\}}$ $=2b\sqrt{\{a^2-ax\}}$, (since $x^2+y^2=a^2$), $\therefore y=\frac{ax-b^2-2b\sqrt{\{a^2-ax\}}}{a}$

Put $\sqrt{\{a^2-ax\}}=z$, or $x=\frac{a^2-z^2}{a}$, then $y=\frac{a^2-z^2-b^2-2bz}{a}$ $=\frac{a^2-(z+b)^2}{a}\;\therefore x^2+y^2=a^2=\frac{2a^4-2a^2z^2+z^4-2a^2(z+b)^2+(z+b)^4}{a^2}$

or by reduction, $(a^2 - z^2)^2 - 2a^2(z + b)^2 + (z + b)^4 = 0$, or $(a^2 - z^2)^2 - 2(a^2 - z^2)(z + b)^2 + (z + b)^4 = 2z^2(z + b)^2$, or by extracting the square root we have $a^2 - z^2 - (z+b)^2 = 2(z+b)\sqrt{2}$, or $a^2 - b^2 = 2z^2 + 2bz + 2(z + b)\sqrt{2} = (2 + \sqrt{2}) \times (z^2 + b^2)$;

$\therefore z^2 + bz = \dfrac{a^2 - b^2}{2 + \sqrt{2}} = \dfrac{2 - \sqrt{2}}{2} \times \{a^2 - b^2\}$; or by quadratics,

we have $z = \dfrac{\sqrt{\{(4 - 2\sqrt{2}).(a^2 - b^2) + b^2\}} - b}{2}$;

hence $x = \dfrac{a^2 - z^2}{a}$ and $y = \dfrac{a^2 - (z+b)^2}{a}$ become known.

Otherwise, Geometrically. Let CH denote the given difference of the lines drawn from the acute angles to the centre of the inscribed circle, then draw the indefinite line AH so as to make the angle $AHO = \frac{1}{8}P$, (where P = two right-angles) then with C as a centre and the given hypotenuse as radius, describe an arc cutting AH in A, and join AC, then draw BC through C so as to make the angle BCO = ACO and let fall the perpendicular AB from A to BC, then ABC is the triangle sought.

For, draw AO bisecting the angle BAC and meeting CH produced in O, then evidently O is the centre of the inscribed circle; but since B = a right-angle $= \frac{1}{2}P$, $A + C$ is also $= \frac{1}{2}P$ $\therefore OAC + OCA = \frac{1}{2}(A + C) = \frac{1}{4}P$, and $AOC = P - \frac{1}{4}P = \frac{3}{4}P$ $\therefore$ in the triangle AOH we have $AHO + HAO = \frac{1}{4}P$, but $AHO = \frac{1}{8}P$ $\therefore$ $HAO = \frac{1}{8}P$ and the triangle AOH is isosceles, and CH = the given difference of the line CO and AO as required.—See calculation, p. 286.

PROBLEM LXVI.

Having given the sides of a quadrilateral which has two of its opposite angles equal; determine its area.

Let ABCD denote the quadrilateral whose four sides are given, and which has the angle B = D; join AC and draw the perpendiculars AF, AF′ to DC, BC. The triangles ADF, ABF′ are evidently similar, hence AD : AB : : DF : BF′ : : AE : AF $\therefore$ BF = AB.DF ÷ AD, and AF′ = AF.AB ÷ AD, hence if A = the area we have $\dfrac{AF}{AD} \times \dfrac{AD.DC + AB.BC}{2} = A$. Put $AB = a$, $BC = b$, $CD = c$, $AD = d$, $AF = p$, $DF = x$; then we have $\dfrac{p}{d} \times \dfrac{cd + ab}{2} = A$ (1),

also $AC^2 = AD^2 + DC^2 - 2DC.DF$, and $AC^2 = AB^2 + BC^2 - 2BC.BF'$ $\therefore a^2 + b^2 - 2b.\ BF = d^2 + c^2 - 2cx$,

or (since $BF = \dfrac{ax}{d}$), $a^2 + b^2 - \dfrac{2abx}{d} = d^2 + c^2 - 2cx$. hence we have

$x = \dfrac{(d^2 + c^2 - (a^2 + b^2))d}{2(cd - ab)}$ $\therefore p = \sqrt{\{d^2 - x^2\}}$ becomes known, and thence A becomes known by (1).

PROBLEM LXXXVIII.

Having given the base of a plane triangle $= 2a$, the perpendicular $= a$, and the sum of the cubes of its other two sides equal to three times the cube of the base; to determine the sides.*

Let $x + d$, $x - d$, denote the sides, then $(x + d)^3 (x - d)^3 = 3 \times (2a)^3 = 24a^3$, or by reduction $x^3 + 3xd^2 = 12$ (1), also by the common rule for the area of a triangle when the three sides are given, (Hutton's Mathematics vol. 1. p. 405), we have $\sqrt{\{(x^2 - a^2) \cdot (a^2 - d^2)\}} =$ the area of the triangle, but $\frac{1}{2}(a \times 2a) = a^2$ (Hutton p. 403), = the area, hence results the equation $(x^2 - a^2) \cdot (a^2 - d^2) = a^4$, (2); by (2) $d^2 = \frac{2a^4 - a^2x^2}{a^2 - x^2}$ which substituted in (1) gives $x^5 + 2a^2x^3 - 12a^3x^2 - 6a^4x + 12a^5 = 0$, this equation is satisfied by putting $x = 2a$, hence $d^2 = \frac{2a^4 - a^2x^2}{a^2 - x^2} = \frac{2a^2}{3} = \frac{6a^2}{9}$, or $d = \frac{a}{3}\sqrt{6}$, $\therefore x + d = a(2 + \frac{1}{3}\sqrt{6})$, $x - d = a(2 - \frac{1}{3}\sqrt{6})$ are the sides sought.

Remark, that the solution of question 51 page 46, might have been much simplified by the method used in the solution of this question. For by denoting the sides by $x + \frac{1}{2}c$, $x - \frac{1}{2}c$ we have the area of the triangle $= \sqrt{\{(x^2 - \frac{1}{4}b^2) \cdot (\frac{1}{4}b^2 - \frac{1}{4}c^2)\}} = a^2$ (per question) and $x^2 = \frac{4a^4}{b^2 - c^2} + \frac{1}{4}b^2$, or $x = \sqrt{\left\{\frac{4a^4}{b^2 - c^2} + \frac{b^2}{4}\right\}}$, hence the sides come out the same as in the solution cited.

Questions* of this kind may be made as follows: Let the base of a triangle be $= b$; the sum of the other two sides $= mb$, and the sum of their cubes $= nb^3$; Then the difference of the sides will be $= b\sqrt{((4n - m^3) \div 3m)}$; where $(4n - m^3) \div 3m$ may be any positive number with the following limitations: Since, from the nature of the question, m must always be greater than unity, it follows that n must be greater than $\frac{1}{4}$, and because $b^3 \times (\frac{1}{4}m^3 + \frac{3}{4}m^3)$ is the greatest limit of the sum of the cubes of the sides, it is evident that n must be less than $\frac{1}{4}m^3 + \frac{3}{4}m^3$.

The two least whole numbers for m and n (unity excepted) that will answer the question: Let $m=2$, and $n=3$; then the difference of the sides will be $2b \div \sqrt{6}$, and the sides themselves $b + b \div \sqrt{6}$ and $b - b \div \sqrt{6}$, and the perpendicular $= \frac{1}{2}b$ as given by the question. For by (*Trig. p.*) $b : 2b :: b \div \sqrt{6} : 4b \div \sqrt{6} =$ twice the distance of the perpendicular from the middle of the base: hence the greater segment $= \frac{1}{2}b + 2b \div \sqrt{6}$; therefore $(b + b \div \sqrt{6})^2 - (\frac{1}{2}b + 2b \div \sqrt{6})^2 = \frac{1}{4}b^2$ the square of the perpendicular. In this manner, when the values of m and n are chosen within the above limits, the perpendicular, and thence the area may be determined.

If $m = 3$, then the least whole number for n will be 7; and the three sides of the triangle will be b and $1\frac{1}{3}b$, and $1\frac{2}{3}b$, and the triangle is right angled.

Let BC be the given base $= 2a$; take a line M a mean proportional between BC and $\frac{1}{6}$BC; on BC describe a triangle such, that the side AB = BC + M, and AC = BC — M, and the thing is done.

For let fall the perpendicular. It is well known that $AB^3 + AC^3 = 2BC^3 + 6M^2 \times BC$; and by construction, $6M^2 = BC^2$, therefore $AB^3 + AC^3 = 3BC^3$, one of the conditions. Again, BC : AB + AC (2BC) : : AB — AC (2M) : 4M, hence $BK = \frac{1}{2}BC + 2M$; and (by Euc. I. 47,) $AK^2 = AB^2 - BK^2 = (BC + M)^2 - (\frac{1}{2}BC + 2M)^2 = \frac{3}{4}BC^2 - 3M^2 = \frac{1}{4}BC^2$ or $AK = \frac{1}{2}BC = a$, the given perpendicular.

†Let the given base BC be bisected, in G by the perpendicular NM; upon which take $GE = \frac{1}{3}GB$, GH = GC, HL = 2GB = BC; upon the diameter LE describe a circle cutting HA parallel to BC in A, the vertex of the required triangle ABC.

For it is evident that $GL = \frac{3}{2}BC$, $GE = \frac{1}{6}BC$, $HE = \frac{2}{3}BC$, and HA or $GK = BC\sqrt{\frac{2}{3}}$ (by Euc. VI. 13.). Also $BK = BC \times (\frac{1}{2} + \sqrt{\frac{2}{3}})$, and $CK = BC \times (\sqrt{\frac{2}{3}} - \frac{1}{2})$. Therefore $BA^2 = BC^2 \times (\frac{7}{6} + \sqrt{\frac{2}{3}})$ (Euc. I. 47,) and $BA = \frac{1}{2}BC \times (2 + \sqrt{\frac{2}{3}})$. In like manner, $AC = \frac{1}{2}BC \times (2 - \sqrt{\frac{2}{3}})$. Whence the whole is manifest, for $(2 + \sqrt{\frac{2}{3}})^3 + (9 - \sqrt{\frac{2}{3}})^3 \div 8 = 3$.

Cor. 1. GK = BA — AC. *Cor.* 2. BA + AC = 2BC.

Cor. 3. If CQ be perpendicular to BC, the cubes on BQ and QC are together equal to seven times the cube on BC. Also BQ + QC = 3BC, and $BQ - QC = \frac{1}{3}BC$.

‡Draw BC = the given base, which bisect in G; make $GK = BC\sqrt{\frac{2}{3}}$, and at K erect the perpendicular KA = BG; join AB, AC, and ABC will be the required triangle.

For (by Euc. I. 47,) $AB^2 = BK^2 + AK^2 = BG^2 + 2BG \times GK + GK^2 + BG^2 = BC^2 \times (\frac{7}{6} + \sqrt{\frac{2}{3}})$, or $AB = BC\sqrt{(\frac{7}{6} + \sqrt{\frac{2}{3}})} = BC \times (1 + \sqrt{\frac{1}{6}})$. Also, $AC^2 = CK^2 + KA^2 = BG^2 - 2BG \times GK + GK^2 + BG^2 = BC^2 \times (\frac{7}{6} - \sqrt{\frac{2}{3}})$, or $AC = BC \times (1 - \sqrt{\frac{1}{6}})$. Now it is obvious that if AB and AC be each cubed and added together, all the terms except the first and third in each will destroy each other, viz. $AB^3 + AC^3 = BC^3 \times (2 + 1) = 3BC^3$.

Let ABC be the proposed triangle, in which $BC = 2a$, and $AD = a$, $ED = x$. Then $BD = a + x$, and $ED = a - x$, or $-(x - a)$ we have $AB = \sqrt{(2a^2 + 2ax + x^2)}$, $AE = \sqrt{(2a^2 - 2ax + x^2)}$. And by the question $(2a^2 + 2ax + x^2)^{\frac{3}{2}} + (2a^2 - 2ax + x^2)^{\frac{3}{2}} = 24a^3$. Let $2a^2 + 2ax + x^2 = m$, and $2a^2 - 2ax + x^2 = n$,

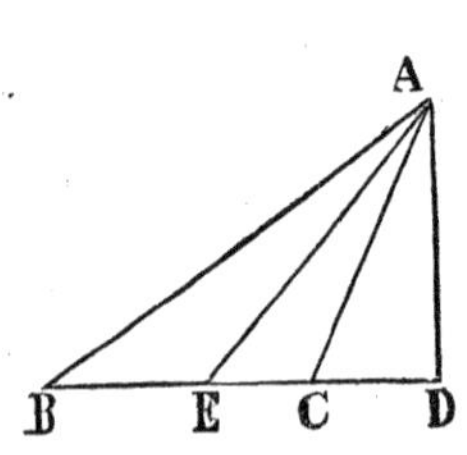

Then $m^{\frac{3}{2}} + n^{\frac{3}{2}} = 24a^3$. And by squaring, $m^3 + n^3 + 2m^{\frac{3}{2}} n^{\frac{3}{2}} = 576a^6$ Or $4m^3n^3 = (576a^6 - m^3 - n^3)^2$. Where, substituting the above values of m and n, the equation reduces to this; viz. $x = a\sqrt{\frac{8}{3}}$. Whence BD $= (1 + \sqrt{\frac{8}{3}})a$, and ED $= (1 - \sqrt{\frac{8}{3}})a$. Which, being negative, shows that the perpendicular falls on the base produced. Therefore AB $= a\sqrt{(\frac{24}{3} + \frac{4}{3}\sqrt{6})}$, and AE $= a\sqrt{\frac{24}{3} - \frac{4}{3}\sqrt{6}}$). And this, by extracting the roots, gives AB $= a(2 + \frac{1}{3}\sqrt{6})$, AE $= a(2 - \frac{1}{3}\sqrt{6})$, which are the two sides required. See page 40.

PROBLEM LXXXIX.

Having given the base of a plane triangle (15), its area, (45), and the ratio of its other two sides as 2 to 3, it is required to determine the lengths of these two sides.

Let ABC be the proposed triangle. Let AB $= a$, CD $= \frac{45}{7\frac{1}{2}} = b$, and AD $= x$, therefore BD $= a - x$; also the ratio AC : CB :: 3 : 2, or $m : n$; then AC$^2 = b^2 + x^2$, and BC$^2 = b^2 + (a - x)^2$ $\therefore b^2 + x^2 : b^2 + (a-x)^2 :: m^2 : n^2$. Whence, we have $m^2b^2 + m^2a^2 - 2m^2ax + m^2x^2 = n^2b^2 + n^2x^2$, or $(m^2 - n^2)\,x^2 - 2m^2\,ax = (n^2 - m^2)\,b^2 - m^2a^2$; therefore, by solving this quadratic and substituting the values of a, b, m, and n, the numeral value of x may be determined, and hence those AC and BC.

PROBLEM XC.

Given one angle, a side adjacent to the said angle, and the difference of the other two sides, to determine the triangle.

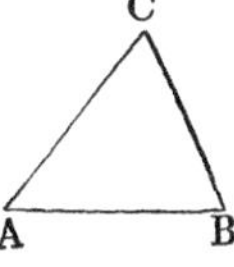

Put $a =$ AB; $d =$ the difference of the sides AC, BC; $c =$ cosine $\angle$A; and $x =$ AC; then BC $= x \pm d$, and (by Geom.) AB^2+AC^2 $- 2c$.AB.AC $=$ BC^2, that is $a^2 + x^2 - 2acx = x^2 \pm 2dx + d^2$; therefore $x = (a^2 - d^2) \div (2ac \pm 2d) =$ AC.

PROBLEM XCI.

Given one side, the difference between the square of the other side, and the square of the base, and the difference of the segments of the base, made by a perpendicular from the vertical angle, of a plain triangle, to determine the triangle.

Put $a =$ the given sides $b^2 =$ the difference of the squares of the other two, $d =$ the difference of the segments, $x =$ the base, and $y =$ the other side; then $y^2 - a^2 = \pm dx$ (by Euc. III. 36,) and $x^2 - y^2 = \pm b^2$, from which equations we get $x^2 \mp dx = a^2 \pm b^2$, an equation which exhibits the four cases of this problem.

Observation I.—Since by the conditions of the question, d can never be greater than x, it is evident that the least value of $x^2 - dx = a^2 - b^2$ is 0; or that, in this case, b must not be given greater than a.

Observation II.—Since $x^2 + dx$ can never be less than $2d^2$, this is the least value of $a^2 - b^2$ in that case.

PROBLEM XCII.

A gentleman has a garden in the form of an equilateral triangle, the sides whereof are each 50 feet: at each corner of the garden stands a tower; the height of A is 30 feet, that of B 34 feet, and that of C 28 feet. At what distance from the bottom of each of these towers must a ladder be placed that it may just reach the top of each tower, and what will be the length of the ladder, the ground of the garden being horizontal?

Observation.—Had the height of A been 38, B 42, and C 45, and the distance from A to B = $50l$ B to C = 40, and from C to A = 47 feet, the operation would have been more difficult.—The length of a ladder in this case would have been 49·552, and hence, the distances would have been found as in my key to Hutton's Mathematics.

Admit P to be the point sought, from which let fall the perpendiculars PG and PF, and produce PF to K; let AD BE, and CH represent the three towers, then the angles DAP, EBP, and HCP will each be right angles, and lines drawn from D to P, from E to P, or from H to P will be equal to each other, and equal to the length of the ladder.

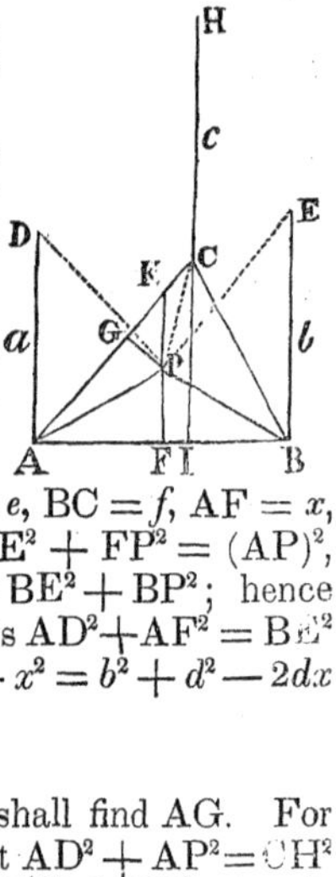

$AD = a$, $BE = b$, $CH = c$, $AB = d$ $AC = e$, $BC = f$, $AF = x$, $AG = y$, $PF = z$, $GP = v$. First, we have $AE^2 + FP^2 = (AP)^2$, and $BF^2 + FP^2 = BP^2$; But $AD^2 + AP^2 = BE^2 + BP^2$; hence $AD^2 + AF^2 + FP^2 = BE^2 + BF^2 + FP^2$, that is $AD^2 + AF^2 = BE^2 + BF^2$; let $AF = x$, $BF = d - x$; hence $a^2 + x^2 = b^2 + d^2 - 2dx + x^2$, then $x = \dfrac{b^2 + d^2 - a^2}{2d} = AF$.

Proceeding exactly by the same process, we shall find AG. For $AG^2 + GP^2 = AP^2$, and $GC^2 + GP^2 = PC^2$; But $AD^2 + AP^2 = CH^2 + PC^2$, hence $AD^2 + AG^2 + GP^2 = CH^2 + GC^2 + GP^2$, that is $AD^2 + AG^2 = CH^2 + GC^2$; Let $AG = y$, $GC = e - y$, hence $a^2 + y^2 = c^2 + y^2 - 2ey + e^2$, from which $y = (c^2 + e^2 - a^2) \div 2e = AG$

Again, because the 3 sides of the $\triangle$ ABC are given, the segments, AI and IB are found thus, $AB : AC + BC :: AC - BC : AI - IB$; Then $\frac{1}{2}AB + \frac{1}{2}(AI - IB) = AI = \frac{1}{2}d + \dfrac{e^2 - f^2}{2d} = \dfrac{d^2 + e^2 - f^2}{2d}$; put this quantity $= r$, then $\sqrt{(AC^2 - AI^2)} = \sqrt{(e^2 - r^2)} = IC$.

Now because FK is parallel to IC, the angle AKF = angle ACI; and the angles AFK, AIC and KGP are ⅃ angles: these triangles are similar AI : IC : : AF : FK, $r : \sqrt{(e^2 - r^2)} : : x : (x \div r) \times \sqrt{(e^2 - r^2)}$; put this quantity $= s$. Again AI : AC : : AF : AK; or $r : e : : x : (ex \div r)$. Now AG $= y$, therefore AK — AG $= (ex \div r) - y =$ GK, and KP = FK — PF $= s - z$. But AF^2 + EP^2 = AG^2 + GP^2, viz. $x^2 + z^2 = y^2 + v^2$, and GP^2 + GK^2 = KP^2, viz. $v^2 + [(ex \div r) - y]^2 = s^2 - 2sz + z^2$, from the first of these equations $v^2 = x^2 + z^2 - y^2$; and from the second $v^2 = s^2 - 2sz + z^2 - [(ex \div r) - y]^2$. Hence $x^2 + z^2 - y^2 = s^2 - 2sz + z^2 - [(ex \div r) - y]^2$; and $z = (s^2 + y^2 - x^2 - [(ex \div r) - y]^2) \div 2s$.

And lastly, because AF^2 + FP^2 = AP^2, and that $\sqrt{}$ (AP^2 + AD^2) = the length of the ladder, it follows, that $\sqrt{}$ (AF^2 + FP^2 + AD^2) = the length sought. Now AF $= x$, FP $= z$, and AD $= a$, all of which quantities are known. If $d = 50$, $e = 47$, $f = 40$; $a = 38$, $b = 42$, and $c = 45$, then will AF $= x =$ 28,2; AG $= y =$ 29,68085; AI $= r =$ 31.09; FK $= s =$ 31.971365, GK $= (ex \div r) - y =$ 12.95022180, FP $= z =$ 14.7033573: And $\sqrt{}$ (AF^2 + FP^2 + AD^2) $= \sqrt{(x^2 + z^2 + a^2)} = \sqrt{2455.4287158914} = 49.55228$ the length of the ladder.

PROBLEM XCIII.

Given the four sides of a quadrilateral figure, two of which are parallel; determine its area. (See Diagram to Problem 97, p. 73.)

Let ABCB denote the quadrilateral having AB parallel to CD and its four sides given; suppose DC ⦣ AB then through A draw AE parallel to BC meeting DC in E, then AE = BC, AB = EC (Geom. p. 84.) ∴ the three sides AD, DE=DC—AB, AE=BC of the triangle ADE are all known and thence its altitude AF=the latitude of its quadrilateral is easily found by known methods, and consequently we find the area of the quadrilateral $= \frac{1}{2}$AF × (AB + DC) becomes known as required.

PROBLEM XCIV.

A quadrilateral figure has two of its sides parallel; determine its area, when its altitude, one of the parallel sides and the two adjacent angles are given.

(See fig. last question.) Let ABCD be the quadrilateral, in which the side AB, CD are parallel, and CD together with the angles at C, D and the altitude AF are known; then by making the same construction as in the last question we have the triangle ADE whose angles are all known, for the angle D is given, and the angle AED=BCE (Geom. p. .) ∴ the remaining angle is easily found; also the perpendicular is known, hence the sides of the triangle are easily found by (trig. p. .) Hence the sides AB=CD — DE becomes known, and then the area $= \frac{1}{2}$AF. (AB + CD) becomes known also.

PROBLEM XCV.

The lengths of two lines that bisect the acute angles of a right angled plane triangle being 40 and 50 respectively, it is required to determine the three sides of the triangle.†

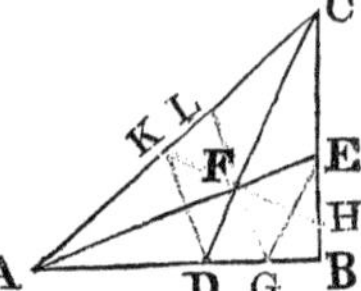

Let ABC be the proposed triangle, AE $= a$, and DC $= b$, the two given lines. Also let x and y represent the sine and cosine ABC respectively: then by (trig. p. .) we have $\sqrt{\frac{1+y}{2}} = cos.$ BAE, and $\sqrt{\frac{1+x}{2}} = cos.$ BCD. Also $x : b :: \sqrt{\frac{1+x}{2}} : \frac{b}{x}\sqrt{\frac{1+x}{2}} =$ AC, and $y : a ::$ $\sqrt{\frac{1+y}{2}} : \frac{a}{y}\sqrt{\frac{1+y}{2}} =$ AC : Whence $\frac{b}{x}\sqrt{\frac{1+x}{2}} = \frac{a}{y}\sqrt{\frac{1+y}{2}}$ or $\sqrt{\frac{1+x}{1+y}} = \frac{ax}{by}$. Again by trig. p. , $sin.\ A^* = \frac{2tan\frac{1}{2}A}{1+tan^2\frac{1}{2}A}$, and $cos.\ A = \frac{1-tan^2\frac{1}{2}A}{1+tan^2\frac{1}{2}A}$; Putting, therefore, tan BAE $= t$, and substituting $x = \frac{2t}{1+t^2}$, and $y = \frac{1-t^2}{1+t^2}$, We have $\frac{1+t}{\sqrt{2}} = \frac{2at}{2(1-t^2)}$, or $b(1+t)(1-t^2) = 2at\sqrt{2}$, or $t^3 + t^2 + (\frac{1+2a\sqrt{2}}{b})\, t = 1$.

Which is a cubic equation, whence the value of t may be determined: viz. the tangent of the angle BAE : and hence, also, the angles BDC and BEA become known, and consequently the sides AB $= 35.80737$, BC $= 47.40728$ and AC $= 59.41143$, as required.

† Make AE$=a$, CD$=b$, $s =$ sine of $\angle$AFD $=$ CFE $= 45°$, $y =$ $tang.$ $\angle$ EAB or EAC, $1 \div x =$ its cosine, and $y \div x$ of course $=$ its sine; also $(1-y^2) \div 2y = tang.$ ACB, and $(s + sy) \div x =$ sin. D $=$ cos. DCB or DCA. Now, (by trig. p ,) as $1 : a :: \frac{1}{x} : \frac{a}{x} =$ AB, and $1 : b :: \frac{s+sy}{x} : \frac{bs+bsy}{x} =$ CB. Hence $tang.$ $\angle ACB = \frac{a}{bs} \times \frac{1}{1+y} = \frac{1-y^2}{2y}$, from whence by a cubic equation the angles are found and consequently the sides as before.

Otherwise, let ABC be the right angled triangle, AE, CD the two given lines bisecting the two acute angles and intersecting one another in F. Draw LFG perpendicular to AE and KFH perpendicular to CD; and join EG, KD. (By Geom. p. .) That each of the angles at F is equal to half a right angle, and consequently that FL

* A denotes any angle; but in this example it is put for the angle BAC See. trig. by Dr. Day, President of Yale College.

$=$ FG $=$ FE, and FK $=$ FH $=$ FD. Put AE $= a$, CD $= b$, FL $=$ FG $=$ FE $= x$, and FK $=$ FH $=$ FD $= y$; then AF $= a - x$ CF $= b - y$, EG $= x\sqrt{2}$, and DK $= y\sqrt{2}$. By similar triangles, AF . FD : : AE : EG, and CF : FL :: CD : DK, which give these two equations, viz. $ay = (a - x) \cdot x\sqrt{2}$, and $bx = (b - y) \times y\sqrt{2}$; from the former $y = (a - x) \cdot x\sqrt{2} \div a$, which substituted in the latter, gives, $(a - x)^2 \cdot 2x\sqrt{2} = (a - 2x) \cdot ab$, a cubic equation by which x will be found as before.

PROBLEM XCVI.

Given the triangle ABC, AB=24, and BC=37.44; together with the segment of the base, DE=16.80; made by the two lines BD and BE drawn from the vertical angle to the base; the angle ABE $=$ CBD $= 90°$. To determine the triangle

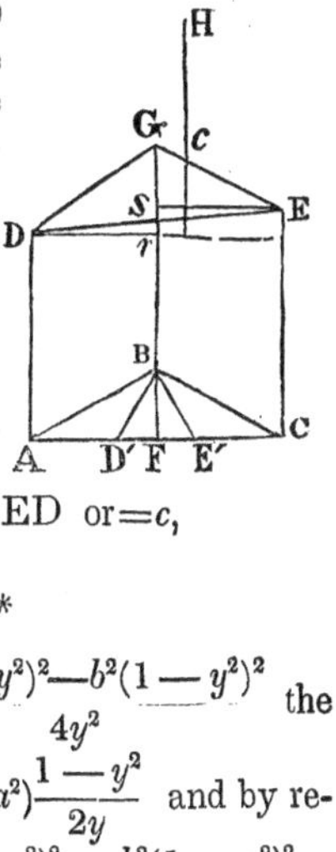

Let ABC represent the triangle, BE and BD the perpendicular to BA and BC, and ED the given distance, instead of the number on the question it will be more convenient to take the numbers 50, 78, and 35, or 5, 7.8 and 3.5, which are in the same ratio with the others.

Let AB $= 5 = a$, BC $= 7.8 = b$, ED $= 3.5 = c$. and BF $= x$. Then AF $= \sqrt{\{b^2 - x^2\}}$, CF $= \sqrt{\{a^2 - x^2\}}$; by similar triangles CF : FB : : FB : FE, or, EF $= \frac{FB^2}{CF} = \frac{x^2}{\sqrt{\{a - x^2\}}}$. and

$$FD = \frac{FB^2}{AF} = \frac{x^2}{\sqrt{\{b^2 - x^2\}}}$$ but EF $+$ FD $=$ ED or$=c$,

consequently, $$\frac{x^2}{\sqrt{\{b^2 - x^2\}}} + \frac{x^2}{\sqrt{\{a^2 - x^2\}}} = c.*$$

Which is made rational by putting $x^2 = \frac{a^2(1+y^2)^2 - b^2(1 - y^2)^2}{4y^2}$ the segments are $\sqrt{(b^2 - a^2)}\frac{1 + y^2}{2y}$ and $\sqrt{(b^2 - a^2)}\frac{1 - y^2}{2y}$ and by reduction we get the final equation $a^2.(1 + y^2)^2 - b^2(1 - y^2)^2 = c\sqrt{(b^2 - a^2)}\, y\,(1 - y^4.)$

* In numbers, a being equal to 5, $b = 7.8$ and $c = 3.5$. then we have

$$\frac{x^2}{\sqrt{\{60.84 - x^2\}}} + \frac{x^2}{\sqrt{\{25 - x^2\}}} = 3.5,$$ by a few trials we easily find $x = 3$. Hence 5 : 24 :: 3 : 14.4 $=$ BF.

Whence AF $= \sqrt{\{a^2 - x^2\}} = 19.2$, FC $= \sqrt{\{b^2 - x^2\}} = 34.56$; consequently, $19.2 + 34.56 = 53.76 =$ the base AC; and EF $= \frac{x^2}{\sqrt{\{a^2 - x^2\}}} = 6$; and FD $= \frac{x^2}{\sqrt{\{b^2 - x^2\}}} = 10.8$. Hence BD $=$ 15 6 and BE $= 18$, and the area $=$ AC $\times \frac{1}{2}$BF $= 387.072$ feet.

Let AB $= c'$; BC $= a$; DE $= b$; the angle A $= \varphi$, C$=\varphi'$. Draw the perpendicular BF to the base AC; then it is evident that it will fall within the triangle DBE, and that the angle EBF $=$ A $= \varphi$; also, that the angle DBF $=$ C $= \varphi'$; but BF $=$ AB $\times$ sin. A $= c'$ sin. φ; and EF$=$ BF $\times$ tan. FBE $= c'$ sin. φ tan φ; and DF $= c'$ sin. φ tan. φ $\therefore$ c' sin. $\varphi \times$ (tan. $\varphi +$ tan. φ') $= b$ (1); also c' sin. $\varphi = a$ sin. φ' (2); it is easy by (2) to eliminate tan. $\varphi' =$ from (1) whence there would be had an equation in terms of φ and known quantities; but it appears to me better to use (1) and (2) as they stand, by assuming in (2) a certain value for φ and then by (2) calculate φ' substitute these values of φ, φ' in (1)and if they satisfy it, φ was rightly assumed, if not then by the usual methods of trial and error φ, can be found to any degree of accuracy desired; and thence every thing else becomes known also; remembering always that the sines and tangents are to be taken to the radius (1); also that each of the angles, φ, φ' must be less than 90°, and indeed it is evident that their sum cannot exceed 90° supposing the point D . E not to lie in AC produced.

Again put DF $= x$, BF $= y$; the right-angled triangle DBC gives $\frac{\text{BF}^4}{\text{DF}^2} = \text{CB}^2 - \text{BF}^2$, or $\frac{y^4}{x^2} = a^2 - y^2$, $\therefore y^4 + x^2y^2 = a^2x^2$ (3); $\therefore y^2 = \frac{1}{2}\{x\sqrt{(4a^2+x^2)} - x^2\}$; similarly by the right-angled triangle ABE, $y^4 + (b-x)^2y^2 = c'^2(b-x)^2$ (4); $\therefore y^2 = \frac{1}{2}\{(b-x)\sqrt{\{4c'^2 + (b-x)^2\}} - (b-x)^2\}$; put these values of y^2 equal to each other and there results the equation $x\sqrt{\{4a^2+x^2\}} - x^2 = (b-x) \times \sqrt{\{4c'^2 + (b-x)^2\}} - (b-x)^2$ (5); x can be found by (5) by the usual methods and the problem will be solved as before.

By constructing the curve whose equation is (3) and the curve whose equation is (4) we shall find the vertex B of the triangle ABC at their intersection; remembering that in these equations x is to be reckoned on DE from D towards E, and that y is to be drawn through the extremity of x at right angles to DE above it. The point B can also be found by polar equations; for put DB $= r$, BE $= r'$; then r tan BDC $= r$ cot. $\varphi' = a$, or $r = a$ tan. φ', (6); $r' = c'$ tan. φ (7); construct then (6) and (7), and the vertex B will be given by their intersection as before. It appears to me, however, that the solution by (1) and (2) will be more easy in practice than any of the other methods which I have mentioned.

Otherwise. Let AB $= 50 = a$, BC $= 78 = b$, ED $= 35 = c$ and the base AC $= x$. Then AC : AB $+$ BC :: BC $-$ AB : FC$-$AF, or FC $-$ AF $= \frac{b^2 - a^2}{x}$, and AF $+$ FC $= x$. Hence FC $= \frac{x^2 + b^2 - a^2}{2x}$ or, $\frac{x^2+n}{2x}$, and AF $= \frac{x^2 - n}{2x}$ BF^2 $=$ AB$^2 -$ AF^2 $=$

$a^2 - \frac{\{a^2-n^2\}}{4x^2} = \frac{4x^2a^2-(x^2-n)^2}{4x^2}$. Again CF : BF : : BF : FE $= \frac{BF^2}{CF} = \frac{4x^2a^2-\{x^2-n\}^2}{2x \times \{x^2+n\}}$, $FD = \frac{BF^2}{AF} = \frac{4x^2a^2-\{x^2-n\}^2}{2x^2 \times \{x^2-n\}}$ but $EF + FD = c$, whence $\frac{4x^2a^2-(x^2-n)^2}{2x \times (x^2+n)} + \frac{4x^2a^2-(x^2-n)^2}{2x \times \{x^2-n\}} = c$, Or $\frac{1}{2x \times \{x^2+n\}} + \frac{1}{2x \times \{x^2-n\}} = \frac{c}{4x^2a^2 - \{x^2-n\}^2}$. Or $\frac{x}{x^4-n^2} = \frac{c}{4x^2a^2-\{x^2-n\}^2}$. By clearing it of fractions $4x^3a^2 - x^5 + 2x^3n^2 - n^2x^2 = cx^4 - cn^2$.

Or $x^5 + cx^4 - (2n + 4a^2)\,x^3 + n^2\,x - cn^2 = 0$; in number $x^5 + 35\,x^4 - 17168\,x^3 + 12845056\,x - 449576960 = 0$.

By rule given in Young's Algebra, p. 212, just published by Carey and Lea, Philadelphia, we find $x = 112$; Therefore 50 : 24 : : 112 : 53.76 = AC and $AF = \frac{x^2-n}{2x} = 40$, and $CF = 72$, $BF = 30$.

Whence $EF = 12.5$ and $FD = 22.5$; consequently $EB = 32.5$, $BD = 37.5$ and $\frac{24 + 32.5}{50} = 15.6 - EB$ and $\frac{24 \times 37.5}{50} = 18 = BD$, and the area $= 387.072$ feet as before.

PROBLEM XCVII.

A triangular field ABC whose sides are given, is to be divided into two parts in the ratio of 2 : 1, by a fence passing across from a given point D, in AC to BC. Determine its length.

Let ABC denote the triangular field whose sides are all given, D the given point in AC, and suppose that AD is not greater than DC it is evident that the division line will meet BC at some point between B and C. Join BD, and

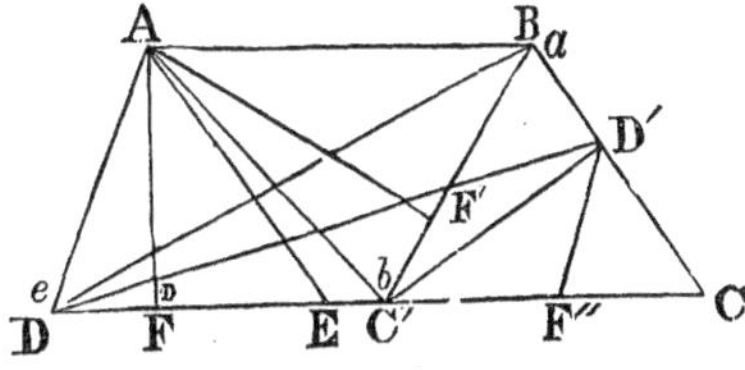

through A draw AE parallel to BD meeting BC produced in E, join DE then the triangle DEB=the triangle ADB, for they have the same base DB and the same altitude (By Geom. p.) hence the triangle ABC=the triangle DEC, then take EF : CF : : 2 : 1; join DF, which represents the fence both in length and direction; for the triangle DEF, DFC having the same altitude are as EF : FC : : 2 : 1, but the triangle EDF=the quadrilateral ABFD; hence the quadrilateral ABFD : triangle DFC : : 2 : 1, as was required.

The calculation is easy, for since the three sides of the triangle

ABC are given, the angle C is easily found. Also, since DB is parallel to AE, we have CD : AD : : CB : EB (Legendre 196.) Hence EB is found ∴ EF $=\frac{2}{3}$EC is found also, or FC$=\frac{1}{3}$EC is found; then in the triangle DFC we have the two sides CD and CF and their contained angle C, whence BF is easily found by (trig. p. 120,) also the angle FDC becomes known, whence DF is found both in length and direction.

Remark. If AB is greater than DC the same construction will hold provided the point F does not fall further from C than the point B; but should F lie beyond B, then draw the division line from D to AB in a similar manner to what has been done above, and the position and length of the fence become known as before. It may also be observed that the above process will serve if two parts of the piece are in the general proportion of $m : n$, where m and n are any two given numbers.

PROBLEM XCVIII.

Having given the sides, $a = 6$, $b = 4$, $c = 5$, and $d = 3$, of a trapezium, inscribed in a circle, to determine the diameter of the circle.*

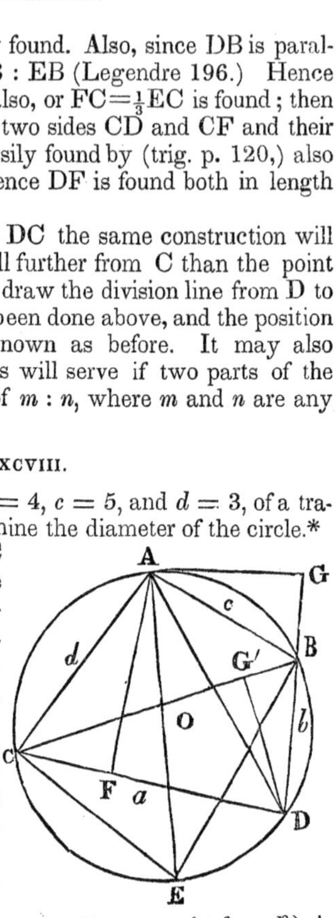

Let ACE denote the circle, ABDC the inscribed trapezium; draw the diameter AE, join EC, EB, the angles ACE, ABE, being each in a semicircle are right angles, (*Geom. p.* 52.) ∴ CE $= \sqrt{\{AE^2 - AC^2\}}$, EB $= \sqrt{\{AE^2 - AB^2\}}$, (*Geom p.* 35.) Join CB, then AB · CE + AC · EB = AE · CB (*Geom. p.* 233) put AB $= a$, AC $= b$, CD $= c$, DB $= d$, AE $= 2r$, then the equation becomes $a\sqrt{\{4r^2-b^2\}}+b\sqrt{\{4r^2-a^2\}} = 2r \cdot$ CB, in like manner we have, $c\sqrt{\{4r^2-d^2\}}+d\sqrt{\{4r^2-c^2\}} = 2r\cdot$ CB, $\therefore a\sqrt{\{4r^2-b^2\}}+b\sqrt{\{4r^2-a^2\}} = c\sqrt{\{4r^2-d^2\}} + d\sqrt{\{4r^2-c^2\}}$ (1), in the same manner we find $b\sqrt{\{4r^2-c^2\}}+c\sqrt{\{4r^2-b^2\}} = d\sqrt{\{4r^2-a^2\}}+a\sqrt{\{4r^2-d^2\}}$, or $b\sqrt{\{4r^2-c^2\}} - a\sqrt{\{4r^2-b^2\}} = d\sqrt{\{4r^2-a^2\}} - c\sqrt{\{4r^2-b^2\}}$, (2,).

By squaring (1) and reducing we have $2(a^2+b^2-c^2-d^2)r^2 - (a^2b^2-c^2d^2) = cd\sqrt{\{4r^2-d^2\}}\times\sqrt{\{4r^2-c^2\}} - ab\sqrt{\{4r^2-b^2\}}\times\sqrt{\{4r^2-a^2\}}$ (3), and by squaring we have (2), $4b^2r^2-b^2c^2-2ab\sqrt{\{4r^2-c^2\}}\times\sqrt{\{4r^2-d^2\}}+4a^2r^2-a^2d^2 = 4r^2d^2-a^2d^2-2dc\sqrt{\{4r^2-a^2\}}\times\sqrt{\{4r^2-b^2\}}+4c^2r^2-c^2b^2$, and reducing $2(a^2+b^2-c^2-d^2)r^2 = ab\sqrt{\{4r^2-c^2\}}\times\sqrt{\{4r^2-d^2\}} - dc\sqrt{\{4r^2-a^2\}}\times\sqrt{\{4r^2-b^2\}}$ (4), eliminating $\sqrt{\{4r^2-d^2\}}\times\sqrt{\{4r^2-c^2\}}$, by (3), & (4), we have $2(ab-cd)\times(a^2+b^2-c^2-d^2)r^2 - (a^2b^2-c^2d^2)ab = -(a^2b^2-c^2d^2)\times\sqrt{\{4r^2-a^2\}}\times\sqrt{\{4r^2-b^2\}}$, or by rejecting the factor $ab - cd$, $2(a^2+b^2-c^2-d^2)r^2-(ab+cd)ab = -(ab+cd)\times\sqrt{\{4r^2-a^2\}}\times\sqrt{\{4r^2-b^2\}}$ (5) by squaring (5) $4(a^2+b^2-c^2-d^2)^2r^4-4ab(ab+cd)\cdot(a^2+b^2-c^2-d^2)r^2+a^2b^2(ab+cd)^2=(ab+cd)^2\times(4r^2-a^2)\times(4r^2-b^2)=(ab+cd)^2\times(16r^4-4r^2(a^2+b^2)+a^2b^2)$, or $4(a^2+b^2-c^2-d^2)^2r^2-4ab(ab+cd)(a^2+b^2-c^2-d^2)=(ab+cd)^2\times(16r^2-4(a^2+b^2))$

Or $(\{a^2+b^2-c^2-d^2)^2-4(ab+cd)^2\}r^2=\{ab(a^2+b^2-c^2-d^2)-(ab+cd)\times(a^2+b^2)\times(ab+cd)=-\{ab(c^2+d^2)+cd(a^2+b^2)\}\times\{ab+cd\}$ or there results $\{4(ab+cd)^2-(a^2+b^2-c^2-d^2)^2\}r^2=\{a^2b(c+d^2)+cd(a^2+b^2)\}\times(ab+cd)$ consequently we have

$$r=\sqrt{\left\{\frac{\{ab(c^2+d^2)+cd(a^2+b^2)\}\times(ab+cd)}{\{4(ab+cd)^2-(a^2+b^2-c^2-d^2)^2\}}\right\}}$$ Or we have $$r=\sqrt{\left\{\frac{\{ab+cd\}\times\{ac+bd\}\times\{ad+bc\}}{4\times\{ab+cd\}^2-\{a^2+b^2-c^2-d^2\}^2}\right\}},\ (6).$$ Again we have $4\times\{ab+cd\}^2-\{a^2+b^2-c^2-d^2\}^2=\{a^2+b^2+2ab-c^2-d^2+2cd\}\times\{c^2+d^2+2cd-a^2-b^2+2ab\}=\{(a+b)^2-(c-d)^2\}\times\{(c+d)^2-(a-b)^2\}=\{a+b+c-d\}\times(a+b+d-c)\times(a+c+d-b)\times(b+c+d-a)$; put $a+b+c+d=2s$, then $a+b+c-d=2(s-d)$, $a+b+d-c=2(s-c)$, $a+c+d-b=2(s-b)$, and $b+c+d-a=2(s-a)$, consequently we have

$$r=\frac{1}{4}\sqrt{\left\{\frac{(ab+cd)\times(ac+bd)\times(ad+bc)}{(s-a)\times(s-b)\times(s-c)\times(s-d)}\right\}}\ (7).$$ as required.

Cor. If $d=0$ the trapezium becomes a triangle and (7) reduces to $$r=\frac{1}{4}\times\frac{abc}{\sqrt{((s-a)\times(s-b)\times(s-c)\times s)}}$$ = the radius of the circle which will circumscribe the triangle.

* Put the angle CDB $= x$ by (trig. p. 100), $a^2+b^2-2ab.\cos.x = BC^2 = c^2+d^2-2cd.\cos.(180^\circ-x)=c^2+d^2+2cd.\cos.x$ Hence $\cos.x=\dfrac{a^2+b^2-c^2-d^2}{2(ab+cd)}$, The area of DCB $= ab.\sin.x =$ CB. DG But BC $=\sqrt{(a^2+b^2-2ab.\cos.x)}=m$. Hence DG $=\dfrac{ab.\sin.x}{m}$ By (Euc. VI. C) $ab = D\times DG$ (the diameter of the circumscribing circle, $ab=\dfrac{ab.\sin.x}{m}\times D$, $\therefore D=\dfrac{m}{\sin.x}=m.\cos.c.x$.

PROBLEM XCIX.

In a given circle inscribe an equilateral triangle; and within this triangle describe a circle, &c.; then if r = radius of the first circle, find the sum of the areas of all the circles and all the triangles ad infinitum.

It is easy to see that the radius of the first circle is twice that of the second, the second twice that of the third, and so on indefinitely, $\therefore$ the radii are r, $\frac{1}{2}r$, $\frac{1}{4}r$, $\frac{1}{8}r$, and so on. Put $p=3.14159265$ &c.

then the areas of the circles are pr^2, $\frac{1}{4}pr^2$, $\frac{1}{16}pr^2$, $\frac{1}{64}pr^2$, &c., (*Geom. p.* 291.) hence $pr^2+\frac{1}{4}pr^2+$ &c. $=pr^2(1+\frac{1}{4}+\frac{1}{16}+$ &c. ad infinitum) = the sum of the circles required. Put this sum $=s$, now $1+\frac{1}{4}+\frac{1}{16}+$ &c. is a decreasing geometrical progression whose ratio of decrease $=\frac{1}{4}$, hence by the common rule for finding the sum of such a series we have $1+\frac{1}{4}+\frac{1}{16}+$ &c. $=1\div 1-\frac{1}{4}=1\div\frac{3}{4}=\frac{4}{3}$, $\therefore s=\frac{4}{3}pr^2=$ the sum of all the circles, as required.

It is also evident that the sides of the first triangle are twice those of the second, and so on as before. By putting $a=b=c$ and $s=\frac{3}{2}a$ in the cor. to the solution of problem 98, there results $r=a\div\sqrt{3}$ and $a=r\sqrt{3}=$ the expression for the side of the equilateral triangle inscribed in the circle whose radius $=r$; by the common rule for the area of the equilateral triangle we have $(a^2\sqrt{3})\div 4=$ the area of the first triangle, and $\{(\frac{1}{2}a)^2\}\times\frac{1}{4}\sqrt{3}=$ that of the second and $\{(\frac{1}{4}a)^2\}\times\frac{1}{4}\sqrt{3}=$ that of the third, and so on. Let s' denote the sum of these areas and we have $s'=\dfrac{a^2\sqrt{3}}{4}\left\{1+\dfrac{1}{4}+\dfrac{1}{16}+\text{\&c.}\right\}=\dfrac{a^2}{\sqrt{3}}=r^2\sqrt{3}$. *Ans.*

If $a=10$ feet, then $s=57.735$ square feet very nearly.

REMARK. If within any triangle we inscribe another by joining the middle of its sides, and within this second triangle we inscribe another by similar means, and so on, the sum of the triangles so formed together with the first is easily found in an analagous manner to the methods used in the above solution. For let s denote the area of the first triangle, then it is evident that $\frac{1}{4}s$ is the area of the second, $\frac{1}{16}s$ that of the third and so on, let s denote the sum of these areas continued ad infinitum; then $s'=s(1+\frac{1}{4}+\frac{1}{16}+$ &c.) ad infinitum, $=4s\div 3$ as required.

PROBLEM C.

If from any point within an equilateral triangle perpendiculars be drawn to the three sides, their sum is equal to a perpendicular drawn from one of the angles on the opposite side. Required proof.

From the point within the triangle draw straight lines to all the angles of the triangle, and they will evidently divide it into three triangles, whose bases are all equal to each other, being each one of the sides of the equilateral triangle. Let $a=$ one of the sides of the equilateral triangle, $p=$ the altitude of the triangle, then (*Geom. p.* 176.) if $A=$ its area we have $A=\frac{1}{2}ap$; also let x, y, z denote the perpendiculars from the point within to the sides of the triangle, then $\frac{1}{2}(ax+ay+az)=$ the sum of the three triangles into which the triangle was divided $=A$ $\therefore\frac{1}{2}a(x+y+z)=\frac{1}{2}ap$, and $x+y+z=p$ as was to be proved.

PROBLEM CI.

Given the four sides of a quadrilateral inscribed in a circle, to find the diagonals.

Let ABDC (see fig. to prob. 98.) be the inscribed quadrilateral, having AD, CB for its diagonals, put AB$=a$, AC$=b$, CD$=c$, BD $=d$, CB$=x$, AD$=y$. Now (*Geom. p.*) $xy=ac+bd$, (1), also (*Geom. p.*) $ab+cd : ad+bc :: y : x$, (2), or $ab+cd : ad+bc :: xy : x^2$, $\therefore$ by (1) $x=\sqrt{\left\{\frac{(ad+bc)\times(ac+bd)}{ab+cd}\right\}}$ hence $y=\sqrt{\left\{\frac{(ab+cd)\times(ac+bd)}{ad+bc}\right\}}$ which are the diagonals.

PROBLEM CII.

If a, b, c, d, be the four sides of a quadrilateral, inscribed in a circle and $s=a+b+c+d$, it is required to prove that the area $=\sqrt{\{(\frac{1}{2}s-a)(\frac{1}{2}s-b)(\frac{1}{2}s-c)(\frac{1}{2}s-d)\}}$

Let ABDC (see fig. to prob. 98.) be the quadrilateral inscribed in the circle ACE, from the angle A draw the perpendiculars AF, AG to the sides CD, BD, respectively, let A denote the area of the trapezium, then $\frac{1}{2}$(AF·CD) $=$ the area of the triangle ACD and $\frac{1}{2}$(AG·BD) that of the triangle ABD, (*Geom. p.* 176.) but these triangles make up the trapezium $\therefore$ A$=\frac{1}{2}$(AF·CD$+$AG·BD.) Now the two angles ACF, ABD when added make two right angles (*Geom. p.* 130) also ABD$+$ABG $=$ two right angles (*Geom. p.* 28.) $\therefore$ the angle ACF$=$ABG, and ACE being acute ABD is obtuse, and the perpendicular AG falls without the triangle ABD; hence (*Geom. p.* 191, 192.) $AD^2=AC^2+CD^2-2CD\cdot CF$, and $AD^2=AB^2+BD^2+2BD\cdot BG$, whence $AB^2+BD^2+2BD\cdot BG=AC^2+CD^2-2CD\cdot CF$, or $BD\cdot BG+CD\cdot CF=\frac{1}{2}(AC^2+CD^2-AB^2-BD^2)$ (1), put AB $=a$, AC $=b$, CD $=c$, BD$=d$, and (1) becomes $BG\times d+CF\times c=\frac{1}{2}(b^2+c^2-a^2-d^2)$, (2). Now since the angles at F and G are right they are equal (*Geom. p.* 27.), and since ACF $=$ ABG the triangles ACF, ABG are equiangular, (*Geom. p.* 74.), $\therefore$ similar and AC : CF :: AB : BG, AC : AB :: AF : AG (*Geom. p.* 202.), or $BG=CF\cdot AB\div AC=\frac{CF\times a}{b}$ & $AG=\frac{AF\times a}{b}$, hence $A=\frac{AF}{2b}(cb+ad)$ (3), also (2) gives $\frac{CF}{b}=\frac{b^2+c^2-a^2-d^2}{2(cb+ad)}$, but $CF=\sqrt{(AC^2-AF^2)}=\sqrt{(b^2-AF^2)}$, hence $\sqrt{(1-\frac{AF^2}{b^2})}=\frac{b^2+c^2-a^2-d^2}{2(cb+ad)}$, or $AF^2\div b^2=1-\left\{\frac{b^2+c^2-a^2-d^2}{2(bc+ad)}\right\}^2=\frac{4(bc+ad)^2-(b^2+c^2-a^2-d^2)^2}{4(bc+ad)^2}$, Or $\frac{AF}{b}=\frac{\sqrt{\{4(bc+ad)^2-(b^2+c^2-a^2-d^2)^2\}}}{2(bc+ad)}$, this value when substituted in (3) gives $A=\frac{\sqrt{\{4(bc+ad)^2-(b^2+c^2-a^2-d^2)^2\}}}{4}$, (4)

Now $4(bc+ad)^2-(b^2+c^2-a^2-d^2)^2=(b^2+c^2+2bc-a^2-d^2+2ad)\cdot(a^2+d^2+2ad-b^2-c^2+2bc)=(\{b+c\}^2-\{a-d\}^2)(\{a+d\}^2-\{b-c\}^2)=\{a+b+c-d\}\cdot\{b+c+d-a\}\cdot\{a+d+b-c\}\cdot\{a+d+c-b\}=$(since $s=a+b+c+d$)$=\{s-2a\}\{s-2b\}\cdot\{s-2c\}\cdot\{s-2d\}$, hence (4) becomes by substitution $A=\sqrt{(}(\frac{1}{2}s-a)\cdot(\frac{1}{2}s-b)\cdot(\frac{1}{2}s-c)\cdot(\frac{1}{2}s-d)$ as required.

Otherwise. It has been proved (*Geom. p.*) that the product of the three sides of any plane triangle = its surface multiplied by twice the diameter of its circumscribed circle; hence (supposing the same notation as in problems 98, 101, and the present problem,) $AC\cdot CD\cdot AD=$the area of the triangle $ACD\times 4r$ and $AB\cdot BD\cdot AD$ =the area of the triangle $ABD\times 4r$ or (since the area of the two triangles=the area of the trapezium=A), by addition $(AC\cdot CD+AB\cdot BD)\times AD=(bc+ad)\times AD=4Ar$, or substituting the value of $AD=y=\sqrt{\left\{\frac{(ab+cd)\times(ac+bd)}{ad+bc}\right\}}$ as found in prob. 101. we have $A=\sqrt{(}(ab+cd)\times(ac+bd)\times(ad+bc))\div 4r=$ (by (7) found in the solution of problem 98.) $(\sqrt{(S-a)\cdot(S-b)\cdot(S-c)(S-d)})$ which agrees with the result found above, for $S=\frac{1}{2}s$.

Cor. If one of the sides (as (d) for example = 0,) the trapezium becomes a triangle and the area = $\sqrt{\{S(S-a)\cdot(S-b)\cdot(S-c)\}}$ which agrees with the common rule for the area of a triangle when the three sides are given.

PROBLEM CIII.

Find the side of a square inscribed in a circular segment, which is contained by a chord, and one third part of the whole circumference.

Let PAQ denote the given circle, apply the radius from A to B, then from B to C, and join AC, and the arc ABC=one third of the circumference (*Geom. p.*). Let EDFG be the inscribed square having DE parallel to AC, draw the radius OB perpendicular to AC and it will be perpendicular to DE (*Geom. p.* 65.) hence (supposing that OB meets DE in z, and AC in y,) y and z are the middle points of AC, and DE; again because the chord AB=the radius AO and that Ay is perpendicular to BO, BO is bisected in y, (*Geom. p.* 47) or $Oy=OB\div 2$. Join OE and let r=the radius OB and $EG=yz=x$, then $Ez=\frac{1}{2}x$, $Oy=\frac{1}{2}r$, $Oz=\frac{1}{2}r+x$, then since by the right angled triangle OzE we have $OE^2=Oz^2+Ez^2$ ∴ by substitution $(\frac{1}{2}r+x)^2+\frac{1}{4}x^2=r^2$ or by reduction $\frac{5}{4}x^2+rx=\frac{3}{4}r^2$, or $x^2+\frac{4}{5}rx=\frac{3}{5}r^2$, which by quadratics we have $x=\frac{1}{5}r(\sqrt{19}-2)=.47178r$ nearly. If $r=100$, then $x=47.178$ feet as required.

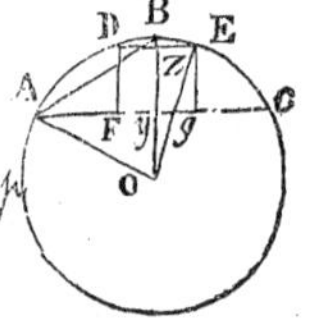

(104.) Having given the area and its four sides respectively of a trapezium, to determine the length of the greatest diagonal.

Put AB $=a$, BC $=b$, CD $=c$, AD $=d$, angle BAD $=x$, angle BCD $=y$, and the double area $=e$. Then by Young's trigonometry, page 47, Amer. Edition. $a^2+d^2-2ad\cos x=\text{BD}^2=b^2+c^2-2bc\cos y$, therefore $2ad\cos x-2bc\cos y=a^2+d^2-b^2-c^2$.

Again, double the area of the triangle BAD $=ad\sin x$, and double the area of the triangle BCD $=bc\sin y$; therefore, putting $2r=a^2+d^2-b^2-c^2$, we have $ad\sin x+bc\sin y=e$ and $ad\cos x-bc\cos y=r$. Hence $a^2d^2\sin^2 x=(e-bc\sin y)^2$, and $a^2d^2\cos^2 x=(r+bc\cos y)^2$, therefore by addition, and remarking that $\sin^2 x+\cos^2 x=1$, and $\sin^2 y+\cos^2 y=1$, $a^2d^2=(e-bc\sin y)^2+(r+bc\cos y)^2=e^2-2ebc\sin y+r^2+2rbc\cos y+b^2c^2$,

$$\text{or, } r\cos y-e\sin y=\frac{a^2d^2-b^2c^2-e^2-r^2}{2bc}=m.$$

Whence $r\cos y-m=e\sin y$;
or; $r^2\cos^2 y-2rm\cos y+m^2=e^2\sin^2 y=e^2-e^2\cos^2 y$, therefore $(r^2+e^2)\cos^2 y-2r\,m\cos y=e^2-m^2$, and, dividing by r^2+e^2 and completing the square,

$$\cos y=\frac{1}{r^2+e^2}\times\left\{e\sqrt{(r^2+e^2-m^2)}+rm\right\},$$

and the diagonal BD $=\sqrt{(b^2+c^2-2bc\cos y)}$.

Or since $(bc\sin y)^2=(e-ad\sin x)^2$ and $(bc\cos y)^2=(r-ad\cos x)^2$, we have as before

$$e\sin x+r\cos x=\frac{a^2d^2-b^2c^2+e^2+r^2}{2ad}=n,$$

$$\text{therefore } \cos x=\frac{1}{r^2+e^2}\times\{r\sqrt{(r^2+e^2-n^2)}+rn\},$$

and the diagonal BD is then $=\sqrt{(a^2+d^2-2ad\cos x)}$.

Here it may be remarked, that when the value of e is such as to make either m^2 or n^2 greater than r^2+e^2, the part under the radical becomes negative, and consequently the problem does not then admit of a solution. Therefore the limit of possibility, or the case in which the area is the greatest possible, will be when m^2 and n^2 are each equal to r^2+e^2. The values of m and n are then equal, but have a different sign, and the above expressions for $\cos y$ and $\cos x$ give $\cos y=\frac{rm}{r^2+e^2}$ or $\frac{r}{\sqrt{(r^2+e^2)}}$; $\cos x=\frac{rn}{r^2+e^2}$ or $\frac{-r}{\sqrt{(r^2+e^2)}}$; therefore $\cos y$ is $=-\cos x$, and consequently the one angle is the supplement of the other, and the trapezium is inscribed in a circle. Again because $m=-n$, $\frac{a^2d^2-b^2c^2-e^2-r^2}{2bc}=-\frac{a^2d^2-b^2c^2+e^2+r^2}{2ad}$; therefore by reduction $e^2=(ad+bc)^2-r^2$, or $e=\sqrt{\{(ad+bc)^2-r^2\}}$

Hence, when the four sides are given, if the double area be greater than $\sqrt{\{(ad+bc)^2-r^2\}}$ the problem is impossible.

The common rule for finding the area of a trapezium capable of being inscribed in a circle, when the four sides are known, may be deduced immediately from the above expression for e, by considering that $(ad+bc)^2-r^2$ is $=(ad+bc+r)(ad+bc-r)=\frac{1}{4}(2ad+2bc+a^2+d^2-b^2-c^2)(2ad+2bc-a^2-d^2+b_2+c^2)=\frac{1}{4}(a+d+c-b)(a+d-c+b)(b+c+a-d)(b+c+d-a)$.

(105.) A dodecaedron is a solid composed of twelve regular pentagonal pyramids, whose vertices meet in the centre of the circumscribing sphere, and the bases of the pyramids form the superfices of the dodecaedron. Now suppose a dodecaedron having the side of each pentagon composing the superficies thereof 8 inches, and supposing every two of its composing pyramids to be hollowed out in the form of the greatest hemisphere, cylinder, cube, cone, triangular pyramid, and square pyramid: What will the remainder of the dodecaedron weigh after having been hollowed or scooped out as above described, supposing each cubic foot of the matter of which it is composed to weigh 60lb?

The dodeceadron weighs, after all the twelve cavities are cut out, 1746.6646016 inches = 1.0108013 feet = solidity remaining; which at 60lb. a foot, weighs 60.6480771b.

(106.) In gauging a spheroidical ale cask, I found the diameter of one head to measure 18.1 inches, that of the other 16, the bung diameter 20, and the distance between the two heads 20.6 inches; also, by the cask lying a little obliquely, I observed that the liquor just rose to, or touched the upper extremities of the two heads. Having noted these dimensions, I was informed that there were in the cask a ball of iron weighing 60lb. another ball of lead weighing 90lb. and a cube box, a foot square. What quantity of liquor was in the cask?

	INCHES.
The spheroid's greatest distance from bung to head	12.054
The lesser distance	8.547
The content of the cask in ale gallons	20.763
The iron ball equal to cubic inches	217.048
The leaden ball	219.717
The cask's vacuity	117.814
The box emerged	1723.000
The sum, cubic inches	2277.579
which are equal to ale gallons	8.076
which deducted from the whole content leaves	12.687

ale gallons, the true quantity of liquor remaining in the cask.

CHAPTER II.

CONSTRUCTION OF ALGEBRAICAL EXPRESSIONS.

(13.) Having, in the preceding chapter, given several examples of the algebraical method of solving problems of geometry, it will be proper now to show how the algebraical may be converted into geometrical solutions. We shall commence with the construction of rational expressions. The simplest of these are such as denote lines; they are necessarily of one dimension, and are called linear expressions; they may always be reduced to one or other of the forms

$x = a - b + c - d +$ &c. $x = \frac{ab}{c}$, $x = \frac{a^2}{c}$, in which a, b, c, &c. represent lines of known length, or rather they express the number of linear units contained in these known lengths.

The construction of the first of these expressions, when put under the form $x = a + c +$, &c. $-$ $(b + d +$, &c.$)$ is obvious. All that is necessary is to draw a line equal to the sum of the lines, a, c, &c and to take from it another, equal to the sum of the lines, b, d, &c the remainder being the line represented by x.

The construction of $x = \frac{ab}{c}$ is reduced to the finding, geometrically, a fourth proportional to the three given lines, c, a, b, for the above expression reduces to the proportion $c : a :: b : x$.

The expression $x^2 = \frac{a^2}{c}$ requires us to find a third proportional to two given lines, c, a, since $c : a :: a : x$.

(14.) Let us now proceed to more complicated expressions.

1. Suppose we had to construct the expression $x = \frac{2abc}{3de}$; then, decomposing it into factors, in order to apply the foregoing elementary constructions, we have $x = \frac{2ab}{3d} \times \frac{c}{e}$. The first factor represents a fourth proportional to the three lines $3d$, $2a$, and b; hence, constructing this line and calling it m, the proposed expression becomes $x = \frac{mc}{e}$, which represents a fourth proportional to the three known lines, e, m, and c.

2. Let $x = \frac{2a^3 b^2 c}{3d^2 f^2 g}$ be the expression proposed, then, putting it under the form, $x = \frac{2a^2}{3d} \times \frac{a}{d} \times \frac{b}{f} \times \frac{b}{f} \times \frac{c}{g}$ we shall have first to construct the fourth proportional $\frac{2a^2}{3d}$ to the three lines, $3d$, $2a$, and a. Calling the line thus found m, the proposed expression becomes, $x = \frac{ma}{d} \times \frac{b}{f} \times \frac{b}{f} \times \frac{c}{g}$, and we have now to construct the fourth proportional, $\frac{ma}{d}$, to the three lines d, m, and a. Calling it m', we have $x = \frac{m' b}{f} \times \frac{b}{f} \times \frac{c}{g}$ Constructing in like manner the fourth proportional, $\frac{m' b}{f}$, and calling it m'', x becomes, $x = \frac{m'' b}{f} \times \frac{c}{g}$ and this is constructed, as in the last example, so that the line x will be constructed after finding five fourth proportionals. And it is obvious that in every such expression the construction will require the aid of as many fourth proportionals as are equal to the sum of the exponents of the letters in the denominator.

After these examples the construction of such compound expressions as $x = \frac{a^2}{c} + \frac{be}{d} + \frac{e^3 f^2 g^2}{k^2 l^2 m^2} - \frac{n^3}{p^2}$, &c. can present no difficulty.

(15.) Before proceeding further, it should be remarked, that every algebraical expression, admitting of geometrical construction, must have its terms all of the same dimension, that is, each term must be either of one dimension, and thus represent a line; or, secondly, each must be of two dimensions, and so represent a surface; or, lastly, each must have three dimensions, and denote a solid. It is plain that if this uniformity of dimension does not belong to all the component terms of an algebraical expression, that such an expression involves a geometrical absurdity, for we can in nowise combine a line with a surface, or a surface with a solid. Nevertheless, it often happens that an expression really admitting of construction does appear under this unsuitable form, but such a result can arise only from the *linear* unit having been represented in the calculation by the *numeral* unit, 1, thus causing every term into which it entered as a factor to appear of lower dimensions than the other terms. Whenever, therefore, for convenience of calculation, the linear unit is so represented, the result should be made *homogeneous*, by introducing it and its powers into the defective terms. Thus, if we happened to have such a result as

$x = ab$, then calling l the linear unit, we should change it into the homogeneous equation $lx = ab \therefore x = \frac{ab}{l}$, showing that the line x is a fourth proportional to the lines l, a, and b.

In like manner, if the result were $x = abc$, we should change it into $l^2x = abc$, whence $x = \frac{abc}{l^2} = \frac{ab}{l} \times \frac{c}{l}$, which expression we have already seen how to construct.

1. Let now the expression to be constructed be a compound fraction, such as $x = \frac{a^3 + 3bc - a}{b + 2c + 3}$, to admit of geometrical representation, both numerator and denominator of this fraction must be homogeneous; and to represent a line, x, the terms in the numerator must be one dimension higher than those in the denominator; so that introducing the linear unit, l, the expression to be constructed must be $x = \frac{a^3 + 3lbc - l^2 a}{lb + 2lc + 3l^2}$ that is, $x = \frac{a^3 + 3lbc - l^2a}{l(b + 2c + 3l)} = \frac{a^3}{lk} + \frac{3bc}{k} - \frac{la}{k}$, where k is put for the sum of the lines $b + 2c + 3l$; hence the problem is reduced to the construction of simple fractional expressions, such as have been considered in art. (14.)

2. As another example of this kind, let there be proposed the expression $x = \frac{2a^3 - 3a^2 b + b^2c}{a^2 - 2ab + b^2}$ This may be constructed as the preceding, if we can represent the denominator as a single product, and this we may do, by putting $b^2 = va$, or by determining v so that $v = \frac{b^2}{a}$, for then the expression becomes $x = \frac{2a^3 - 3a^2 b + b^2 c}{a(a - 2b + v)} = \frac{2a^2}{k} - \frac{3ab}{k} + \frac{vc}{k}$ where k is put for the line $a - 2b + v$.

3. Again, let $x = \frac{abcd + efgh}{mnp + qrs}$ be taken. To reduce the denominator to a single product, put $qr = vm \therefore v = \frac{qr}{m}$ is a known line, and the denominator becomes $m(np + vs)$, it remains then to reduce $np + vs$ to a single product. Put, then, $vs = wn \therefore w = \frac{vs}{n}$ is a known line, and the proposed expression becomes finally $x = \frac{abcd + efgh}{mn(p + w)} = \frac{abcd}{mnk} + \frac{efgh}{mnk}$ where k is put for the $p + w$.

(16.) We now proceed to consider *irrational expressions.* These may always be reduced to one or other of the following simple forms, viz, $x = \sqrt{ab}$, $x = \sqrt{\{a^2 + b^2\}}$, $x = \sqrt{\{a^2 - b^2\}}$, we shall, therefore, begin by constructing these elementary expressions.

From the first, $x = \sqrt{ab}$, we deduce $x^2 = ab$ $\therefore$ $a : x :: x : b$, therefore x is determined by finding, geometrically, a mean proportional between the given lines, a, b, (*Geom. p.* 136.)

From the second expression, $x = \sqrt{\{a^2 + b^2\}}$, we have $x^2 = a^2 + b^2$, so that x is the hypothenuse of a right angled triangle, of which the sides are a and b. (*Geom. p.* 58.)

The last expression, $x = \sqrt{\{a^2 - b^2\}}$, when put under the form $x = \sqrt{\{(a+b)(a-b)\}}$, represents a mean proportional between the two lines $a + b$ and $a - b$.

(1.) As a first example, let $x = \sqrt{\{a^2 - b^2 + c^2 - d^2 + e^2 - \&c.\}}$ be proposed. Put $m = \sqrt{\{a^2 - b^2\}}$, and construct this line; then $m^2 = a^2 - b^2$, and $x = \sqrt{\{m^2 + c^2 - d^2 + e^2 - \&c.\}}$ Put now $n = \sqrt{\{m^2 + c^2\}}$, and construct the line n; then, since $n^2 = m^2 + c^2$, and $x = \sqrt{\{n^2 - d^2 + e^2 - \&c.\}}$

This series of constructions being continued we shall at length have but two squares under the radical, and the construction of this last expression will be the line sought. In the same way may any numerical surd be accurately represented by a line first assuming some fixed length for unity, for any number may be decomposed into square numbers, thus $\sqrt{7} = \sqrt{\{2^2 + 2^2 - 1\}}$; $\sqrt{11} = \sqrt{\{3^2 + 1 + 1\}}$ and $\sqrt{13} = \sqrt{\{3^2 + 2^2\}}$; $\sqrt{43} = \sqrt{\{6^2 + 3^2 - 1 - 1.\}}$

2. Let the expression to be constructed be $x = \sqrt{\{a^2 + 3bc.\}}$ Put $3bc = v^2$ $\therefore$ $v = \sqrt{3bc}$ is a known line, and the expression is reduced to $x = \sqrt{\{a^2 + v^2.\}}$ Or the same expression may be constructed by putting $3bc = au$ $\therefore$ $x = \sqrt{\{a(a+u)\}}$ $\therefore$ x is a mean proportional between a and $a + u$.

3. Let $x + \sqrt{\left\{\frac{ab^2}{c} - de\right\}}$; then, putting $\frac{b^2}{c} = m$ and $de = an$, we have $x = \sqrt{\{a(m-n)\}}$ $\therefore$ x is a mean between a and $m - n$.

4. As a last example, let $x = \sqrt{\left\{\frac{a^3 - 2b^2c + 3b^3}{a \quad a - b}\right\}}$; then, putting it under the form $x = \sqrt{\left\{\frac{b^2(\frac{a^3}{b^2} - 2c + 3b)}{a - b}\right\}}$ we can first construct the line $\frac{a^3}{b^2} - 2c + 3b$, then $\frac{b^2}{a-b}$, and, calling these m and n, we shall have, lastly, to construct $x = \sqrt{mn}$.

We shall leave the student to point out the constructions of the following expressions, viz.

$$x = a + \sqrt{\left\{\frac{ab^2 + cd^2}{b + c}\right\}}. \qquad x = \sqrt{\{ac - fg + eh + mn\}}.$$

$$x = \frac{\sqrt{\{a^2b + a^2c\}}.}{\sqrt{\{3d + 2c}} \qquad x = \sqrt{\left(a + \frac{bc}{d^3}\right)}.$$

SECTION II.

CHAPTER I.

ON THE POINT AND THE STRAIGHT LINE.

(1.) In the preceding section we have endeavoured to show the use of algebra, when combined with geometry, in the solution of problems. In the remaining part of the present treatise we shall proceed in a manner more strictly analytical, dispensing with the truths of geometry, except a few of the simplest kind, and depending upon analytical expressions, as well for the establishment of theorems, as for the solution of problems; that is, as well for the determination of the form and properties, as of the magnitude of geometrical quantity. It is this extended application of the principles of analysis that, strictly speaking, constitutes the science of Analytical Geometry.

On the Equation of a Point.

(2.) Let AX, AY, be two assumed straight lines, intersecting, in any angle, at A; and let P be a point in the same plane, whose position it is required to determine relatively to these assumed lines.

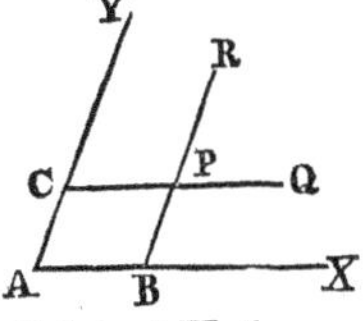

Let the lines PC, PB, be drawn respectively parallel to the lines AX, AY; then, if the lengths of the former be known, it is obvious that the position of the proposed point will be easily determined. It will be situated at the intersection of two lines, CQ, BR, drawn, the one parallel to AX, from a point, C, in AY, the distance of which from A is the given length, BP, and the other parallel to AY, from a point, B, in AX, the distance of which from A is the given length, CP.

The two lines AX, AY, in reference to which the position of the point is to be determined, are called *axes*, and their point of intersection, A, is called their *origin*. The distance, AB, is denominated the *abscissa* of the point; P, and BP, or its equal, AC, is called the *ordinate* of the same point; hence the axis AX is distinguished from the axis AY by the name *axis of abscissas*, the latter being called *axis of ordinates*.

The abscissa and ordinate of a point, when spoken of together, are, for the sake of brevity, called the *coordinates* of the point, and, for a like reason, the two axes are referred to as *axes of coordinates*. An abscissa is generally denoted by the letter x, and an ordinate by the letter y; and often, for shortness, the axis of abscissas is called the *axis of* x, and the axis of ordinates the *axis of* y.

We have just seen that a point becomes determinable when its coordinates, x and y, are known; it follows, therefore, that, in order to this determination, we need only have the two equations

$$x=a,\ y=b,$$

in which a and b are given. These equations are, therefore, called *the equations of a point.*

(3.) It is of importance to remark, that not only the absolute values of a and b must be given, in order to fix the position of a point, but also the signs of these quantities. If the axes are produced through the origin to X′ and Y′, it is obvious that the abscissas reckoned in the direction AX′ ought not to have the same sign as those taken in the opposite direction, AX, nor should the ordinates taken in the direction AY′ have the same sign as those taken in the opposite direction, AY; for, if there were no distinction in this respect, the position of a point, as determined by its equations, would be ambiguous. Thus the equations of the point P would equally belong to the points P′, P″, P‴, provided the absolute lengths of the coordinates of each were respectively equal to those of P. All this ambiguity is, however, avoided, by denoting the axes in one direction +, and in the opposite direction—. Hence then, regarding the abscissas to the right of the origin, A, as positive, those to the left will be negative. In like manner, considering the ordinates above the origin as positive, those below it will be negative.

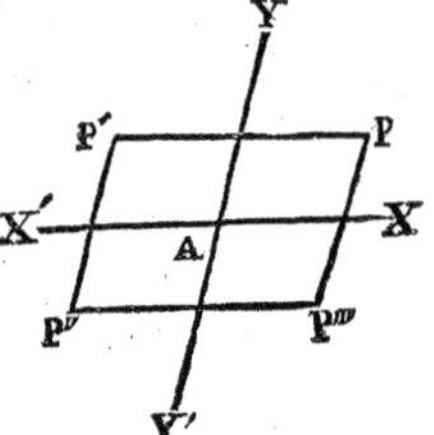

We thus have for the point P the equations $x=a,\ y=b,$ for the point P′, $x=-a,\ y=b,$ for the point P″, $x=-a,\ y=-b,$ and for the point P‴, $x=a,\ y=-b.$

If the point be situated on the axe AX, the equation $y=b$ becomes $y=o$, so that the equations $x=a, y=o$, characterise a point on the axis of abscissas, at the distance a from the origin.

If the point be on the axe AY, then $x=a$ becomes $x=o$, hence the equations $x=o,\ y=b,$ characterise a point on the axis of ordinates, at the distance, b, from the origin.

And lastly, if the point be common to both axes, that is, if it be at the origin, its position will be expressed by the equations.

$$x=o,\ y=o.$$

A point is said to be given when its coordinates are given, and, instead of referring to it as the point whose coordinates are x, y, it is more briefly and more usually designated as the point $(x, y.)$

On the Equation of the Straight Line.

(4.) Let it now be required to determine the equation of a straight line, or, in other words, to find an analytical expression by which it may be characterized.

Let MN be any straight line, and, in the same plane with it, let two straight lines AX, AY, be taken for axes of coordinates, and, for greater simplicity, let their origin, A, be upon the proposed line.

From any two points, D, B, in AB, let DC, BE, parallel to AY, be drawn, then will AC, CD, be the coordinates of the point D, and AE, EB, the coordinates of the point B, and they form this proportion, viz. AC : CD :: AE : EB

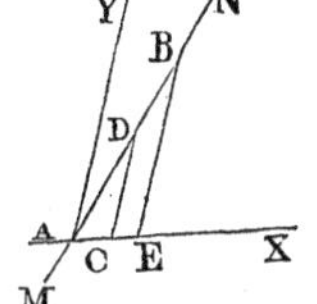

$$\therefore \frac{CD}{AC} = \frac{EB}{AE}$$

hence ***each abscissas is to its ordinate in the same constant ratio*** for every point taken in the proposed line. If, therefore, we represent this ratio by a, which will obviously be an abstract number, there will always exist this relation between any ordinate and its abscissa, viz. the former equal to a times the latter: thus $CD = a$ AC, $BE = a$ AE, &c. that is, agreeably to the notation already established, we shall have for every point in the line MN the relation

$$y = ax \quad . \quad . \quad . \quad . \quad (1,)$$

intimating, that whatever abscissa we take, a times that abscissa will be the value of the corresponding ordinate. Knowing, therefore, the constant a, we may, by giving arbitrary values to x, determine as many points in the line as we please, and, consequently, the line itself; equation 1 is hence called *the equation of the straight line*, MN.

(5.) If the axes of coordinates are rectangular, then the ratio $\frac{CD}{AC}$ or a expresses the tangent of the angle DAC, which the proposed line makes with the axis of abscissas.

If this angle is obtuse, the tangent will be negative; hence, if AN (next fig.) be the position of the proposed line, in reference to the rectangular axes, AX, AY, the equation is $y = -ax$ . . . . (2.) where it must be observed that the sign — applies only to the number a, and not to x, the abscissa, for the sign of this depends upon its direction from A, the abscissas of every point in the proposed line, which is below AX, being positive, while for every point in the portion above the abscissa is negative. Thus the abscissa AC, of the point P, is positive, for it is to the right of A; but the abscissa, AC′, of the point P′ being on the opposite direction, is negative. As regards the ordinates, it is plain, both from equation (2), and from the diagram, that to positive abscissas belong negative ordinates, and to negative abscissas, positive ordinates.

(6.) When the axes of reference are oblique, the coefficient a may still be represented by trigonometrical quantities,

for $\frac{\text{CD}}{\text{AC}}$, or a, is the same as $\frac{\text{sin. DAC}}{\text{sin. ADC}}$; therefore, since the angle ADC is equal to the angle YAD, if we represent the angle NAX, which the proposed line makes with the axis of x by α, and the inclination, YAX, of the axes themselves by β, we shall have

$$\frac{\text{sin. DAC}}{\text{sin. ADC}}=\frac{\text{sin. }\alpha}{\text{sin. }(\beta-\alpha)}$$

and for the equation of the line AN, $y=\frac{\text{sin. }\alpha}{\text{sin. }(\beta-\alpha)}x$

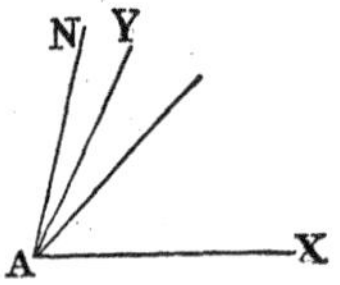

In this equation the coefficient of x will obviously be negative, when $\alpha > \beta$, that is, when AN takes the position in the annexed diagram to the left of the axis of y. We see, therefore, that, whether the axes be rectangular or oblique, the coefficient of x, in the equation of a straight line passing through their origin, will be positive, if the portion of this line situated above the axis of x lie to the right of the axis of y, but the same coefficient will be negative, if it lie to the left.

(7.) The equation to the straight line, which has just been exhibited, applies only when the line passes through their origin. Let us now suppose that this restriction is removed, and that the proposed line takes the position LM, cutting the axes in C and B.

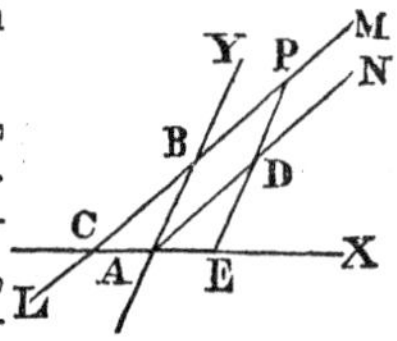

Let AN be parallel to LM, and from any point, P, in the latter, let the ordinate, PDE, be drawn. Then, since AB=DP, it follows that any ordinate, PE, is equal to AB, plus the ordinate ED, of that point, D, in AN, which has the same abscissa, AE, as the point P. Now this latter ordinate is always expressed by the equation $y=ax$, as we have already seen; consequently, if we put b for AB, the ordinate of the proposed line at the origin, we shall have for every point in LM this relation between the coordinates, viz. $y=ax+b$, this, therefore, is *the equation of the straight line in general.*

With regard to the sign of a, its changes have already been examined; and as to the sign of the ordinate b, we know that it will be positive so long as LM cuts AY above the origin, and negative when the intersection is below it. It may, however, be satisfactory to the student to have here exhibited the form of the equation for every possible position of the proposed line.

1. Let the line, LM take the position shown in the annexed diagram, cutting the axis of x to left of the origin, and the axis of y above it, then a and b are both positive, and the equation is

$$y = +ax + b.$$

2. Next, let the proposed line cut the axes on the opposite sides of the origin, as here represented, then a will still be positive, but b will be negative; the equation, therefore, in this position of the line is

$$y = +ax - b.$$

3. Thirdly, let the line cut the axis of y above, and the axis of x to the right of the origin, then a becomes negative, and b positive, in this case, therefore the equation is

$$y = -ax + b.$$

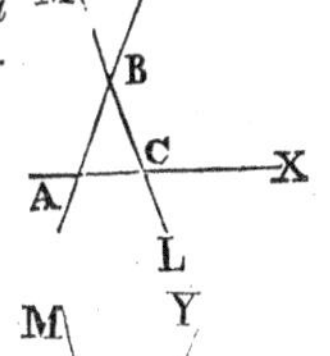

4. Lastly, let the axis of y be cut below, and the axis of x to the left of the origin, then both a and b will be negative, so that the equation becomes

$$y = -ax - b.$$

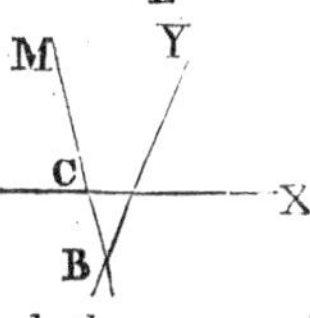

It thus appears that when both axes are intersected, the proposed line may take four different positions analytically represented by four distinct equations.

There remain two other positions to be considered, viz. those in which the line is parallel to one of the axes. If it be parallel to the axis of abscissas, as it cannot then form an angle therewith $\alpha = 0$, and therefore (6) $a = 0$, so that this position is characterized by the equation

$$y = 0x \pm b, \text{ or } y = \pm b,$$

intimating that, whatever abscissa be taken, the value of the ordinate remains the same.

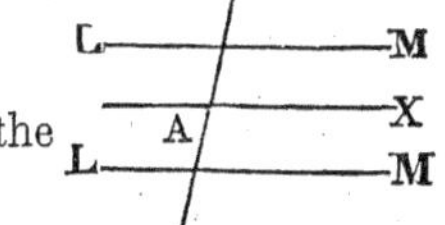

When the line is parallel to the axis of ordinates, substitute for a, in the general equation, its equal $\frac{b}{c}$, putting c for AC, the distance of the intersection, with the axis of x from the origin, b representing, as usual, the distance of the intersection with the axis of y, in the present case infinite, we thus have

$$y = \frac{b}{c}x + b, \quad \therefore x = \frac{1}{\frac{b}{c}}\,y - \frac{b}{\frac{b}{c}}$$

which, since b is infinite, is the same as $x = 0\,y \mp c$, or $x = \mp c$, an equation which indicates that whatever be the ordinate, the abscissas are constantly equal to c. From the preceding discussion it follows, that the general equation $y = ax + b$ comprehends in it all those that can characterize straight lines, whatever be their position in reference to two assumed axes, any how inclined to each other, each particular position being denoted by the particular values given to a and b. As these quantities remain the same, while the coordinates x and y vary in value for every point in the same line, the former are called *constants*, and the latter *variables*.

Since two constants enter into the general equation of a straight line, as many particular values as are given to them, so many particular lines will be represented. They may, therefore, take such values as will render the line whose equation is expressed, subject to certain proposed conditions, provided such conditions are possible. Thus we may suppose such values given to the constants, that the line represented must of necessity pass through two given points, or only one of the constants may be fixed, and of such value, that all the lines represented by the equation shall pass through a proposed point. We have already seen, that, if the origin of the axes were the proposed point, the constant, b, must take the particular value $b = 0$, the equation $y = ax$ representing all lines subject to this condition. Let us now proceed to a few determinations of this kind, and, as we are at liberty to assume any angle of inclination for the axes of reference, we shall, in general, for greater simplicity, assume them rectangular, excepting only in a few cases, to be hereafter pointed out, where oblique axes may be more advantageously employed.

PROBLEM I.

(8.) To find the equation of a straight line passing through a given point.

Let us denote the coordinates of the given point by x' and y'; then, since the general equation for every point in the required line is

$$y = ax + b \;\ldots\; (1),$$

it follows that for the particular point in question we must have the relation $y' = ax' + b \therefore b = y' - ax' \ldots (2)$, hence, substituting this value of b in (1,) we have

$$y - y' = a\,(x - x'), \text{ or } y = a\,(x - x') + y',$$

which is the equation sought, and characterizes every straight line that can be drawn through the point (x', y'). By comparing this with the general equation of the straight line, we find that the ordinate at the origin is $y' - ax'$.

If the given point were on the axis of x, then $y' = 0$, and the equation would be $y = a(x - x')$.

If it were on the axis of y, then $x' = 0$ and the equation would become $y - y' = ax$, or $y = ax + y'$.

PROBLEM II.

(9.) To find the equation to the straight line, which passes through two given points.

Representing the given points by $(x', y',)$ and $(x'', y'',)$ we have to subject the equation $y - y' = a(x - x')$ of a line passing through one of the points to the additional condition $y' - y'' = a(x' - x'')$, which equation as x', x'', y', y'', are all given, determines for a the particular value $a = \frac{y' - y''}{x' - x''}$; substituting, therefore, this value of a in the former equation, the analytical representation of the required line is $y - y' = \frac{y' - y''}{x' - x''}(x - x')$, in which equation all are constants, except x and y, the variable abscissa and ordinate of the line.

By writing the equation thus, $y = \frac{y' - y''}{x' - x''}x + \frac{x'y'' - y'x''}{x' - x''}$ which, however, is less simple than the former, its identity with the equation $y = ax + b$, for the particular case in question, is more distinctly seen, as it shows at once the value of the tangent a, and of the ordinate b at the origin. If, for instance, the coordinates of one of the given points $(x', y',)$ be 4 and 6, and those of the other point $(x'', y'',)$ 3 and 5, then the equation of the line passing through them is $y = 1x + 2$; therefore 1 being the trigonometrical tangent of the angle made by the line with the axis of x, this angle must be 45°, and the ordinate at the origin is 2.

If $(x', y',)$ is on the axis of x, $y' = 0$, and the equation is

$$y = \frac{y''}{x'' - x'}(x - x').$$

If it is on the axis of y, then $x' = 0$, and the equation is

$$y - y' = \frac{y'' - y'}{x''}x.$$

And lastly, if it be on both axes, that is, at the origin, then $x' = 0$, $y' = 0$, and the equation becomes $y = \frac{y''}{x''}x$.

PROBLEM III.

(10.) To find the equation of the straight line which passes through a given point, and is also parallel to a given straight line.

Representing as before the given point by (x', y'), we have for every line passing through it the general equation $y - y' = a(x - x')$, and among these lines we are required to distinguish that which is parallel to the given line, or, in other words, that which makes a given angle with the axis of abscissas; putting a' for the tangent of this angle, the equation of the line sought will be $y - y' = a'(x - x')$.

PROBLEM IV.

To determine the point where two given straight lines intersect.

Referring both lines to the same axis, let their equations be $y = ax + b$, $y = a'x + b'$, then, since at the point of intersection the ordinate is the same for both lines, we must have for this particular point $ax + b = a'x + b'$, whence we obtain for the coordinates of the intersection $x = \frac{b' - b}{a - a'}$, and $y = \frac{ab' - a'b}{a - a'}$

If we suppose $a = a'$, the expressions for x and y become infinite, as they evidently ought to do; since the lines, being in that case parallel, can meet only at an infinite distance.

If $b = b'$, then $x = 0$, and $y = \frac{a - a'}{a - a'} b = b$, showing that the ordinate at the origin belongs to the point of intersection.

PROBLEM V.

(11.) To find the expression for the angle of intersection of two given straight lines.

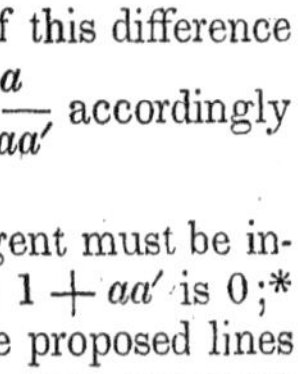

Let the two lines be A'B and CD, P being their point of intersection; then it is required to find an expression for the angle A'PC.

Let the equations of A'B and CD be respectively, $y = ax + b$, and $y = a'x + b'$; then a will be the tangent of the angle PA'X, and a' the tangent of the angle PCX, the lines being referred to the rectangular axes, AX, AY. Now the angle A'PC is equal to the difference of the angles PCX, PA'X; hence, calling the tangent of this difference v, we have (*Trig. p.*) $v = \frac{a - a'}{1 + aa'}$, or $v = \frac{a' - a}{1 + aa'}$ accordingly as the angle is to the left or right of A'P.

If the angle of intersection be a right angle, its tangent must be infinite, which the expression for v becomes, only when $1 + aa'$ is 0;* this condition, therefore, is necessary, in order that the proposed lines may be perpendicular to each other; so that, in this case, we must have $aa' = -1$, or $a = -\frac{1}{a'}$.

* Or, since the cotangent of a right angle is 0, and that the cotangent of an angle is the reciprocal of the tangent, we must have $\frac{1}{v}$, or $\frac{1 + aa'}{a - a'} = 0 \therefore 1 + aa' = 0$.

It follows from this, that, if $y = a'x + b'$ be the equation of a given straight line, then will $y = -\frac{1}{a'}x + b$ be the equation of a line perpendicular to it. These perpendiculars may be innumerable. If we fix one of them by the condition that it may pass through a given point, $(x', y',)$ then (*Prob.* 1) it will be characterized by the equation

$$y - y' = -\frac{1}{a'}(x - x').$$

When $(x', y',)$ is on the axis of abscissas, $y' = 0$, and the equation is

$$y = -\frac{1}{a'}(x - x').$$

When it is on the axis of ordinates, then $x' = 0$, and, therefore, the equation is $y - y' = -\frac{1}{a'}x$. And when the given point is at the origin x' and y', being both 0, the equation is, $y = -\frac{1}{a'}x$

(12.) If we wish for the sine or cosine of the angle of inclination of two lines, instead of the tangent, they may be obtained thus. By trigonometry $\sin.(\alpha - \alpha') = \sin.\alpha \cos.\alpha' - \sin.\alpha' \cos.\alpha$, in which formula, if α, α', represent the angles, whose tangents are a, a', we shall have, $\sin.\alpha = \frac{a}{\sqrt{\{1 + a^2\}}}$, $\cos.\alpha = \frac{1}{\sqrt{\{1 + a^2\}}}$;

$\sin.\alpha' = \frac{a'}{\sqrt{\{1 + a'^2\}}}$, $\cos.\alpha' = \frac{1}{\sqrt{\{1 + a'^2\}}}$ hence by substitution,

$$\sin.(\alpha - \alpha') = \frac{a - a'}{\sqrt{(1 + a^2)(1 + a'^2)}}$$

In like manner, from the expression

$$\cos.(\alpha - \alpha') = \cos.\alpha \cos.\alpha' + \sin.\alpha' \sin.\alpha,$$

we get, $\cos.(\alpha - \alpha') = \frac{aa' + 1}{\sqrt{(1 + a^2)(1 + a'^2)}}$.

PROBLEM VI.

(13.) To find the equation of the straight line which passes through a given point, and which makes a given angle with a given straight line.

Let the given line be represented by the equation, $y = ax + b$; then, because the required line passes through a given point, $(x', y',)$ its equation will take the form $y - y' = a'(x - x')$; and we have to determine a', so that these lines may intersect in a given angle. Put v for the tangent of this angle, then (*art.* 11)

$$v = \frac{\pm(a - a')}{1 + aa'}, \therefore a' = \frac{a - v}{1 + av}, \text{ or } \frac{a + v}{1 - av}$$

hence the equation of the line sought is

$$y - y' = \frac{a - v}{1 + av}(x - x'), \text{ or } y - y' = \frac{a + v}{1 - av}(x - x').$$

Two lines, therefore, may be drawn through the given point, fulfilling the required condition; the one forming the proposed angle, to the left, and the other forming it to the right, of the given line, conformably to the two expressions for its tangent, v.

Thus, in the annexed diagrams, the equation of PC forming the given angle to the left of the given line, AB, is

$$y - y' = \frac{a - v}{1 + av}(x - x');$$

but the equation of PC′, forming an equal angle to the right, is $y - y' = \dfrac{a + v}{1 - av}(x - x')$.

PROBLEM VII.

(14.) To find the analytical expression for the distance of two given points.

Let the given points be M, $(x', y',)$ and N, $(x'', y''.)$ Draw Mp parallel to AX; then the axes being rectangular, the distance, MN, will be $\sqrt{\{Mp^2 + Np^2\}}$; but $Mp = x'' - x'$, and $Np = y'' - y'$, therefore the expression for the distance, D, is

$$D = \sqrt{\{(x'' - x')^2 + (y'' - y')^2\}}.$$

If one of the points, as (x', y'), is at the origin, then $x' = 0$, and $y' = 0$; therefore, $D = \sqrt{\{x''^2 + y''^2}$.

The expression for D would have been much less simple, if we had chosen oblique axes of coordinates; for, if the angle MpN had been oblique, we should have had (*Trig. p.* ,)*

$$MN = \sqrt{\{Mp^2 + Np^2 - 2Mp \cdot Np \cdot \cos. Mp\,N\}};$$

or, putting A for M′pN we have $\cos. MpN = -\cos. A$,

$$\therefore D = \sqrt{\{(x'' - x')^2 + (y'' - y')^2 + 2(x'' - x')(y'' - y')\cos. A\}}.$$

If $(x', y',)$ is at the origin, $D = \sqrt{\{x''^2 + y''^2 + 2x''y'' \cos. A\}}$.

PROBLEM VIII.

(15.) To find the analytical expression for the distance of a point from a line.

Let N be the point, and BC the line, then it is required to find the length of the perpendicular NM. If the coordinates of M be represented by x, y, and those of N by x', y', we have by the preceding problem $MN = \sqrt{\{(x' - x)^2 + (y' - y)^2\}}$ in which the values of $x' - x$ and $y' - y$, must be determined by means of the equations to the lines BC, MN.

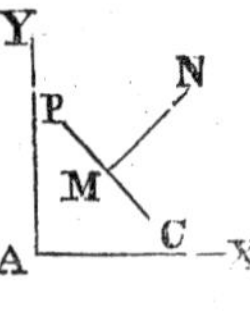

* The Trigonometry usually referred to, is that of Dr. Gregory.

Taking for the equation of the former $y = ax + b$ . . . (1,) that of the latter will be (*art.* 11,) $y - y' = -\frac{1}{a}(x - x')$ . . . . (2.)

In order to obtain from these equations the values sought, in the simplest manner, put equation (1) under this form:

$$y - y' = a(x - x') - y' + ax' + b \quad . \ . \ . \ . \quad (3.)$$

by subtracting y' from one side, and its equal, $y' + ax' - ax'$, from the other, in order that the equations (2) and (3) may both contain the unknowns $x - x'$ and $y - y'$, then the determination of the values of these becomes easy. Subtract (2) from (3) and the result is

$$0 = (a + \frac{1}{a})(x - x') - y' + ax' + b,$$

$$\therefore x - x' = \frac{a(y' - ax' - b)}{a^2 + 1}$$

and by substitution in equation (3)

$$y - y' = -\frac{y' - ax' - b}{a^2 + 1}$$

These expressions for $x - x'$ and $y - y'$, by changing the signs prefixed to them, represent $x' - x$ and $y' - y$; but, as these latter enter into the expression for MN only in the second power, it is indifferent what signs are prefixed; we have, therefore, by substitution,

$$\mathrm{MN} = \frac{\sqrt{(a^2 + 1)(y' - ax' - b)^2}}{(a^2 + 1)} = \frac{y' - ax' - b}{\sqrt{(a^2 + 1)}}$$

If the point N be at the origin, then $x' = 0$ and $y' = 0$; therefore

$$\mathrm{MN} = \frac{-b}{\sqrt{(a^2 + 1)}}$$

expresses the distance of the proposed line from the origin.

If the proposed line pass through the origin, then $b = 0$, and the value of the perpendicular upon it from $(x', y',)$ is $\mathrm{P} = \frac{y' - ax'}{\sqrt{(a^2 + 1)}}$

(16.) By means of this last expression, we can immediately arrive independently of any trigonometrical property, at the values of the sine and cosine of the angle formed by the intersection of two given straight lines.

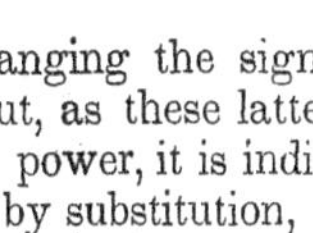

For let AM, AM′, through the origin, A, be parallel to any two given intersecting straight lines, thus forming an angle at A, equal to the angle of intersection. Take any point, M′, on one of the lines, and, representing its coordinates by $(x', y',)$ the value of the perpendicular from this point to the other line will be $\mathrm{M'M} = \frac{y' - ax'}{\sqrt{(a^2 + 1)}}$, a being the tangent of the angle MAX. Now, if we represent the tangent of the angle M′AX by a', we shall have at the point M′, y', $= a'x'$; hence, by substitution

$$\mathrm{M'M} = \frac{a'x' - ax'}{\sqrt{(a^2+1)}}$$

Now M′M is the sine, and AM the cosine, of the angle M′AM tc radius AM′, whose length is expressed by the equation

$$\mathrm{AM'} = \sqrt{\{x'^2 + y'^2\}} = \sqrt{\{x'^2 + a'^2 x'^2\}} = x'\sqrt{\{1 + a'^2\}};$$

hence, calling this radius R, we have $x' = \dfrac{\mathrm{R}}{\sqrt{(1+a'^2)}}$

$$\therefore \mathrm{M'M} = \frac{\mathrm{R}\,(a' - a)}{\sqrt{\{(a^2+1)\,(a'^2+1)\}}} = \text{sine} \angle \mathrm{A};$$

and, subtracting the square of this from R^2, and then taking the square root, we get $\mathrm{AM} = \dfrac{\mathrm{R}\,(aa'+1)}{\sqrt{\{(a^2+1)\,(a'^2+1)\}}} = \text{cosine} \angle \mathrm{A}$;

and these expressions are identical with those given at (12), where the radius, R, of the tables is unity.

The preceding problems contain every useful particular relative to the straight line. They should be attentively studied by the student, who before closing this chapter, should be fully prepared to state the conditions under which any straight line is drawn, when its equation is given, or on the contrary, when the conditions are given to write the equation. We shall now devote a short chapter to the solution of a few problems, in which the principles here established will find their application.

CHAPTER II.

PROBLEMS IN WHICH THE EQUATION OF THE STRAIGHT LINE IS EMPLOYED.

PROBLEM I.

(17.) It is required to determine whether the perpendiculars drawn from the vertices to the opposite sides of a triangle meet in a point.

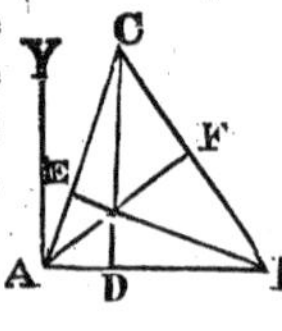

Let the perpendiculars, AF, BE, CD, from the vertices to the opposite sides of the triangle, be drawn and assume AB, AY, for rectangular axes of reference. Let (x', y') represent the given point, C, and for AB put c.

Then the equation of AC, passing through the origin and the given point, C, is $y = \dfrac{y'}{x'}\,x$; and the equation of BC passing through the two given points, B,C, the former $(c, 0)$ being on the axes of x, is $y = \frac{y'}{x' - c}(x - c)$.

Now BE, AF, being respectively perpendicular to AC, BC, and passing each through a given point on the axes of x, their equations are, of BE, $y = -\frac{x'}{y'}(x - c)$; and of AF, $y = -\frac{x' - c}{y'}x$.

At the point where these intersect, the ordinates must be equal, so that at this point $\frac{x'}{y'}(x - c) = \frac{x' - c}{y'}x$, whence, $x = x'$; that is, x, the abscissa of the intersection of BE, AF, is equal to x', the abscissa of the point C; hence, the perpendicular, CD, passes through that intersection.

PROBLEM II.

(18.) It is required to determine whether perpendiculars from the middle of each side of a triangle meet in a point.

Let M, M′, M″, mark the middle points of the sides of the triangle ABC. Let P be the point where two of the perpendiculars, MP, M′P, meet, and, as before, take the rectangular axes, AB, AY. Represent the point C by (x', y'), and the base, AB, by c; then the point M′ will be $(\frac{1}{2}c, 0)$, and the points M″, M, having their ordinates parallel to the ordinate of C, will obviously be $(\frac{1}{2}x', \frac{1}{2}y')$, and $(\frac{1}{2}c + \frac{1}{2}x', \frac{1}{2}y')$, respectively, for the triangles AM″ m', M′Mm will be equal.

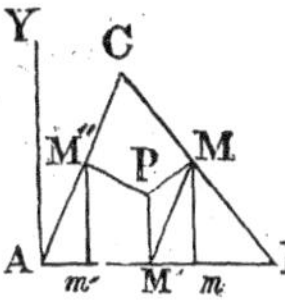

Now the equation of AC, passing through the origin and the point, $(x', y',)$ is $y = \frac{y'}{x'}x$, and that of BC, through the points $(c, 0)$ and $(x', y',)$ is $y = \frac{y}{x' - c}(x - c)$.

Now PM″, PM, being respectively perpendicular to these, and, at the same time, passing the one through the given point M″, and the other through the given point M, we have for the equation of PM″, $y - \frac{1}{2}y' = -\frac{x'}{y'}(x - \frac{1}{2}x')$ and for the equation of PM,

$$y - \frac{1}{2}y = \frac{c - x'}{y'}\{x - \frac{1}{2}(c + x').\}$$

Now, if these two lines meet in the perpendicular from M′, they must necessarily have a common ordinate for the abscissa, A M′, or $\frac{1}{2}c$, otherwise the ordinates will be different. Substituting this value of x, in the equation, for PM″, we find $y = \frac{1}{2}y' + \frac{x'^2 - cx'}{2y'}$ and, making the same substitution in the equation for PM, there results

$$y = \tfrac{1}{2}y' + \frac{x'^2 - cx'}{2y'}$$

The two ordinates are, therefore, identical, and thus the three perpendiculars meet in a point.

PROBLEM III.

(19.) It is required to determine whether the straight lines drawn from the vertices of a triangle to bisect the opposite sides, meet in a point.

Let the lines, CM, BM$'$, AM$''$, bisect the opposite sides of the triangle, CAB; and, let us, in this case, employ the oblique axes, CM, MB. Since AM$'$ is half AC, a parallel to CM, from the point M$'$, will bisect AM, and be equal to half CM. In like manner, a parallel to CM, from the point M$''$, will bisect MB, and be also equal to half CM, so that the coordinates of the points M$'$, M$''$, are numerically equal.

Now, if $(x', 0)$ represent the point B, $(-x', 0)$ will represent the point A; and, if (x'', y'') denote the point M$''$, $(-x'', y'')$ will denote the point M$'$; hence the equation of AM$''$, passing through the points $(-x', 0)$ and (x'', y''), is $y = \frac{y''}{x''+x'}(x + x')$; and the equation of BM$'$, passing through the points $(x', 0,)$ and $(-x'', y'',)$ is,

$$y = \frac{y''}{-x''-x'}(x - x').$$

Now, in order that these lines may intersect on the axis, MC, the ordinates of both at the origin must be the same, and this they evidently are, for the ordinate corresponding to $x = 0$ is, in both equations, $y = \frac{y''}{x''+x'}x' = \text{PM}.$

Cor. Since $x'' = \frac{1}{2}x'$, and $y'' = \frac{1}{2}\text{CM}$, it follows that

$$\text{PM} = \frac{\frac{1}{2}\text{CM}}{\frac{3}{2}x'}x' = \tfrac{1}{3}\text{CM}.$$

PROBLEM IV.

(20.) To determine whether the lines bisecting the three angles of a triangle meet in a point.

Let AF, CD, BE, bisect the three angles of the triangle CAB; and, as before, let the oblique lines, CD, DB, be taken for axes of reference; then $-$DA will be the abscissa of the point A, and DB will be the abscissa of the point B; consequently the equation of AF will be $y = a(x + \text{AD})$ . . . . . . (1.); and the equation of BE, $y = a'(x - \text{DB})$ . . . (2).

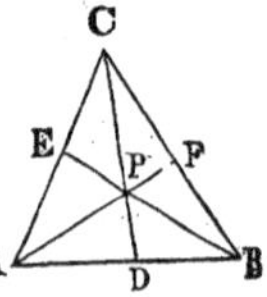

Now it appears from (*art.* 6,) that, as the axes are oblique, the value of a will be $\frac{\sin. \text{FAB}}{\sin. (\text{CDB} - \text{FAB})} = \frac{\sin. \frac{1}{2}\text{A}}{\sin. \frac{1}{2}(\text{A} + \text{C})} = \frac{\sin. \frac{1}{2}\text{A}}{\cos. \frac{1}{2}\text{B}}$ In like manner, the value of a' will be $a' = -\frac{\sin. \frac{1}{2}\text{B}}{\cos. \frac{1}{2}\text{A}}$ hence, when $x = 0$, the equations (1) and (2) become, by these substitutions, $y = \frac{\sin. \frac{1}{2}\text{A}}{\cos. \frac{1}{2}\text{B}}\text{AD}$ . . . (3), and $y = \frac{\sin. \frac{1}{2}\text{B}}{\cos. \frac{1}{2}\text{A}}\text{DB}$ . . . (4); and it now remains to inquire whether these two expressions for y are identical.

For AD, in equation (3,) substitute its equal, viz. $\text{AD} = \frac{\sin. \text{B}}{\sin. \text{A}}\text{DB}$, and the equation becomes, $y = \frac{\sin. \frac{1}{2}\text{A}. \sin. \text{B}}{\cos. \frac{1}{2}\text{B}. \sin. \text{A}}\text{DB}$, which is identical with equation (4.) (*Trig. p.*)* Hence the three lines meet in a point.

PROBLEM V.

(21.) To express the area of a triangle in terms of the coordinates of two of its angular points.

Let the triangle, BAC, be proposed; and let the rectangular axes originate at the point A; let x', y', be the coordinates of B, and x'', y'', the coordinates of C; then we have for the equation of BC, passing through both these points,

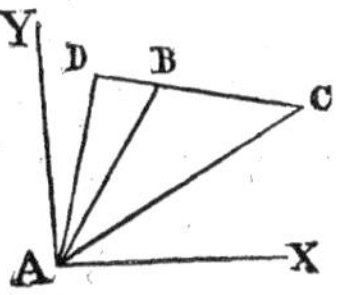

$$y = \frac{y' - y''}{x' - x''}x + \frac{x'y'' - y'x''}{x' - x''}$$

and for the length of the perpendicular AD, upon this line, from the origin, we have (*art.* 15), the expression $\text{P} = \frac{-b}{\sqrt{\{a^2 + 1\}}}$ substituting here for a and b, their values as exhibited in the foregoing equation, viz. $a = \frac{y' - y''}{x' - x''}$, $b = \frac{x'y'' - y'x''}{x' - x''}$ and the expression for P becomes, after reduction, $\text{P} = \frac{-(x'y'' - y'x'')}{\sqrt{\{(x' - x'')^2 + (y' - y'')^2\}}}$;

Now the denominator of this fraction expresses the distance of the point (x'', y''), from the point $(x', y',)$ that is, it denotes the line BC; so that $\text{BC} \times \text{P} = -(x'y'' - y'x'')$; consequently $\frac{1}{2}\{\text{BC} \times \text{P}\} = -\frac{1}{2}\{x'y'' - y'x''\}$ = area of the triangle.

The foregoing exercises on the equation of the straight line may suffice for the present; further applications will repeatedly occur in the succeeding chapters.

* For, at art. 18, p. 43, we have $\sin. \frac{1}{2}\text{B} = \frac{\sin. \text{B}}{2\cos. \frac{1}{2}\text{B}}$, and $\cos. \frac{1}{2}\text{A} = \frac{\sin. \text{A}}{2\sin. \frac{1}{2}\text{A}}$

(22.) At the commencement of the present section it was shown that a straight line, whatever its position, might be represented by a simple indeterminate equation; and it will be proper, before we proceed to the circle, to show, conversely, that every simple indeterminate equation containing two variables is the analytical representation of some straight line.

Let $My = Nx + P$ be any simple indeterminate equation, containing the two variable quantities, x and y, then we have $y = \frac{N}{M}x + \frac{P}{N}$, or putting for simplicity A for $\frac{N}{M}$, and B for $\frac{P}{N}$

$$y = Ax + B \quad . \; . \; . \; . \quad (1).$$

Now, draw any two straight lines, X′AX XAY′, intersecting at A, and make AB = B, $AC = \frac{B}{A}$, and through the points C, B, draw the straight line CBL, which will be the geometrical representation of the proposed equation. For the equation of this line, in reference to the axes, AX, AY, is $y = \frac{BA}{AC}x + AB$ . . . (2); but by the construction $\frac{BA}{AC} = B \div \frac{B}{A} = A$, also AB = B, therefore the equations (1) and (2) are identical, each, therefore, is the analytical representation of the line CBL.

(23.) The line which any equation represents, or in which the variable point (x, y) is always found, is called the locus of that equation, or of the point (x, y). Hence the locus of a simple equation containing two variables is a straight line.

When the equation is given, and it is required to construct the locus, it will be sufficient to determine two points in it, since the locus will be the straight line passing through them. Now the two points most easily found are those where the locus intersects the axes. The abscissa of the one point will be furnished by the proposed equation, by making therein $y = 0$, and the ordinate of the other, by making $x = 0$.

Let, for example, the locus of the equation $2y = 3x - 5$ be required.

Making $y = 0$, there results for x the value $x = \frac{5}{3}$; and, making $x = 0$, we have $y = -\frac{5}{2}$; therefore having assumed the axes AX, AY, on the former take $AC = \frac{5}{3}$; and, on the latter, take $AB = \frac{5}{2}$; then the straight line, BCL, drawn through the points B, C, will be the locus sought.

This method of determining the locus can, however, be applied only when the equation is of the form $y = ax + b$; for, if it were of the form $y = ax$, then b being 0, the locus would pass through the origin, so that its intersection with the axes would furnish but one point; another, therefore, must be found, before we can determine the line; for this purpose, we may give to x any particular value, and this, with the resulting value of y, will be the coordinates of another point.

It is obvious that of every point in the locus of the indeterminate equation $y = ax + b$, the coordinates exhibit a geometrical solution; and, as an infinite number of these points may be taken, the locus supplies all the infinite solutions of the equation. If, therefore, any other indeterminate equation were susceptible of solutions that belong also to the former equation, and if the loci of both equations were to be constructed on the same axes, these common solutions would be geometrically represented by so many points being common to both loci.

CHAPTER III.

ON THE CIRCLE.

(24.) Let r represent the radius of a circle, the centre of which is O. In the same plane as the cicrle, assume any rectangular axes, AX, AY; and let it be required to determine the equation of its circumference, or the analytical representation thereof, in reference to the assumed axes.

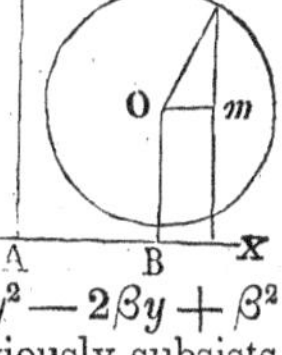

Let the coordinates AB, OB, of the centre, be represented by α, β; while the coordinates of any point, P, in the circumference, are denoted by the variables x, y; then, drawing the radius, OP, and Om, parallel to the axis of x, we shall have O$m = x - \alpha$, and P$m = y - \beta$; consequently, since O$m^2 +$ P$m^2 =$ OP^2, we have $(x-\alpha)^2 + (y-\beta)^2 = r^2$ or $x^2 - 2\alpha x + \alpha^2 + y^2 - 2\beta y + \beta^2 = r^2$ . . . (1), which is the equation sought, and obviously subsists for every point, P or $(x, y,)$ taken in the curve. For brevity, the equation is usually called *the equation of the circle*, the circumference, however, and not the enclosed surface, is to be understood.

Y P O m A B X

If the origin of the axes be assumed on the circumference, as at A, in the annexed diagram, the equation will be more simple in form; for then, if AO be drawn, we shall have $AB^2 + BO^2 = AO^2$, that is, $\alpha^2 + \beta^2 = r^2$; so that equation (1) reduces to $x^2 - 2\alpha x + y^2 - 2\beta y = 0$,
or $x^2 + y^2 - 2(\alpha x + \beta y) = 0$, ... (2), the equation of the circle, when the origin is on the circumference.

If, in this case, the axis of x pass through the centre, then $\alpha = r$, and $\beta = 0$, and equation (2) becomes $x^2 + y^2 - 2rx = 0$, or $y^2 = (2r - x)x$ . . (3).

But, if the axis of y pass through the centre, then $\alpha = 0$, and $\beta = r$, and equation (2) takes the form

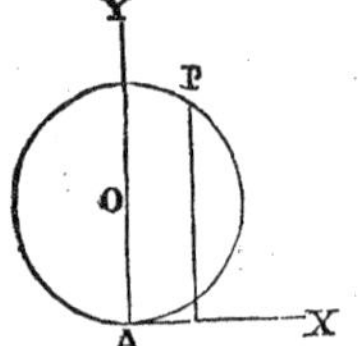

$$x^2 + y^2 - 2ry = 0 \quad . \; . \; . \; . \quad (4).$$

When the axes originate at the centre, the form of the equation is still more simple; for, as in this case, both α, and β are 0, equation (1) becomes $x^2 + y^2 = r^2$ . . . . (5), and this form, on account of its simplicity, is most generally employed.

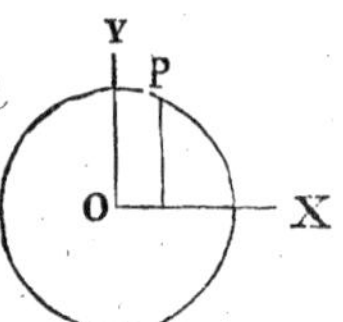

Equation (3) may obviously be converted into this proportion, viz.

$$x : y :: y : 2r - x;$$

that is, a perpendicular, PM, from any point in the circumference, to a diameter, AD, is a mean between the parts AM, MD, into which the diameter is divided by it.

If in the foregoing cases the angle at A had been oblique, instead of right, the several equations would have been more complicated; equation (1) would then have taken the form (*art.* 14,)

$$(x - \alpha)^2 + (y - \beta)^2 + 2(x - \alpha)(y - \beta) \cos. A = r^2,$$

and the simplest form of this equation, viz. that corresponding to equation (5), above, the axes originating at the centre, would be

$$x^2 + y^2 + 2xy \cos. A = r^2.$$

We shall now proceed to the solution of some problems relating to the circle, always referring the curve, for the sake of simplicity, to rectangular axes.

PROBLEM I.

(25.) To find the equation of the tangent at a point in the circumference of a circle.

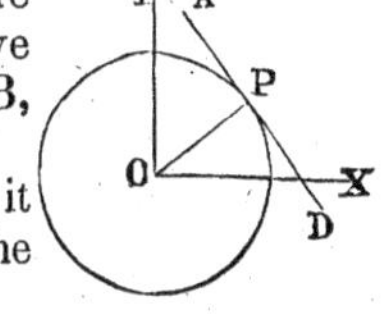

Let the rectangular axes originate at the centre and let (x', y') represent the point, P; then we have to find the equation of the straight line, AB, which touches the circumference in this point.

Let the radius, OP, be drawn: then since it passes through the point P, and is drawn from the origin, the equation of OP is $y=\frac{y'}{x'}x$.

Now a tangent is perpendicular to the radius at the point of contact, consequently, we have merely to express the equation to the line drawn through $(x', y',)$ and perpendicular to that represented by the foregoing equation, the equation of the tangent is, therefore,

$y-y'=-\frac{x'}{y'}(x-x')$, which, by reduction becomes

$$yy'+xx'=y'^2+x'^2, \text{ or } =r^2.$$

The first form of the equation is that most frequently employed.

The equation of the tangent may also readily be determined, independently of the geometrical property referred to above, by a mode of investigation which is applicable to all curves whatever. Thus,

Let us first consider a secant to the curve, that is, a line cutting it in two points, $(x', y',)$ and (x'', y'').

The equation of this secant is $y-y'=\frac{y'-y''}{x'-x''}(x-x') \ldots (1)$;

and, as in the present instance, both points are in the circumference of a circle, we must have $x'^2+y'^2=r^2 \ldots (2)$, and $x''^2+y''^2=r^2 \ldots (3)$, equation (3), subtracted from (2), gives

$$y'^2-y''^2=-(x'^2-x''^2);$$

that is, $(y'+y'')(y'-y'')=-(x'+x'')(x'-x'')$;

whence $\frac{y'-y''}{x'-x''}=-\frac{x'+x''}{y'+y''}$, consequently, by substitution, equation (1) becomes $y-y'=-\frac{x'+x''}{y'+y''}(x-x') \ldots (4)$

If now we suppose that the points through which this secant passes coincide, it will then become a tangent; we have only, therefore, to put in equation (4) $x'=x''$ and $y'=y''$, and there results for the tangent the equation $y-y'=-\frac{x'}{y'}(x-x')$, as before found.

Since the equation of the line drawn from the origin through the point (x', y') is $y=\frac{y'}{x'}x$, it follows, from the foregoing equation, which is obviously that of a perpendicular to this, through the point $(x', y',)$ that *the tangent through any point is perpendicular to the radius at that point*, a property which was assumed in the preceding investigation.

PROBLEM II.

(26.) To draw a tangent to a circle from a given point without it.

Let $(a, b,)$ characterize the given point, P, when referred to rectangular axes originating at the centre, then it is required to find through what point, $(x', y',)$ on the circumference the line must pass to be a tangent.

The point $(x', y',)$ being on the circumference, there must exist the relation $x'^2 + y'^2 = r^2 \ . \ . \ . \ (1)$. Also, since the point $(a, b,)$ is on the tangent, we must have, by substituting its coordinates for x and y, in the equation of the tangent, the relation $ax' + by' = r^2 \ . \ . \ (2)$.

Now, from these two equations, we may determine the unknown coordinates x' and y'; and it is obvious that we shall arrive at two systems of values, for the first of the equations above is of the second degree; we may infer, therefore, that there will be two points in the curve, to each of which a tangent may be drawn from the given point. As the analytical representation of the coordinates of these points will be rather complicated, instead of obtaining them from the preceding equations, we shall determine the points geometrically; and to do this, we shall merely have to construct the locus of the simple equation $ax + by = r^2$, in reference to the axes of the circle, for this locus must of necessity intersect the curve in the two points sought, since, by virtue of equations (1) and (2), the coordinates x', y', belong both to this locus and to the circle (*see art.* 23.)

For $y = 0$, the value of x is $\frac{r^2}{a}$, and for $x = 0$, the value of y is $\frac{r^2}{b}$ therefore, on AX take $AC = \frac{r^2}{a}$, and on AY take $AB = \frac{r^2}{b}$, then the straight line, BC, drawn through the points B, C, will intersect the circumference in the required points, M, M′. The value of x for $y = 0$, that is to say, AC, being $\frac{r^2}{a}$, an expression independent of b, must be the same, whatever value b may take, that is, whatever be the length of the perpendicular, NP; we may infer, therefore, that in whatever point of this perpendicular P be situated, the chord joining the points of contact of tangents drawn from it, will always intersect the axis of x in the same point, C. This property may be thus expressed.

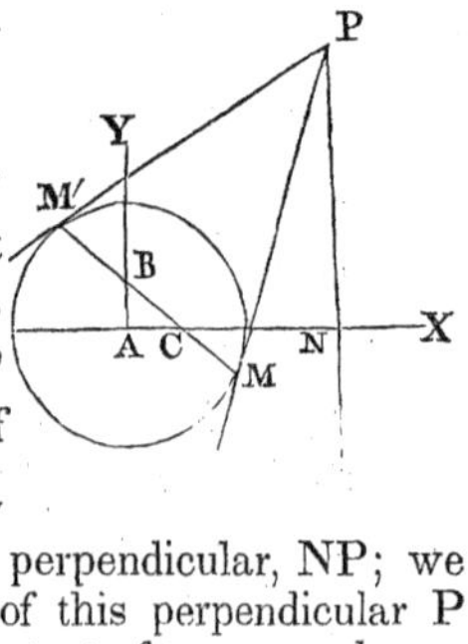

If from any number of points in a straight line tangents be drawn to a circle, the chords joining each pair of tangents will all intersect the perpendicular, from the centre to the line, in the same point.

If the proposed line cut the circle, the intersection of the chords will lie without the circle; for as then $a \angle r$, $x = \frac{r^2}{a}$ must exceed r; but, if the line be wholly without the circle, the intersection will obviously be within it.

If instead of constructing the locus of the equation $ax + by = r^2$, we had actually solved the equations (1) and (2), we should have arrived at the following expressions for the coordinates of the points of contact, viz. $y' = \frac{br^2}{a^2 + b^2} \pm \frac{ar}{a^2 + b^2} \sqrt{\{a^2 + b^2 - r^2\}}$ . . . . . (3)

and $x' = \frac{ar^2}{a^2 + b^2} \mp \frac{br}{a^2 + b^2} \sqrt{\{a^2 + b^2 - r^2\}}$ . . . . . (4).

By means of these coordinates we can find the equation of the line passing through the two points to which they belong, and this equation we shall find to be $ax + by = r^2$, as above. For the equation of a line passing through two points is of the form

$$y - y' = a'(x - x') \quad . . . . . (1),$$

in which the coefficient a' is equal to the difference of the ordinates of the points divided by the difference of the abscissas; in the present case, the difference of the ordinates is $+\frac{2ar}{a^2 + b^2} \sqrt{\{a^2 + b^2 - r^2,\}}$

and the difference of the abscissas, $-\frac{2br}{a^2 + b^2} \sqrt{\{a^2 + b^2 - r^2\}}$;

so that, dividing the first of these expressions by the second, we have for the value of a', $a' = -\frac{a}{b}$ and, consequently equation (1) is

$y - y' = -\frac{a}{b}(x - x')$, or $by + ax = by' + ax'$; but, by the conditions of the problem, $ax' + by' = r^2$, and $\therefore ax + by = r^2$, the equation required.

PROBLEM III.

(27.) A circle and a point being given, it is required to draw from the point a straight line through the circle, so that the part intercepted by the circumference may be of a given length.

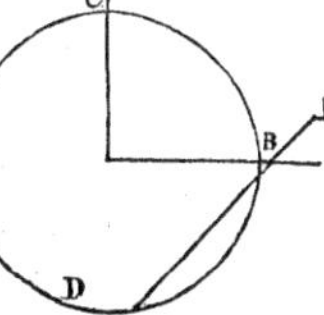

Let ACBD be the given circle, and P the given point, the coordinates of which are x', y', the axes being as before. Let the unknown distance of the point (x', y') from one of the points of intersection be represented by z, we shall then have the following equations, viz.

$x^2+y^2=r$ . . (1) $y-y'=a(x-x')$ . . (2) and $z^2=(x-x')^2+(y-y')^2$ . . (3). the first representing the given circle; the second, a straight line through the given point; and the third, the square of the unknown distance.

Now at the points where the line and circle intersect, the same co-ordinates will belong to each, so at these intersections the values of x and y will be the same, in each of the above equations.

Hence, substituting for $(y-y')^2$, in equation (3), its value $a^2(x-x')^2$, in equation (2), we have $z^2=(x-x')^2(1+a^2)$,

from which we get $x-x'=\dfrac{z}{\sqrt{\{1+a^2\}}}$ and $y-y'=\dfrac{az}{\sqrt{\{1+a^2\}}}$

consequently $x=x'+\dfrac{z}{\sqrt{\{1+a^2\}}}$, and $y=y'+\dfrac{az}{\sqrt{\{1+a^2\}}}$.

Substituting these values of x and y, in equation (1,) it becomes, after reduction, $z^2+\dfrac{2(x'+ay')}{\sqrt{\{1+a^2\}}}z+x'^2+y'^2-r^2=0$ . . . . (4).

The roots of this quadratic are

$$z=\frac{x'+ay'}{\sqrt{\{1+a^2\}}}\pm\frac{1}{\sqrt{\{1+a^2\}}}\sqrt{\{(r^2-x'^2)a^2+2ay'x'+r^2-y'^2\}}.$$

Since the difference of these two values of z must express the given length, we have, by calling it $2m$,

$$2m=\frac{2}{\sqrt{\{1+a^2\}}}\sqrt{\{(r^2-x'^2)a^2+2ax'y'+r^2-y'^2\}}.$$

If now we square this expression, we shall obtain, after reduction, the quadratic $(x'^2+m^2-r^2)a^2-2x'y'a+y'^2+m^2-r^2=0$ . . . (5), in which all the quantities are known, except a; this, therefore, may now be determined, and thence the required line $y-y'=a(x-x')$ drawn, and it is plain that, as the solution of the above quadratic will give two values for a, two lines may be drawn from P, fufilling the proposed condition.

If the given point be upon the axis of x, then $y'=0$, and equation (5), becomes $(x'^2+m^2-r^2)a^2+m^2-r^2=0$, so that the expression for a, which in this case is, $a=\sqrt{\dfrac{r^2-m^2}{x'^2+m^2-r^2}}$ . . . (6), will be most simple, if we choose for the axis of abscissas a line from the centre through the given point.

Let us now actually construct the line from its equation $y=a(x-x')$, which equation (2) becomes, when the axis of x passes through (x', y'). The numerator $\sqrt{\{r^2-m^2\}}$ of the coefficient a, in this equation, obviously expresses the side of a right-angled triangle, of which the hypothenuse is r, and the other side m. Let, then, this line be constructed, and call it p. The denominator, also, of the fraction a being $\sqrt{\{x'^2+m^2-r^2\}}$, or $\sqrt{\{x'^2-(r^2-m^2)\}}$, expresses the side of a right-angled triangle of which the hypothenuse is x', and the other side $\sqrt{\{r^2-m^2\}}$, the line just constructed. If then this second line be also constructed, and represented by q, the equation to the required line will be $y=\frac{p}{q}(x-x')$.

Take, therefore, on OP, the distance, $PF=q$, and on a perpendicular at the extremity, take $FG=p$, then the straight line, GE′, through P, will be drawn as required; for the trigonometrical tangent of the angle FPG will be $p \div q$.

If $m=0$, the proposed line will be a tangent to the circle, and, in that case, equation (6) becomes $a=r\div\sqrt{\{x'^2-r^2\}}$. This, therefore, is the value of the trigonometrical tangent, which the line from the given point to the centre must make with another line drawn from the same point, in order that this latter may touch the circle, hence we have an analytical solution to prob. 2.

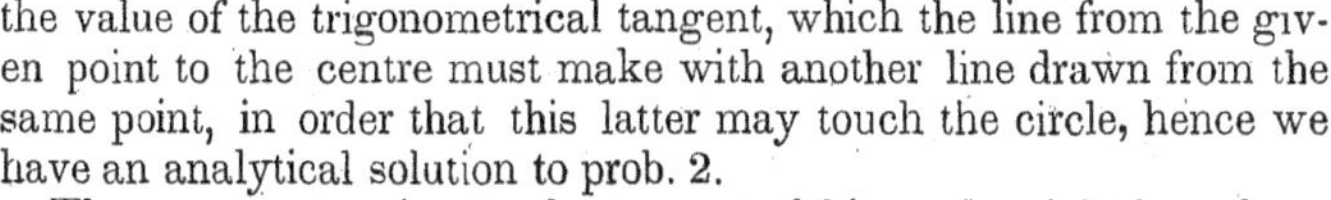

The same expression for the tangent of this angle might have been readily derived from equations (3) and (4), in problem 2. For, since a tangent, PC, is perpendicular to the radius, OC, it follows that the angle, OCp, included by this radius and the ordinate of the point C, is equal to the angle OPC; so that the trigonometrical tangent of this angle will be expressed by dividing the abscissa, Op, of the point C by the ordinate, pC, the axis of x being supposed to pass through P. Now, on this supposition, b, in the equations referred to, is 0, so that the coordinates of C become

$y'=\frac{r}{a}\sqrt{\{a^2-r^2\}}$, $x'=\frac{r^2}{a}$, $\therefore \frac{x'}{y'}=\frac{r}{\sqrt{\{a^2-r^2\}}}$ which expression is the same as that above the abscissa of the point, P, being represented by a in this expression, and by x' in the former.

From equation (4) two well-known theorems may be easily deduced, for representing the roots of that equation by z' z'', we have, by the theory of equations, (*Alg. p.* 177), $z'z''$, or $PE \times PE'=x'^2+y'^2-r^2$; and as the values of x', y', and r, are quite independent of m, this equation subsists for every position of PE′; hence, when P is within the circle, we infer that chords intersecting each other in P are divided, so that the rectangle of the parts of each is the same; and, when P is without the circle, we conclude that all lines drawn therefrom, and terminated by the concave part of the circumference, are so divided by the opposite part, that the rectangle of the whole line and the external part is the same in each.

PROBLEM IV.

(28.) To find the coordinates of the points of intersection of two circumferences.

Let the radii of the two circles be r, r', and the distance of their centres, d. Let the rectangular axes originate at the centre of that circle, whose radius is r, and let it pass through the centre of the other circle; then the equation of the former circle will be $y^2+x^2=r^2$ (1), and the equation of the latter, the coordinates of whose centre is $\alpha=d$, $\beta=0$, will be $y^2+(x-d)^2=r'^2$ . . (2). At the points of intersection, the values of x and y must be the same in both these equations. To determine them, subtract equation (2) from equation (1), and there results $2dx-d^2=r^2-r'^2$, $\therefore x=\dfrac{r^2-r'^2+d^2}{2d}$. Substituting this value of x, in equation (1), we have $y^2=r^2-\left\{\dfrac{r^2-r'^2+d^2}{2d}\right\}^2$

whence $y=\dfrac{1}{2d}\sqrt{\{4d^2r^2-(r^2-r'^2+d^2)^2\}}$ . . . (3).

Now we may observe of this equation that the expression within the brackets is the difference of two squares; it may, therefore, be replaced by two factors, the one denoting the sum, the other, the difference of the roots of these squares, that is by the factors,

$$(2dr+r^2+d^2-r'^2)\ (2dr-r^2-d^2+r'^2).$$

Here again it occurs that each factor is the difference of two squares, the first being $(r+d)^2-r'^2$, and the second, $r'^2-(r-d)^2$; hence, by decomposing each of these into factors, equation (3) is finally reduced to

$$y=\frac{1}{2d}\sqrt{\{(r+r'+d)(r+d-r')(r+r'-d)(r'+d-r)\}}.(4).$$

Since y has here two values, numerically the same but with contrary signs, it follows that *the line joining the centres of two intersecting circles bisects at right angles the line joining the intersections.*

The form under which we have just exhibited the expression for y is very convenient for the examination of the circumstances of the problem, which examination will lead us to the theorems relative to intersecting circles already established in Elements of Geometry.

As the first factor under the radical is necessarily positive, the whole expression must also be positive, provided that all or only one of the remaining factors are likewise positive. Now two of the remaining factors, at least must be positive; for if one, $(r+d-r')$ for instance, be negative, then $(r+d) \angle r'$; consequently $r \angle r'$, and also $d \urcorner r'$, which proves that the other two factors must be positive. There can, therefore, be but two cases to examine, viz. that in which all the factors are positive, and that in which one is negative. In the first case the values of y will be real, in the second, they will be imaginary. In the first case, there must obviously subsist the conditions $r+d \urcorner r'$, $r+r' \urcorner d$, $r'+d \urcorner r$, which prove that *if two circumferences cut, the distance of their centres must be less than the sum, and greater than the difference of the radii.*

In the second case where y becomes imaginary because of a negative factor, we must have one of the conditions $d' < r' - r$, $d > r + r'$, $d < r - r'$; so that *two circumferences can have no point in common, if the distances of the centres be less than the difference or greater than the sum of the radii.*

Lastly, let one of the three last factors be 0, which can happen only when d is equal either to the sum or difference of the radii. In this case, $y = 0$, showing the circumferences have but one point in common, and that this is on the axis of x; so that *two cirumferences touch, when the distance of the centres is equal to the sum or difference of the radii.*

(29.) In addition to the problems here given, a variety of others relating to the circle might be proposed, which would conduct to other properties of this curve. But, as the circle occupies so large a portion of elementary geometry, where its most important properties are unfolded with the utmost simplicity and elegance, it would be superfluous to dwell upon it at any great length here. Indeed, the theorems established in elementary geometry are, for the most part, obtained with less ease and simplicity by analytical processes than by pure geometrical reasoning. This fact the student has, no doubt, had occasion to remark, in some of the foregoing investigations; these however, it would not have been proper wholly to have omitted, on this account; their introduction has not only furnished the student with the means of applying the fundamental principles of analysis, but has, at the same time, given him confidence in those principles, by conducting him to results previously known to be true.

In the remaining sections of this work, which will treat of curves not within the limits of elementary geometry, the great advantage of analysis will be more distinctly seen. Many important properties of these curves will be obtained with the utmost ease and facility, which could not be established by common geometry but by very lengthy and elaborate reasonings.

(30.) We shall terminate this section with one or two problems on *loci*; first, however, showing that, as the general equation of a circle, when referred to rectangular coordinates, is $(x - \alpha)^2 + (y - \beta)^2 = r^2$, or, $x^2 - 2\alpha x + a^2 + y^2 - 2\beta y + \beta^2 - r^2 = 0$, so, conversely, every equation of the second degree of the form $x^2 + y^2 + Ax + By + C = 0$ (1) will be the equation of a circle: this may be proved as follows.

Take any rectangular axes, AX, AY, and find the point 0, whose abscissa is $-\frac{1}{2}A$ and ordinate $-\frac{1}{2}B$; then, from this point as a centre, with a radius equal to $\sqrt{\left\{\frac{A^2 + B^2}{4} - C\right\}}$, describe a circumference; this circumference will be the locus of the proposed equation.

For the equation of this circumference being $(x-\alpha)^2+(y-\beta)^2=r^2$, where, by construction, $\alpha=-\frac{1}{2}A$, $\beta=-\frac{1}{2}B$, and $r^2=\frac{1}{4}(A^2+B^2)-C$, it is the same as $(x+\frac{1}{2}A)^2+(y+\frac{1}{2}B)^2=\frac{1}{4}(A^2+B^2)-C$, which reduces to $x^2+y^2+Ax+By+C=0$; hence the circumference just described is the locus of this equation.

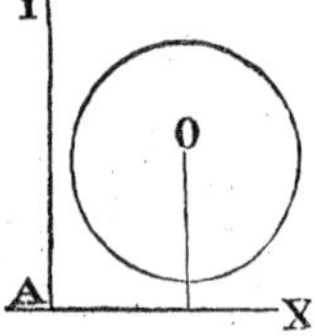

Suppose, for example, it were required to construct the locus of the equation $2x^2+2y^2-3x+4y-1=0$, which reduces to $x^2+y^2-\frac{3}{2}x+2y-\frac{1}{2}=0$. First, then, find a point O, having $\frac{3}{4}$ for its abscissa, and -1 for its ordinate; then from this point, as a centre, with a radius equal to $\frac{1}{4}(\frac{3^2}{2^2}+2^2)+\frac{1}{2}=\sqrt{\frac{33}{16}}=\frac{1}{4}\sqrt{33}$, describe a circumference, which will be the locus sought.

It must be observed that the coefficients of the proposed equation may be related so as to render $\frac{1}{4}(A^2+B^2)=C$, in which case the equation will represent a circle whose radius is 0, that is, merely a point; such is the equation $x^2+y^2-3x-2y+\frac{13}{4}=0$, which represents a point whose coordinates are $x=\frac{3}{2}$ and $y=1$.

The coefficients may also be so related that the equation may have no geometrical representation, as in the equation $x^2+y^2+4x-2y+7=0$, which, for every possible value of x, gives an imaginary value for y; so that no real line or point can be represented by the equation. In such equations, the expression for the square of the radius will always be negative; when, therefore, we say that equation (1) represents a circle, we must be understood as meaning, that no other line can be represented by it; so that, when the locus is not a circle, the geometrical representation is impossible.

We shall now add a few questions, leading to indeterminate equations of the second degree, the loci of which will furnish every geometrical solution.

PROBLEM V.

(31.) Given the base and the sum of the squares of the sides to determine the triangle.

Let AB be the base, and put m for the sum of the squares of the sides, AC, BC.

Let the perpendicular, OY, from the middle of the base, form, with the base, the rectangular axes; then putting a for AO, or OB, and (x, y), for the point C, we shall have the equations $y^2+(x+a)^2=AC^2$. (1) and $y^2+(x-a)^2=BC^2$. (2). Adding these together, $2y^2+2x^2+2a^2=AC^2+BC^2$. (3); therefore, putting m for AC^2+BC^2, $y^2+x^2=\frac{1}{2}(m-2a^2)$ which equation represents a circle, of which the centre is the origin, O, and the radius $\sqrt{(\frac{1}{2}m - a^2)}$; so that, if this circle be described, and lines be drawn from A, B, to any point in its circumference, a triangle having the proposed conditions will always be formed, and thus, when the base and sum of the squares of the sides are constant, the locus of the vertex is a circle.

Since $y^2 + x^2 = OC^2$, it follows, from equation (3), that we have $2AO^2 + 2OC^2 = AC^2 + BC^2$; that is, *in aay triangle, the sum of the squares of the sides is equivalent to twice the squares of half the base, and of the line from the vertex to the middle of the base.* (*Geom. p.* 38.)

PROBLEM VII.

(32.) Given the base and the vertical angle to determine the triangle.

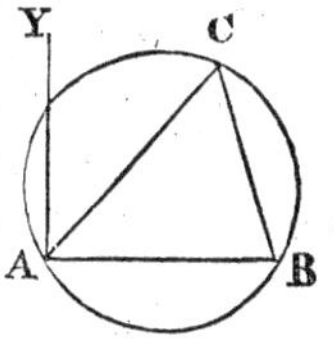

Let b represent the given base, AB, and put v for the tangent of the given angle; then, taking AB, AY, for rectangular axes, we shall have, for the equation of any line, AC, drawn from the origin $y = ax$ .... (1), and the equation of another line, BC, drawn from the point $(b, 0)$, and making the proposed angle with the former line will be

$y = \frac{a+v}{1-av}(x-b)$. (2), (x, y) being the point of intersection, C.

Now, for a, in equation (2), substitute $\frac{y}{x}$, its value in equation (1).

and we thus obtain the equation $y = \dfrac{\frac{y}{x}+v}{1-\frac{y}{x}v}(x-b)$, which reduces to

$y^2 + x^2 - \frac{b}{v}y - bx = 0$, the equation of a circle of which the coordinates of the centre are $\beta = \frac{b}{2v}$, and $\alpha = \frac{b}{2}$, and radius $\sqrt{(\frac{b^2}{4v^2} + \frac{b^2}{4})} = \sqrt{(\beta^2 + \alpha^2)}$.

From the foregoing expression for the radius, it is plain that the circumference passes through the origin; and, since for $y = 0$ the above equation gives $x = b$, it follows that it also passes through B; hence the base of the triangle subtends the arc of which the vertex is the locus.

If the given angle is right, then its tangent, v, is infinite, and therefore $\beta = 0$; so that, in this case, the centre of the circle is at the middle of AB. If the angle be acute, β must be positive, for v will be so in this case; therefore, the centre is situated above the base; but, if the angle be obtuse, then β will be negative, and the centre will be below the base. Hence the segment containing a right angle must be a semicircle, and the segment will be greater or less than a semicircle, according as the angle contained in it is acute or obtuse.

It must be remarked, that no part of the arc below the base belongs to the locus, which we have determined, because equation (2) requires that the angle be formed to the *right* of AC, (*see art.* 13,) fixing the intersection above AB. As, however, there is no restriction of this kind in the problem, we may admit, the proposed angle to be formed to the *left* of AC, in which case the equation of BC will be $y = \frac{a - v}{1 + av}(x - b)$, and the locus, $y^2 + x^2 + \frac{b}{v}y - bx = 0$, representing a circle of the same radius as before, the coordinates of the centre being $\beta = -\frac{b}{2v}$, and $\alpha = \frac{1}{2}b$; so that, in strictness, the locus consists of two equal arcs, situated the one below the other, as in the annexed diagram.

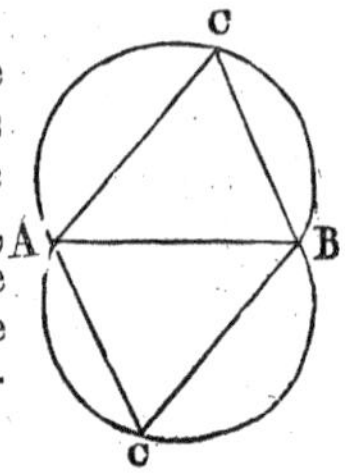

The three problems next following resolve themselves into the preceding.

PROBLEM VII.

(33.) Given the base and vertical angle of a triangle to determine the locus of the intersection of perpendiculars from the angles to the opposite sides.

Let AB be the base, P the intersection of the perpendiculars, AF, BE, on the sides BC, AC; then, since the sum of the angles of a quadrilateral amount to four right angles, the angle P must be the supplement of the angle C, and therefore constant, because C is. Hence we have the base, AB, and vertical angle P of the triangle PAB, to find the locus of P. By the preceding problem, this locus is the arc APB, whose chord, AB, subtends an angle equal to the supplement of C. An equal arc below AB also belongs to the locus.

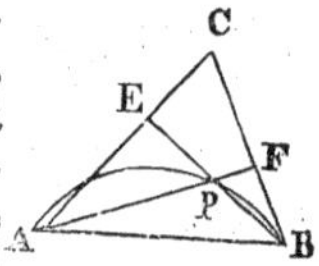

PROBLEM VIII.

(34.) To find the locus of the centre of the inscribed circle when the base and vertical angle of the triangle are given.

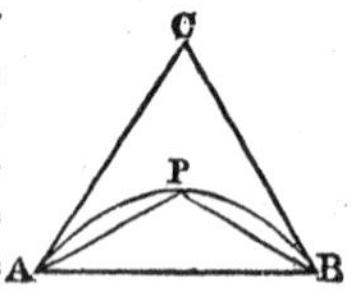

The centre of the inscribed circle is at the intersection of the lines bisecting the angles at the base, and, as the sum of these angles is constant, because the vertical angle is, the half sum must be constant, so that the triangle APB, formed by the given base, AB, and the lines AP, BP, to the centre of the inscribed circle, has the vertical angle P, constant, and equal to two right angles, minus half the sum of the angles at the base of the proposed triangle, that is, to one right angle plus half the given angle; hence the required locus is an arc described on AB, containing this angle. A similar arc below AB, belongs also to the locus.

PROBLEM IX.

(35.) The base and vertical angle of a triangle being given to find the ocus of the intersection of the straight lines, drawn from the angles to the middle of the opposite sides.

By art. (19.) if P be a point of intersection, its distance from M, the middle of the base, AB, will be equal to half its distance from the vertex C.

If, therefore, Pm, Pm', be drawn respectively parallel to AC, BC, we shall have Mm, Mm', each equal to one third of AM or BM, and the angle P equal to the angle C; that is to say, the base, mm', and vertical angle, P, of the triangle, mPm', are constant, the locus of P is, therefore, the arc mPm'. To construct it, we shall have to trisect the base, and to describe upon the middle portion an arc to contain the given angle, or rather, two such arcs, one on each side of mm'.

We shall give another solution to this problem.

Take the rectangular axes, AB, AY, then, since $PM = \frac{1}{3}CM$, the perpendicular from P to the base will be one third of that from C; hence, representing the base by b, the vertex by (x, y), and the point P by (X, Y), we shall have $y = 3Y$; $x = 3X - b$; and, substituting these values of x and y in the locus of (x, y), as represented by its equation in Prob. 6, it becomes $Y^2 + X^2 \mp \frac{b}{3v} Y - bX + \frac{2}{9} b^2 = 0$, the equation of a circle, of which the coordinates of the centre are $\alpha = \frac{b}{2}$, $\beta = \pm \frac{b}{6v}$.

Let $Y = 0$, then the corresponding values of X, given by the solution of the quadratic, are $X = \frac{1}{3}b$; $X = \frac{2}{3}b$, showing that the locus intersects the base in two points, at these distances from A, thus intercepting the middle of three equal portions.

SECTION III.

ON LINES OF THE SECOND ORDER.

(36.) The only curve which we have as yet considered is the circle, whose equation we have found to be of the second degree. Besides this there are three other curves, which, like the circle, are each represented by an equation of the second degree. These three curves we propose in this section to examine, first determining the equation and form of each, and then proceeding to investigate its properties. We shall afterwards show, that every equation of the second degree, with two variables, whatever be its form, can represent no curve, but the circle, or one of these three; and, from this circumstance of their equations being all of the second degree, these four curves are called lines of the second order. Before proceeding to the three new curves, of which we have just spoken, the student should attentively read the following preliminary chapter.

CHAPTER I.

ON THE TRANSFORMATION OF COORDINATES.

(37.) Every equation which characterizes a line, whether straight or curve, will depend for its simplicity upon two circumferences; the relative position of the axes to which the line is referred, and the absolute position of the origin.

This fact has already been observed, as regards the two classes of lines which have hitherto been examined, the straight line and the circle. It was seen in the case of the straight line, that when the axes were oblique, and their origin not upon the line, the equation which characterized it was far less simple than when rectangular axes were employed; the relative position, therefore, of the axes affected the equation. It was moreover observed, that if, in addition to the axes being rectangular, their origin were upon the proposed line, the form of its equation became still more simple. Similar remarks apply to the circle, the equation of which is much more complicated when the axes of reference are oblique, and originate without the curve, than when they are rectangular, and originate at the centre. It may, therefore, readily be conceived, that, with regard to other curves, there may also exist certain positions for the axes, and certain points for their origin, by assuming which, the curve may be susceptible of a more commodious analytical representation, than when the axes and origin are chosen at random. Now the object of this chapter is to show that when a curve, is represented by an equation, in reference to any system of axes, we can always transform that equation into another, which shall equally represent the curve, but in reference to a new system of axes chosen at pleasure. This is called *the transformation of coordinates:* it may consist either in altering the relative position of the axes, without displacing the origin; in removing the origin, without disturbing the relative position of the axes; or, lastly, it may be found necessary to alter both the direction of the axes and the situation of their origin. By means of these transformations, we may often simplify the equation of a curve, and many of its properties, not readily derivable from its equation in one form, may frequently be obtained with great facility by a transformation of it into another, as will be repeatedly seen in the course of the subsequent chapters.

(38.) Let the axes, AX′, AY, be those to which any line, MM′M″, is related by its equation, and let A′X′, A′Y′, be the new axes, to which it is proposed to refer the same line.

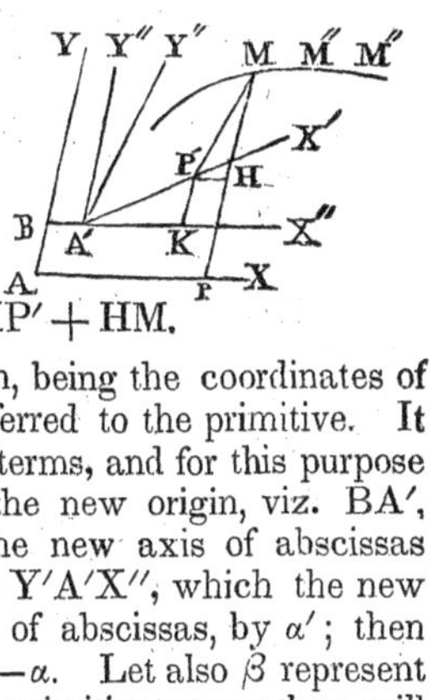

Let the coordinates of any point, M, in the line, relative to the primitive axes, be x and y, and the coordinates of the same point, referred to the new axes, $A'P' = x'$, and $P'M = y'$. Draw A′X″ and P′H each parallel to AX, and P′K parallel to AY, then we shall have $x = AP = BA' + A'K + P'H$, and $y = PM = AB + KP' + HM$.

In these equations BA′, AB, are known, being the coordinates of the origin, A′, of the new axes, when referred to the primitive. It remains, therefore, to determine the other terms, and for this purpose let us represent the known coordinates of the new origin, viz. BA′, AB by a, b; the angle X′A′X″, which the new axis of abscissas makes with the old, by α, and the angle Y′A′X″, which the new axis of ordinates makes with the old axis of abscissas, by α'; then the inclination of the new axes will be $\alpha' - \alpha$. Let also β represent the angle Y″A′X″, the inclination of the primitive axes; then will $\beta - \alpha$ be the angle formed by the new axis of x and old axis of y, and $\beta - \alpha'$ will be the angle formed by the new and old axes of y. Now, by trigonometry, the value of A′K is

$$A'K = \frac{A'P' (\sin. A'P'K)}{\sin. A'KP'} = \frac{x' \sin. (\beta - \alpha)}{\sin. \beta},$$ observing that A′P′K $= Y''A'X'$, or $\beta - \alpha$, and that Y″A′X″, or β, is the supplement of A′KP′. In like manner, for KP′ we have

$$KP' = \frac{A'P' (\sin. P'A'K)}{\sin. A'KP'} = \frac{x' \sin. \alpha}{\sin. \beta}.$$ Also, in the triangle MP′H,

we have, for P′H, $$P'H = \frac{P'M (\sin. P'MH)}{\sin. MHP'} = \frac{y' \sin. (\beta - \alpha')}{\sin. \beta}$$ and for HM, $HM = \frac{P'M\ (\sin. MP'H)}{\sin. MHP'} = \frac{y' \sin. \alpha'}{\sin. \beta}$. Hence, for the values of x and y, we have $x = a + \frac{x' \sin. (\beta - \alpha) + y' \sin. (\beta - \alpha')}{\sin. \beta}$ and $y = b + \frac{x' \sin. \alpha + y' \sin. \alpha'}{\sin. \beta}$. These then are the values which must be substituted in the equation of the curve, when related to the primitive axes, AX, AY, in order to transform it into the equation which the same line must have when referred to the new axes A′X′, A′Y′. The above general expressions become modified in particular cases, the principal of which we shall here exhibit.

1. *When the new axes are parallel to the old.*

In this case, the inclination of the axes remaining unaltered, while the origin is removed, we have $\alpha = 0$, $\beta - \alpha' = 0$; hence the above expressions become $x = a + x'$ and $y = b + y'$ . . . . (1).

2. *When the primitive axes are rectangular, and the new ones oblique.*

Here $\sin. (\beta - \alpha) = \cos. \alpha$, and $\sin. (\beta - \alpha') = \cos. \alpha'$, therefore

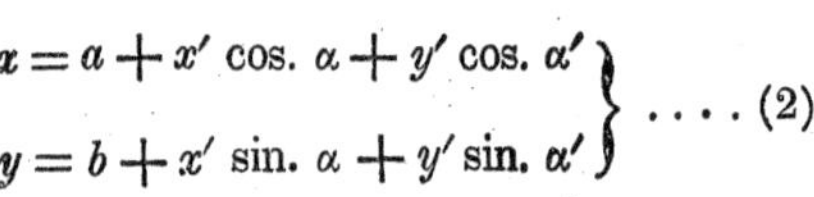

$$\left.\begin{aligned} x &= a + x' \cos. \alpha + y' \cos. \alpha' \\ y &= b + x' \sin. \alpha + y' \sin. \alpha' \end{aligned}\right\} \ldots . (2).$$

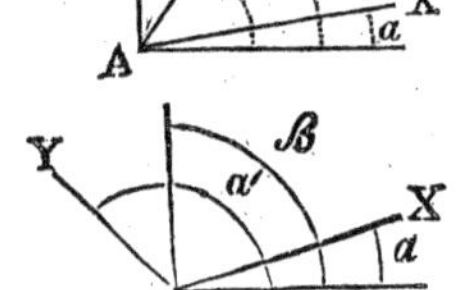

3. *When both systems are rectangular.*

Here $\alpha' = 90° + \alpha$, $\therefore \sin. \alpha' = \cos. \alpha$; also $\sin. (\beta - \alpha') = \sin. (-\alpha) = -\sin. \alpha$; and $\sin. (\beta - \alpha) = \cos. \alpha$;

hence the expressions become, in this case,

$$\left.\begin{aligned} x &= a + x' \cos. \alpha - y' \sin. \alpha \\ y &= b + x' \sin. \alpha + y' \cos. \alpha \end{aligned}\right\} \ldots (3).$$

4. *When the primitive axes are oblique, and the new ones rectangular.*

In this case, $\alpha' = 90° + \alpha \therefore \sin. \alpha' = \cos. \alpha$ also, since the complement of $\beta - \alpha'$ is $90° - (\beta - \alpha') = 180° - (\beta - \alpha)$, we have $\sin. (\beta - \alpha') = -\cos. (\beta - \alpha)$, so that the formulas for this case are

$$\left.\begin{aligned} x &= a + \frac{x' \sin.(\beta - \alpha) - y' \cos.(\beta - \alpha)}{\sin.\beta} \\ y &= b + \frac{x' \sin.\alpha + y' \cos.\alpha}{\sin.\beta} \end{aligned}\right\} (4).$$

In the first of the marginal figures, the two systems of axes are so placed that, $\beta - \alpha$ is less than 90°; its sine and cosine are, therefore, both positive; and the expression for x shows that y' times the latter is to be *subtracted* from x' times the former. In the second figure the two systems are placed so that $\beta - \alpha$ is greater than 90°; its sine, therefore, is positive, as before, but its cosine is negative; and the formula shows that in this case y' times the cosine, is to be taken positively, that is, it must be *added* to x' times the sine. It must be remembered that, in each of the preceding cases, when the transformation is confined merely to the direction of the axes, the origin remaining fixed the terms a, b, become 0. It must further be remarked, that, when the axes of x in the two systems coincide, then $\alpha = 0$, and when the axis of y in both systems are identical, then $\beta = \alpha'$.

Suppose, for example, we wish to pass from an oblique to a rectangular system, the origin and axis of x remaining undisturbed, the formulas (4) will give

$$x = x' - \frac{y' \cos.\beta}{\sin.\beta} = x' - y' \cot.\beta \text{ and } y = \frac{y'}{\sin.\beta} = y' \operatorname{cosec}.\beta.$$

(39.) Throughout the whole of the preceding investigation the angle α, or that which the new axis of x makes with the old, is supposed to be positive, that is, we have uniformly conceived this angle to be situated *above* the primitive axis of x. If, on the contrary, it be supposed negative, or to fall *below* the same axis, then its sine will be negative, but its cosine will continue positive. Hence, in this position of the new axis of x relatively to the old, the preceding formulas will require some modification. We shall, therefore, in order to complete this theory, here present them with the necessary changes.

For the first of the preceding cases the formulas are the same.

For the second they are (2') . . . $\begin{cases} x = a + x' \cos.\alpha + y' \cos.\alpha'. \\ y = b - x' \sin.\alpha + y' \sin.\alpha'. \end{cases}$

For the third, (3') . . . $\begin{cases} x = a + x' \cos.\alpha + y' \sin.\alpha. \\ y = b - x' \sin.\alpha + y' \cos.\alpha. \end{cases}$

For the fourth, (4') . . . $\begin{cases} x = a + \dfrac{x' \sin.(\beta + \alpha) - y' \cos.(\beta + \alpha)}{\sin.\beta} \\ y = b + \dfrac{y' \cos.\alpha - x' \sin.\alpha}{\sin.\beta}; \end{cases}$

or, since the angle $(\beta + \alpha)$ is here the same as α' in figs. 1 and 2, we may write these last expressions thus: α' expressing the inclination of the old axis of y and new axis of x, and $\alpha' - \alpha$ the inclination of the old axes of x and y, (4′) . . . $\begin{cases} x = a + \dfrac{x' \sin. \alpha' - y' \cos. \alpha'.}{\sin. (\alpha' - \alpha).} \\ y = b + \dfrac{y' \cos. \alpha - x' \sin. \alpha.}{\sin. (\alpha' - \alpha).} \end{cases}$

It may, perhaps, be satisfactory to the student to verify this last formula, which is the most complicated, by determining the values of x' and y' from formulas (2); we shall thus have the values of the oblique coordinates in terms of the rectangular, as above. Omitting the constants a and b, in (2), and multiplying the first equation by sin. α', the second by cos. α', and subtracting the latter result from the former, we have, $x \sin. \alpha' - y \cos. \alpha' = x' (\cos. \alpha \sin. \alpha' - \cos. \alpha' \sin. \alpha)$ $= x' \sin. (\alpha' - \alpha) \quad \therefore x' = \dfrac{x \sin. \alpha' - y \cos. \alpha'}{\sin. (\alpha' - \alpha)}$. In like manner, by multiplying the first equation by sin. α, the second by cos. α, and proceeding as before, we get $y' = \dfrac{y \cos. \alpha - x \sin. \alpha}{\sin. (\alpha' - \alpha)}$ values which verify the preceding formulas.

(40.) Some authors employ a different notation in these formulas; thus, instead of using the letters α, α', &c, to denote the angles about A′, they employ the sides which contain them, the primitive axes being denoted by X, Y, and the new ones by X′, Y′, so that [X, X′] is put for α, [Y′, X] for α', &c. By adopting this notation, the first class of formulas, where [X, X′] is positive, will be

1. *When the primitive axes are rectangular, and the new ones oblique.*

$$x = a + x' \cos. [X, X'] + y' \cos. [X, Y']$$
$$y = b + x' \sin. [X, X'] + y' \sin. [X, Y'].$$

2. *When both systems are rectangular.*

$$x = a + x' \cos. [X, X'] - y' \sin. [X, X']$$
$$y = b + x' \sin. [X, X'] + y' \cos. [X, X'].$$

3. *When the primitive axes are oblique, and the new ones rectangular.*

$$x = a + \frac{x' \sin. [X', Y] - y' \cos. [X', Y]}{\sin. [X, Y]}$$
$$y = b + \frac{x' \sin. [X, X'] + y' \cos. [X, X']}{\sin [X, Y]}.$$

The second class of formulas, where [X, X′] is negative, will be

1. *When the primitive axes are rectangular, and the new ones oblique.*

$$x = a + x' \cos. [X, X'] + y' \cos. [X, Y']$$
$$y = b - x' \sin. [X, X'] + y' \sin. [X, Y']$$

2. *When both systems are rectangular.*

$$x = a + x' \cos. [X, X'] + y' \sin. [X, X']$$
$$y = b - x' \sin. [X, X'] + y' \cos. [X, X'].$$

3. *When the primitive axes are oblique, and the new ones rectangular.*

$$x = a + \frac{x' \sin. [X', Y] - y' \cos. [X', Y]}{\sin. [X, Y]}$$

$$y = b + \frac{y' \cos.[X, X'] - x' \sin.[X, X']}{\sin.[X, Y]}$$

Under this form of notation, the angles introduced are more distinctly marked, and therefore more readily recognized, than when they are each represented by a single letter. Still, however, as these formulas are less brief, and, consequently, less commodious in calculation, the form of notation first given is generally adopted in preference.

CHAPTER II.

ON THE ELLIPSE.

Its equation and Properties.

(41.) An ellipse is a curve from any point, P, in which, if straight lines be drawn to two fixed points, F′, F, their sum will always be the same.

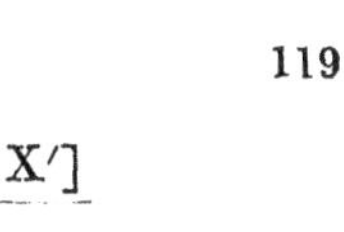

P
F′
F

The points F′, F, are called the foci of the ellipse and the distance F′P or FP of either from a point in the curve, is called the *focal distance* or *radius vector* of that point.

From the definition of an ellipse, the curve may be readily described mechanically; thus, to the two fixed points, F′, F, let the extremities of a cord be fastened, let this cord be stretched into a loop, F′PF, by means of a pencil, P, then the motion of this pencil, still keeping the cord stretched, will evidently describe an ellipse. The cord must obviously exceed in length the distance F′F.

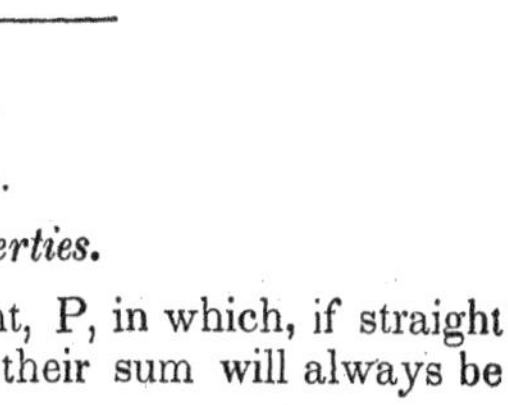

Y
C
P
A
F′
O
F
M
B
X
D

Let us now seek the equation of this curve, and, for this purpose, let us take OX, OY, for rectangular axes, the origin O being placed at the middle of F′F, and let the sum of the distances of any point, P, in the curve from the foci, be represented by 2A. Put also c for OF or OF′; then, if PM be drawn perpendicular to OX, we shall have OM $= x$, PM$=y$ and, consequently, $y^2+(x-c)^2=\text{PF}^2$, (1), $y^2+(x+c)^2=\text{PF}'^2$, (2), also $\text{PF}+\text{PF}'=2\text{A}$, . . (3).

Hence, by addition and subtraction, we get $2y^2+2x^2+2c^2=\text{PF}^2+\text{PF}'^2$, (4), and $4cx=\text{PF}'^2-\text{PF}^2=(\text{PF}'+\text{PF})(\text{PF}'-\text{PF})=2\text{A}(\text{PF}'-\text{PF})$, $\therefore \text{PF}'-\text{PF}=2cx \div \text{A}$; hence, combining this last equation with equation (3) there results $\text{PF}'=\text{A}+\frac{cx}{\text{A}}$ and $\text{PF}=\text{A}-\frac{cx}{\text{A}}$; and, if these values be substituted in equation (4), we have $y^2 + x^2 + c^2 = A^2 + \frac{c^2 x^2}{A^2}$, which finally reduces to

$$A^2 y^2 + (A^2 - c^2) x^2 = A^2 (A^2 - c^2), (5).$$

In this equation, x and y are the coordinates of any point in the curve, and the other terms are all constant; this, therefore, is the equation of the curve.

(42.) Let us now inquire at what points the curve cuts the axes. For this purpose, put $y = 0$, in equation (5), and there results for the abscissa of the point where the curve cuts Ox, $x = \pm A$. We hence learn that x has two values, viz. $x = A$, and $x = -A$, so that the curve cuts the axis of x in two points, B, A, each at the distance, A, from the origin, the one being to the right, the other to the left.

Suppose, now, $x = 0$, in the same equation, and there results the ordinate of the point when the curve cuts the axis of y, $y = \sqrt{\{A^2 - c^2\}}$. Since this value admits of being taken positively or negatively, we infer that the curve cuts the axis of y also in two points, C, D, equi-distant from O, the one above and the other below it. Hence the two chords, AB, CD, are mutually bisected at the point O. As the former chord is represented by 2A, let us denote the latter by 2B, that is, put $2\sqrt{\{A^2 - c^2\}} = 2B$, and then the equation of the ellipse (5), in terms of the two chords AB, CD, when these are taken for axes, assumes the more simple form, $A^2 y^2 + B^2 x^2 = A^2 B^2$ or $y^2 = \frac{B^2}{A^2}(A^2 - x^2)$, (6), and $\therefore y = \pm \frac{B}{A}\sqrt{\{A^2 - x^2\}}$;

$x = \pm \frac{A}{B}\sqrt{\{B^2 - y^2\}}$.

From these expressions for y and x, it appears that for the same value of x, there are two values of y numerically equal, but having contrary signs; hence the chord AB bisects all the chords drawn parallel to CD. In like manner with regard to x; this also has two values numerically equal, but differing in sign for one value of y; therefore, the chord CD bisects all the chords drawn parallel to AB.

Moreover, since, for $x = A$, or $x = -A$, the corresponding value if y is 0, it follows that parallels to the axis of y drawn through the points B, A, of the curve, meet it in no other point, that is, they are tangents to it at that point. In like manner, for $y = B$ or $y = -B$, the corresponding value of x is 0; hence it may in the same way be inferred that parallels to the axis of x through the points C, D, of the curve, are also tangents to it. It appears, likewise, that at no point of the curve can the abscissa exceed A, for when $x > A$, y is imaginary.

(43.) Let us now seek the expression for the distance of any point (x, y) in the ellipse, from the origin, O. We shall merely have to substitute for y^2 in the expression, $D = \sqrt{\{x^2 + y^2\}}$ its value $\frac{B^2}{A^2}(A^2 - x^2)$, as given in equation (6), and the expression sought will be $D = \sqrt\{B^2 + \frac{A^2 - B^2}{A^2} x^2\}$, (7). This expression for the distance will obviously remain the same, whether x, y, be positive or negative. or the one positive and the other negative; hence, those points in the curve, whose coordinates have respectively the same numerical values, however the signs thereof may differ, are equally distant from O, so that if $OM = x$, $PM = y$, and $OM' = -x$, $P'M' = -y$ be the coordinates of two points, P, P', then will OP be equal to OP', and, consequently, the angles M'OP, MOP, are equal; therefore, M, M', being a straight line, PP' must be also a straight line, so that OP' is only the continuation of OP Since, to every abscissa there belongs two ordinates, equal in length, but of different signs, it follows that, if the ordinates of P', P, be produced to p', p these points will be expressed by $(-x, y,)$ $(x, -y)$; they are, therefore, at the same distance from O as the former, and p, p', will, in like manner, be a straight line, and equal to PP'.

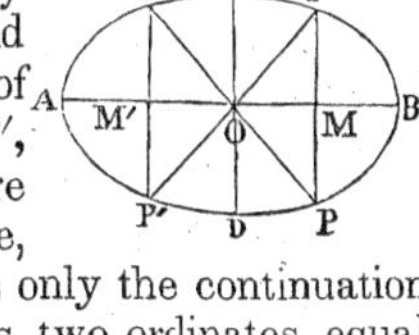

Because every chord passing through O is thus bisected in that point, O is called the *centre* of the curve, and the chords passing through it, *diameters*. It appears, from the above, that diameters PP', pp', which make equal angles, POM, p'OM', with the axis of x, are equal.

Since A is necessarily greater than B (*art.* 42), the preceding expression for D must increase or diminish, accordingly as x increases or diminishes; D will, therefore, be greatest when x is greatest, or equal to A, and it will be least when x is least, or equal to 0; so that OB or OA is the greatest distance of the centre, O, from the curve, and OC or OD is the least distance. Hence, of all the diameters, AB is the greatest and CD the least, and, for this reason, AB is often called the *major diameter*, and CD the *minor diameter*; when spoken of together, they are called the *principal diameters*, or the *principal axes* of the ellipse; consequently, the equation $A^2y^2 + B^2x^2 = A^2 B^2$, (8) is *the equation of the ellipse related to its principal diameters, the origin being at the centre.*

When $A = B$, this equation characterizes a circle, for it then becomes $y^2 + x^2 = A^2$. If we wish for the equation of the ellipse, when the origin of the axis is removed to A, the vertex of the principal diameter, A, it will be obtained by a very simple transformation. We shall merely have to substitute in the primitive equation, $x - A$, for x, (p. 116), because the coordinates of the new origin are $-A$, 0, hence the transformed equation is $A^2 y^2 + B^2 x^2 - 2AB^2 x = 0$, or $y^2 = \frac{B^2}{A^2}(2Ax - x^2)$, (9), *the equation of the ellipse, when the origin of the rectangular axes is at the vertex of the major axis.*

It is sometimes convenient to introduce the quantity, c, denoting the distance of either focus from the centre, into the equation of the ellipse, that is, to substitute for B^2 its equal, $A^2 - c^2$. The quantity, c, is called the eccentricity; hence, from equation (6),

$y^2 = (1 - \frac{c^2}{A^2})(A^2 - x^2)$, or putting e for $\frac{c}{A}$, $y^2 = (1 - e^2)(A^2 - x^2)$, (10), *the equation of the ellipse as a function of the eccentricity.*

We shall now proceed to examine more attentively the foregoing equations, for the purpose of deducing from them the principal properties of the curve.

Properties of the Ellipse as related to its principal Diameters.

(44.) Referring to equation (6), we find that the second form of that equation reduces to $\frac{y^2}{(A + x)\ (A - x)} = \frac{B^2}{A^2}$ which furnishes the proportion $y^2 : (A + x)(A - x) :: B^2 : A^2$. Now $A + x$, $A - x$, are the two portions of the major diameter, into which the ordinate, y, of any point, $(x, y,)$ divides it; hence the above proportion shows that the square of any ordinate is to the product of the parts into which it divides the major diameter, as the square of the minor diameter, is to the square of the major.

Consequently, *the squares of the ordinates are as the products of the parts into which they divide the major diameter.*

If we suppose, in equation (6), the axes of reference to be transposed, that is, x to become y and y to become x, we must then, in the foregoing theorems, substitute *major* diameter for *minor*, and *minor* for *major*; so that the theorems hold good, whichever diameter be taken for the axis of x. They are true, also, of the circle, which the ellipse becomes, when $B = A$; hence, *if a circle be described on a principal diameter of the ellipse, any ordinate in the ellipse will be to the corresponding ordinate in the circle in a constant ratio*, viz. as B to A, or as A to B, accordingly as the major or minor diameter is employed.

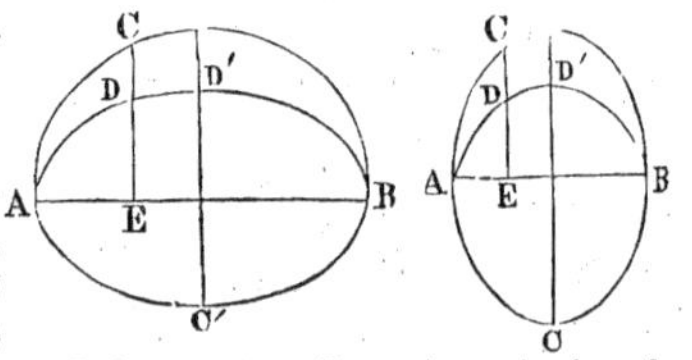

Thus, in the annexed diagrams, $ED : EC :: C'D' : AB$. From equation (10) we have $A^2y^2 = (A^2 - c^2)(A^2 - x^2)$; hence, when $x = c$, $Ay = A^2 - c^2 = B^2$, $\because 2y : 2B :: 2B : 2A$, where $2y$ is the double ordinate through the focus, and is called the *parameter* of the major diameter; it is also sometimes called the *latus rectum.*

From this proportion it appears that *the parameter is a third proportional to the major and minor diameters* Calling the parameter p, we have, therefore, $2Ap = 4B^2$, or dividing by $4A^2$, $\frac{B^2}{A^2} = \frac{p}{2A}$, hence, by substitution, in equation (6), $y^2 = \frac{p}{2A}(A^2 - x^2)$. In like manner, from equation (9), $y^2 = px - \frac{p}{2A}x^2$.

The following problem will conduct us to some other properties.

PROBLEM I.

(45.) To find the expression for an angle inscribed in a semi-ellipse.

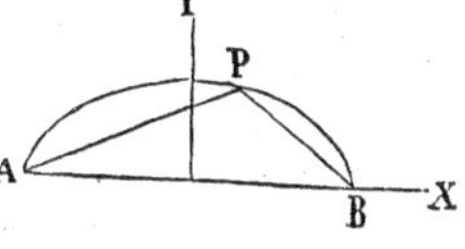

Let APB be a semi-ellipse, and P an angle inscribed in it. Then the equation of AP passing through the point A, whose coordinates are $x = -A$, $y = 0$, is $y = a(x + A)$ $\therefore a = \frac{y}{x+A}$ (1) and the equation of BP passing through the point B, whose coordinates are $x = A$, $y = 0$, is $y = a'(x - A)$ $\therefore a' = \frac{y}{x-A}$ (2), in which equations a, a', represent the trigonometrical tangents of the angles PAX, PBX, respectively.

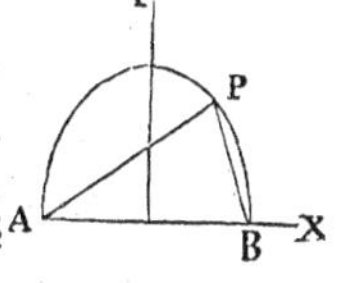

Now the expression for the angle P, which these lines form when they meet, is tan. $P = \frac{a' - a}{1 + aa'}$ which, by substituting for a, a', their values in (1) and (2), becomes, tan. $P = \frac{2Ay}{y^2 + x^2 - A^2}$ (3).

But the lines in question not only meet, but they meet in the curve, the equation of which gives $x^2 - A^2 = -\frac{A^2y^2}{B^2}$ (4); hence, by substitution, (3) becomes, tan. $P = \frac{2Ay}{y^2 - \frac{A^2}{B^2}y^2} = \frac{-2AB^2}{(A^2 - B^2)y}$, (5).

As this expression will remain the same, though x be negative, it follows that there are two points in the semi-ellipse, at which the diameter subtends the same angle, viz. $(x, y,)$ and $(-x, y)$.

If A is greater than B in this expression, that is, if the angle be subtended by the major diameter, the tangent of P is negative, the angle is, consequently, obtuse. Now an obtuse angle increases as its tangent diminishes; when, therefore, the tangent is least, the angle is greatest; and the above expression will evidently be least when its denominator is greatest, that is, when $y = B$. We infer, therefore, that *all the angles subtended by the major diameter are obtuse, and the greatest is that whose vertex coincides with the extremity of the minor diameter.* The expression for this maximum angle is tan. $P = \frac{-2AB}{(A^2 - B^2)}$

If B is greater than A in equation (5), that is, if the angle be subtended by the minor diameter, then the expression for the tangent is positive; the angle is, therefore, acute, and diminishes as its tangent diminishes; it is least, therefore, when y is greatest, that is, when $y = \mathrm{B}$. Hence *all the angles subtended by the minor diameter, are acute, and the least is that whose vertex coincides with the extremity of the major diameter.* The expression for this minimum angle is

$$\tan. \mathrm{P} = \frac{-2\mathrm{BA}}{\mathrm{B}^2 - \mathrm{A}^2}$$

As this expression differs from the former only in its sign, we conclude that *these two angles are supplements of each other.*

From the foregoing theorems it appears, that, if an arc of a circle be made to pass through the extremities of one principal diameter, and a vertex of the other, it will be wholly within the ellipse, if its chord be the major diameter; and wholly without it, if it be the minor diameter. Also, if upon the major diameter there be described a semicircle, or any greater arc, such arc will be entirely without the ellipse; but if, on the contrary, it be described on the minor diameter, it will be entirely within the ellipse.

(46.) Returning to equations (1) and (2) in the preceding problem, we find for their product $aa' = \frac{y^2}{x^2 - \mathrm{A}^2}$, or, substituting for $x^2 - \mathrm{A}^2$, its value in (4,) $aa' = -\frac{\mathrm{B}^2}{\mathrm{A}^2}$. This equation shows that, *the product of the trigonometrical tangents of the two angles, formed by lines drawn from the extremities of a principal diameter to meet in the curve, is constant and equal to* $-\frac{\mathrm{B}^2}{\mathrm{A}^2}$. A representing the half of that diameter which subtends the inscribed angle.

PROBLEM II.

(47.) To find the expression for the distance of any point in the ellipse from the focus. (See Diagram to Art. 41, p. 119.)

Let (x, y) be any given point, P, in the ellipse. Then $\mathrm{FP}^2 = (x - c)^2 + y^2 = x^2 - 2cx + c^2 + y^2$; but, from equation (10), p. 122, we have $y^2 = (1 - e^2)(\mathrm{A}^2 - x^2)$; hence $\mathrm{FP}^2 = x^2 - 2cx + c^2 + (1 - e^2)(\mathrm{A}^2 - x^2) = \mathrm{A}^2 - 2\mathrm{A}ex + e^2 x^2$ $\therefore$ $\mathrm{FP} = \mathrm{A} - ex$ (1); and, since $\mathrm{FP} + \mathrm{F'P}$ must be equal to 2A, we therefore have $\mathrm{F'P} = \mathrm{A} + ex$ (2); hence, (1) and (2) are the expressions sought; and we may conclude that the radius vector of any point in the ellipse is always a rational function of the abscissa of that point.

For the difference of the focal distances, we have $\mathrm{F'P} - \mathrm{FP} = 2ex$; and for their product, $\mathrm{FP} \cdot \mathrm{F'P} = \mathrm{A}^2 - e^2 x^2$.

Properties of the Ellipse when related to its conjugate Diameters.

(48.) The foregoing are the principal properties of the ellipse, which are directly deducible from its equation, when the principal diameters of the curve are taken for axes of coordinates. By referring the curve to other systems of coordinates, we shall obtain equations leading to other properties; we shall, therefore, now inquire what form the equation assumes when the axes of coordinates do not coincide with the principal diameters, but make any angles whatever with them, the origin remaining the same.

It appears from equation (2), p. 116, that in order to transform the coordinates from rectangular to oblique without displacing the origin, we must substitute for x and y in the equation of the curve, the values $x = x \cos. \alpha + y \cos. \alpha'$, and $y = x \sin. \alpha + y \sin. \alpha'$; making, therefore, this substitution in the equation $A^2 y^2 + B^2 x^2 = A^2 B^2$, it becomes

$$\left.\begin{matrix} A^2 \sin.^2 \alpha' \\ B^2 \cos.^2 \alpha' \end{matrix}\right| y^2 + \left.\begin{matrix} 2A^2 \sin. \alpha \sin. \alpha' \\ 2B^2 \cos. \alpha \cos. \alpha' \end{matrix}\right| xy + \left.\begin{matrix} A^2 \sin.^2 \alpha \\ B^2 \cos.^2 \alpha \end{matrix}\right| x^2 = A^2 B^2.$$

Such is the general form of the equation when the oblique coordinates x, y, make any proposed angles, α, α', with the major diameter of the curve. It is obvious that, if the term containing xy were absent from this equation, it would then correspond in form to the primitive equation; and, in order, therefore, that this correspondence may take place, the angles, α, α' must be so related that we may have $A^2 \sin. \alpha \sin. \alpha' + B^2 \cos. \alpha \cos. \alpha' = 0$; or dividing by $\cos. \alpha \cos. \alpha'$, $A^2 \tan. \alpha \tan. \alpha' + B^2 = 0$; the relation, must be such that

$$\tan. \alpha' = \frac{-B^2}{A^2 \tan. \alpha} \quad \ldots \; (1).$$

It hence appears that the transformed equation will have the same form as the primitive, whatever be the angle, α, provided that the other angle, α', be subject, to the condition (1). The angle α, therefore, being arbitrary, it follows that the systems of axes that may be chosen so as to render the transformed equation of the form

$$(A^2 \sin.^2 \alpha' + B^2 \cos.^2 \alpha') y^2 + (A^2 \sin.^2 \alpha + B^2 \cos.^2 \alpha) x^2 = A^2 B^2 \; (2)$$

are limited in number. This equation will be more concisely expressed, and, at the same time, its analogy to the primitive equation more distinctly shown, if we put A'^2 for $\frac{A^2 B^2}{A^2 \sin.^2 \alpha + B^2 \cos.^2 \alpha}$, and B'^2 for $\frac{A^2 B^2}{A^2 \sin.^2 \alpha' + B^2 \cos.^2 \alpha'}$, because then equation (2) may be put under the form $\frac{1}{B'^2} y^2 + \frac{1}{A'^2} x^2 = 1$, as will appear by making the foregoing substitutions for A'^2, B'^2, in this equation: hence, finally, $A'^2 y^2 + B'^2 x^2 = A'^2 B'^2$ (3). Such, then, is the equation of the ellipse related to oblique axes originating at the centre of the curve, the angles α, α', at which these oblique axes are inclined to the primitive axis of abscissas being related to each other, as in equation (1).

(49.) If, in this last equation, we put $x = 0$, the resulting value of y will be the ordinate of the point where the curve meets the axis of y; and, if we put $y = 0$ the resulting value of x will be the abscissa of the point where the curve meets the axis of x; in other words, these particular values of y and x will be the value of those semi-diameters of the ellipse, which have been taken for axes.

Now for $x = 0$, $y = \pm B'$, and for $y = 0$, $x = \pm A'$; hence the semi-diameters in question are A′ and B′.

We may therefore conclude that the equation (3) is not only similar to the equation $A^2y^2 + B^2x^2 = A^2B^2$ in form, but also that in the same manner as the semi-diameters A, B, enter this equation, so do the semi-diameters A′, B′, enter the former.

Equation (3) furnishes the following properties, viz.

1. *Each diameter*, 2A′, 2B′, *bisects all the chords drawn parallel to the other*, as was shown of the principal diameters, 2A, $2B^2$ (42.) Diameters possessing this property are called conjugate diameters, hence the equation $A'^2y^2 + B'^2x^2 = A'^2B'^2$ *is the equation of the ellipse referred to conjugate diameters.*

To distinguish the principal diameters from the other systems of conjugate diameters, they are generally called the *conjugate axes* of the ellipse; the longer is also sometimes called the *transverse* axis, and the shorter its *conjugate*.

1. *Straight lines drawn at the extremities of a diameter*, 2A′, *parallel to its conjugate*, 2B′, *are tangents to the curve* (42.)

3. *The squares of chords drawn parallel to one of two conjugate diameters are as the rectangles of the parts into which they divide the other* (44.)

These properties are established as in the articles referred to.

4. *If any number of parallel chords be drawn, the line which bisects them all will be a straight line passing through the centre.* For if a diameter be drawn parallel to the chords, that which is conjugate to it must bisect the chords, and, therefore, must coincide with the former line. Hence the solution of the two following problems, viz.

1. A diameter being given, to find its conjugate.

Draw a chord parallel to the given diameter, and the line bisecting both will be the diameter sought.

2. To find the centre of an ellipse.

Draw a line to bisect any two parallel chords, and we shall thus have a diameter; bisect this diameter, and the centre will be determined.

(50.) From equation (1) we may infer that *no system of conjugate diameters can be perpendicular to each other, except the principal dia-*

meters. For, that the diameters may be perpendicular to each other, we must have the condition, (11) $\tan. \alpha' = \frac{1}{\tan. \alpha}$, but, by equation (1), this cannot be, unless $A = B$, that is, unless the curve ceases to be an ellipse, and becomes a circle; this teaches us, however, that each system of conjugate diameters in a circle includes a right angle.

As to the principal diameters of the ellipse, the equation $-\frac{}{\tan. \alpha'} = -\frac{B^2}{A^2 \tan. \alpha}$ in which the above condition is implied, does subsist, for then, α being 0, this equation is the same as $\infty = \infty$. Equation (1) moreover shows that if one of the tangents, $\tan. \alpha$, $\tan. \alpha'$, be positive, the other must be negative; consequently, accordingly as the axis of x is *above* or *below* the major diameter of the curve, so will the axis of y be to the *left* or to the *right* of the minor diameter of the curve; for it can make an *obtuse* angle with the major diameter only in the former position, and an *acute* angle only in the latter.

Again, from the same equation, it follows, that *the product of the tangents of the two angles which a system of conjugate diameters make with a principal diameter is constant,* for that equation gives $\tan. \alpha \tan. \alpha' = -\frac{B^2}{A^2}$

But it has been shown (46) that if a, a', be the tangents of the angles which two straight lines drawn from the extremities of a principal diameter to a point in the curve make with that diameter, we must have $a a' = -\frac{B^2}{A^2}$ or $a' = -\frac{B^2}{A^2 a}$. It follows, therefore, that *if from the vertices of a principal diameter two lines be drawn to meet in the curve, the diameters parallel to these will be conjugate,* and conversely, *chords drawn from the vertices of a principal diameter parallel to a system of conjugates meet in the curve.* Art. (50) might obviously have been inferred from this property. Chords drawn from a point in the curve to the extremities of a diameter are called *supplemental chords*

Hence (45) *of all systems of conjugate diameters, those contain the greatest angle which are drawn parallel to the equal supplemental chords from the extremities of the major diameter.* These last diameters are also *equal,* for they form equal angles with the major diameter (43).

When the conjugate diameters, $2A'$, $2B'$, are equal, the equation of the ellipse related to them is $y^2 + x^2 = A'^2$, which corresponds in form to the equation of the circle, and which curve it would characterize, were it not that here the axes of reference are oblique.

(51.) It is obvious, from the condition (1), that if any diameter of an ellipse be represented by the equation $y = ax$, the conjugate thereto will be represented by the equation $y = -\frac{B^2}{A^2 a}x$, the principal diameters being taken for axes. If from the extremities of any diameter, $2A'$, supplement chords be drawn, and they be referred to the semi-conjugates, A', B', as axes, their equation will be $y = m(x + A')$ and $y = m'(x - A')$, since the one chord passes through the point $(-A', 0)$, and the other through the point $(A', 0)$. Consequently, by imitating the steps in art. (46), where the principal axes were employed, we arrive at the analogous property when any system of conjugates are taken for axes, viz. $mm' = -\frac{B'^2}{A'^2}$ in which equation it is to be observed, that the symbols m, m', do not denote the *tangents* of the angles which the chords form with the axis A', but they denote the respective ratios of the sines of the two angles which each chord makes with the axes A', B', these being oblique.

(52.) The several properties of the ellipse, which we have just noticed all immediately flow from the equation of the curve, when referred to the oblique axes, A', B'. If now we return from these to the original rectangular axes, A, B, by a transformation of the equation, other properties will unfold themselves : let us, therefore, effect this transformation.

As we here propose to pass from oblique axes to rectangular, we must substitute for x and y, in the primitive equation, the values in equation (4'), (*art.* 39),

$x = \frac{x \sin. \alpha' - y \cos. \alpha'}{\sin. [A', B']}$, and $y = \frac{y \cos. \alpha - x \sin. \alpha}{\sin. [A', B']}$, the new axis of x being situated *below* the primitive axis. Making, therefore, these substitutions, in equation (3), p. 125, the transformed will be

$$\left.\begin{matrix} A'^2 \cos.^2 \alpha \\ B'^2 \cos.^2 \alpha' \end{matrix}\right| y^2 \left.\begin{matrix} -2A'^2 \sin. \alpha \cos. \alpha \\ -2B'^2 \sin. \alpha' \cos. \alpha' \end{matrix}\right| xy + \left.\begin{matrix} A'^2 \sin.^2 \alpha \\ B'^2 \sin.^2 \alpha' \end{matrix}\right| x^2 = A'^2 B'^2 \sin.^2 [A', B']$$

Now this equation is to be identical with $A^2 y^2 + B^2 x^2 = A^2 B^2$; hence there must exist the following relations, viz.

$A'^2 \cos.^2 \alpha + B'^2 \cos.^2 \alpha' = A^2$ (1). $A'^2 \sin.^2 \alpha + B'^2 \sin.^2 \alpha' = B^2$ (2).

$A'^2 \sin. \alpha \cos. \alpha + B'^2 \sin. \alpha' \cos. \alpha' = 0$ . . . . (3)

$A'^2 B'^2 \sin.^2 [A', B'] = A^2 B^2$ . . . . (4).

By adding together equations (1) and (2), we obtain the property $A'^2 + B'^2 = A^2 + B^2$ (5), that is, *the sum of the squares of any system of conjugate diameters is equal to the sum of the squares of the principal diameters.*

(53.) From equation (4) there results $4A' B' \sin. [A' B'] + 4AB$. (6).

Now the first member of this equation expresses the surface of a parallelogram, of which the adjacent sides are equal to the conjugate diameters, $2A'$, $2B'$, and included angle equal to the angle $[A', B']$;

and the second member of the equation represents the rectangle of the principal diameters, 2A, 2B. Now, if through the extremities of each of the two conjugates, A′B′, C′D′, parallels be drawn to the other, they will be tangents to the ellipse, and the angle O′ will be equal to the angle O, that is, to the angle [A′ B′]; hence the parallelogram formed by these tangents is equal to the rectangle of the principal diameters. Equation (6) therefore expresses this theorem, viz.

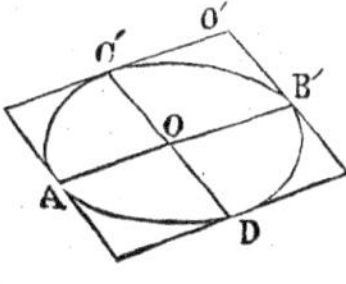

Any parallelogram circumscribing an ellipse, and having its sides parallel to a system of conjugate diameters, is equivalent to the rectan gle of the two axes.

Since systems of conjugate diameters are unlimited in number (48), it follows that an infinite number of circumscribing parallelograms may be found all equal in surface. Of these but one will be a rhombus, viz. that of which the sides are parallel to the equal conjugate diameters; and but one will be a rectangle, viz. that of which the sides are parallel to the principal diameters.

It can be shown, conversely, that if a parallelogram circumscribing an ellipse be equivalent to the rectangle of the principal diameters, its sides must be parallel to a system of conjugate diameters.

For let PR be a circumscribing parallelogram, the sides of which are not parallel to a system of conjugate diameters. Draw a diameter, A′B′, parallel to one of the sides PQ, and let C′D′ be the conjugate to this diameter, and complete the parallelogram P′R′, having its sides parallel to the system of conjugates just drawn. Then, since A′B is parallel to PQ, but does not meet the parallels SP, RQ, it is less than PQ, but it is equal to P′Q′; therefore P′Q′ is less than PQ; and as both parallelograms are between the same parallels, PQ′, S′R; P′R′ must be less than PR, but P′R′ is equivalent to the rectangle of the principal diameters; hence PR is greater than that rectangle. It follows, therefore, that *of all parallelograms circumscribing an ellipse, those about conjugate diameters are least in surface*

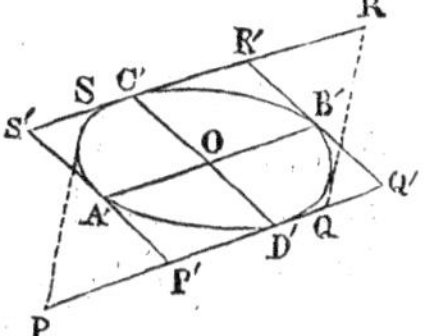

From equation (6), $A'B' = \dfrac{AB}{\sin.\,[A',\,B']}$, and adding twice this to equation (5), and extracting the square root we have

$$\sqrt{(A'^2 + 2A'B' + B'^2)}, \text{ or } A' + B' = \sqrt{(A^2 + B^2 + \frac{2AB}{\sin.[A',B']})}$$

consequently $A' + B'$ is greatest when sin. [A′, B′] is least, that is, when the obtuse angle [A′, B′] is greatest, and $A' + B'$ is least, when sin. [A′, B′] is greatest, that is when [A′, B′] is a right angle; and hence *of all systems of conjugate diameters, the sum of th which are equal is the greatest, and the sum of those which are rectangular is the least.* We shall terminate this division of the present chapter with the following additional problems.

PROBLEM III.

(54.) The axes of an ellipse and the vertex of any diameter being given to find the length of that diameter, and of its conjugate.

Let (x', y') represent the given vertex of the diameter $2A'$, the length of which is required. Then the distance of (x', y') from the centre of the ellipse is $A'^2 = x'^2 + y'^2$; but by the equation of the curve, $y'^2 = B^2 - \frac{B^2}{A^2} x'^2$: hence, by substitution,

$$A'^2 = B^2 + \frac{A^2 - B^2}{A^2} x'^2 = B^2 + \frac{c^2}{A^2} x'^2 = B^2 + e^2 x'^2 \therefore A' = \sqrt{(B^2 + e^2 x'^2)};$$

also since $A'^2 + B'^2 = A^2 + B^2 \therefore B'^2 = A^2 - e^2 x'^2 \therefore B' = \sqrt{(A^2 - e^2 x'^2)}$, and these values of A' and B' are the expressions sought.

If the radii vectores of the point (x', y') be $F'P$, FP, then (47) $F'P \cdot FP = A^2 - e^2 x'^2$. Hence *the product of the radii vectores of any point is equal to the square of the semi-diameter conjugate to that passing through the point,* that is $F'P \cdot FP = B'^2$.

PROBLEM IV.

(55). The axes and the inclination of a system of conjugate diameters being given to determine them in length and direction.

Let A', B' represent the semi-conjugates; then, from equations (5) and (6), p. 128, we have $(A' + B')^2 = A^2 + B^2 + \frac{2AB}{\sin. [A', B']}$, and $(A' - B')^2 = A^2 + B^2 - \frac{2AB}{\sin. [A', B']}$; Consequently, by addition and subtraction,

$$A' = \tfrac{1}{2} \sqrt{\left\{A^2 + B^2 + \frac{2AB}{\sin. [A', B']}\right\}} + \tfrac{1}{2} \sqrt{\left\{A^2 + B^2 - \frac{2AB'}{\sin. [A', B']}\right\}}$$

$$B' = \tfrac{1}{2} \sqrt{\left\{A^2 + B^2 + \frac{2AB}{\sin. [A', B']}\right\}} - \tfrac{1}{2} \sqrt{\left\{A^2 + B^2 - \frac{2AB}{\sin. [A', B']}\right\}}$$

These are the expressions for the lengths of the semi-conjugates. It remains, to determine at what angles these are inclined to the major diameter, $2A$.

We have seen (51) that if the tangent of the angle at which A' is inclined to the major diameter be a, then will $-\frac{B^2}{A^2 a}$ be the tangent of the angle at which B' is inclined to it; hence for the angle $[A', B']$, at which the conjugates themselves are inclined, we have, if we denote its tangent by v, the expression (11),

$v = \frac{a' - a}{1 + aa'} = -\frac{(A^2 a^2 + B^2)}{(A^2 - B^2)a}$. Hence, by reduction, we obtain the quadratic $a^2 + (1 - \frac{B^2}{A^2}) va = -\frac{B^2}{A^2}$ which, solved, gives

$$a = -\frac{1}{2A^2}\{(A^2 - B^2)v \pm \sqrt{\{(A^2 - B^2)^2 v^2 - 4A^2 B^2\}}\}$$

If the given tangent, v, be less than $\frac{2AB}{A^2 - B^2}$, abstracting from the sign of v, the question will be impossible, that is, no system of conjugate diameters can have the proposed inclination, a fact which the above expression for a plainly intimates, and which we previously knew from (45). If v be equal to $-\frac{2AB}{A^2 - B^2}$, then will A', B', be the equal semi-conjugates (45, 50), in which case $a = +B \div A$, consequently, $a' = -B \div A$, the inclinations being supplements of each other.

For the lengths of the equal semi-conjugates we have, from the property (5), p. 128, $2A'^2 = A^2 + B^2$ $\therefore A' = \sqrt{\tfrac{1}{2}(A^2 + B^2)}$.

The geometrical construction of the preceding problem is very simple: On the major diameter, AB, of the ellipse describe a circular arc, capable of containing the proposed angle, if it be obtuse, or its supplement if acute. Then from either of the points, P, P′, in which it intersects the ellipse (45), let chords be drawn to A, B; then diameters drawn parallel to either of these systems of supplemental chords will be conjugate, and will include the given angle. The two values of a in the preceding analytical expression agree with the two systems of conjugates here constructed. If the arc pass through C, it will touch the ellipse at that point, so that P, P′, will coincide, and the conjugates sought will be equal.

Properties of the Tangent to the Ellipse.

(56.) In order to obtain the equation of the tangent to the ellipse, let us first, as in the circle, consider a secant to the curve, or a straight line cutting it in two points, (x', y') and (x'', y'').

The equation of this secant is $y - y' = \frac{y' - y''}{x' - x''}(x - x')$, (1); and as both points are on the curve, there subsists the equations $A'^2 y'^2 + B'^2 x'^2 = A'^2 B'^2$, (2), $A'^2 y''^2 + B'^2 x''^2 = A'^2 B'^2$ . . . (3). Hence, subtracting (3) from (2),

$$A'^2 (y' + y'')(y' - y'') = -B'^2 (x' + x'')(x' - x'')$$

and from this we get $\frac{y' - y''}{x' - x''} = -\frac{B'^2}{A'^2} \cdot \frac{x' + x''}{y' + y''}$ therefore equation (1) becomes, by substitution, $y - y' = -\frac{B'^2}{A'^2} \cdot \frac{x' + x''}{y' + y''} (x - x')$ . . (4).

This, then, is the equation of the secant passing through the two points (x', y'),(x'', y'') of the ellipse, whose equation is $A'^2 y^2 + B'^2 x^2 = A'^2 B'^2$ If now we suppose these two points to coincide, the secant will become a tangent; making, therefore, $x' = x''$, and $y' = y''$; we have $y - y' = -\frac{B'^2}{A'^2} \cdot \frac{x'}{y'} (x - x')$; or more simply, by reduction, $A'^2 y' y + B'^2 x' x = A'^2 B'^2$, (5), the equation of the tangent related to any system of conjugate diameters.

The second form of this equation is very easily retained in the memory, from its resemblance to the equation of the curve; the only difference is that, in the equation of the tangent, $x' x$ occurs instead of x^2, and $y'y$ instead of y^2. Connected with the tangent are several other lines, which it is requisite to consider. Thus, if AB, CD are the principal diameters of an ellipse, and the tangent PR be referred to them as axes, then the distance, MR, of the ordinate of the point of contact from the intersection of the tangent with the axis of x, is called the *subtangent*, the perpendicular, PN, to the tangent from the point of contact, is called the *normal*, and NM, the distance of its intersection with the axis of x from the ordinate, is called the *subnormal*; moreover, in estimating the length of the tangent, we consider only the portion PR, included between the point of contact and the axis of x. Hence on one side of the ordinate of the point of contact are situated the tangent and subtangent, PR and MR; and on the other side the normal and subnormal, PN and MN.

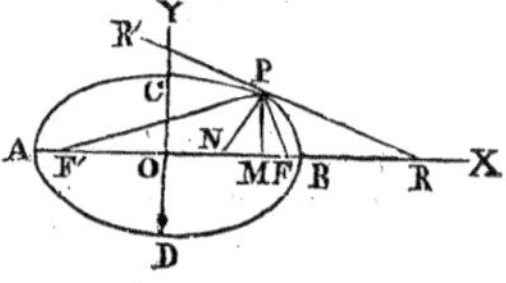

We shall now proceed to deduce the equations of these lines, and to determine analytical expressions for their several lengths.

(57.) For the tangent the equation has already been exhibited when any system of conjugates are employed for axes. For the rectangular system, 2A, 2B, the equation is therefore

$$y - y' = -\frac{B^2}{A^2} \cdot \frac{x'}{y'} (a - x'), \text{ or } A^2 yy' + B^2 x'x = A^2 B^2.$$

The coefficient $-\frac{B^2}{A^2} \cdot \frac{x'}{y'}$ in the first of these forms, which is that most frequently referred to, is the trigonometrical tangent of the angle PRX; it is obviously negative for positive values of x, and positive for negative values of x, since the angle is obtuse in the former case, and acute in the latter. For the normal the equation is immediately deducible from that of the tangent. We shall have merely to characterize a perpendicular to the latter, drawn from the point of contact

(x', y'). The equation of this line is therefore $y - y' = \frac{A^2 y'}{B^2 x'}(x - x')$. The length of the subtangent is easily derived from the equation of the tangent, for supposing in that equation $y = 0$, which is the case at the point R, the resulting value of x will express the length, OR; that is, $x = OR = A^2 \div x'$ and, if from this we deduct x', or OM, we shall have, for the length of the subtangent, $MR = \frac{A^2 - x'^2}{x'}$. In like manner, if, in the equation of the normal, we suppose $y = 0$, which is the case at the point N, the resulting value of x will express the length, ON; and this subtracted from OM, or x', will give the length of the subnormal, MN. We have, therefore, only to express the value of $x' - x$ in the equation of the normal when $y = 0$, which value is $MN = \frac{B^2}{A^2}x'$. From these expressions for the subtangent and subnormal are obtained those for the tangent and normal. Thus, since $PR = \surd \{MR^2 + PM^2\}$ we have by substitution, $PR = \surd \{\frac{(A^2 - x'^2)^2}{x'^2} + y'^2\}$, or because $y'^2 = \frac{B^2}{A^2}(A^2 - x'^2)$ this becomes $PR = \surd \{\frac{(A^2 - x'^2)^2}{x'^2} + \frac{B^2}{A^2}(A^2 - x'^2)\}$. Likewise, since $PN = \surd \{MN^2 + PM^2\}$, we have, by substitution, $PN = \surd \{\frac{B^4}{A^4}x'^2 + \frac{B^2}{A^2}(A^2 - x'^2)\} = \frac{B}{A}\surd \{(\frac{B^2}{A^2} - 1)x'^2 + A^2\}$ $= \frac{B}{A}\surd (A^2 - e^2 x'^2)$.

(58.) For more convenient reference, let us now collect the preceding formulas together.

The equation of the tangent is $y - y' = -\frac{B^2 x'}{A^2 y'}(x - x')$.

The equation of the normal is $y - y' = \frac{A^2 y'}{B^2 x'}(x - x')$.

The length of the tangent is $T = \surd \{\frac{(A^2 - x'^2)^2}{x'^2} + \frac{B^2}{A^2}(A^2 - x'^2)\}$.

The length of the subtangent is $T_{\prime} = \frac{A^2 - x'^2}{x'}$.

The length of the normal is $N = \frac{B}{A}\surd (A^2 - e^2 x'^2)$.

The length of the subnormal is $N_{\prime} = \frac{B^2}{A^2}x'$.

(59.) We may now by aid of these formulas deduce some other properties of the ellipse. And, first, we may remark that, as the expression for the subtangent is independent of B, it remains the same for all the values of B, so that in every ellipse described upon the axis 2A, the subtangent is the same for the same abscissa.

This is true also when $B = A$, that is, when the ellipse becomes a circle; and hence is suggested a method of drawing a tangent to an ellipse from any given point in the curve. Thus, let P be the given point; then, having described a semi-circle on AB, draw P′PM perpendicular to AB, and at P′ draw a tangent, P′R, to the circle; then draw the line PR, and it will be the tangent required; for the points P, P′ have the same subtangent, MR.

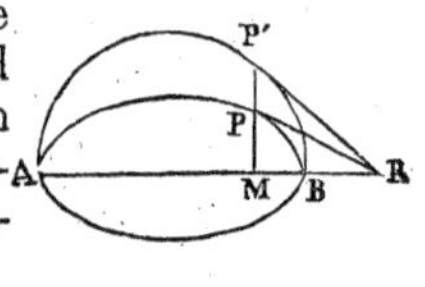

It is further obvious, from the manner in which the expression for the subtangent has been obtained, that if the tangent had been referred to oblique conjugates, as in (56), instead of rectangular, the expression would have preserved the same form, that is, it would have been $T_{\prime} = \frac{A'^2 - x'^2}{x'}$ and this independently of the sign of y'; so that a tangent to the curve through the point (x', y'), and another through the point $(x', -y')$, both meet the axis of x in the same point; and thus from any point without an ellipse two tangents may be drawn to the curve. If the point of contact of one of the tangents drawn from a given point be known, and it is required to find the point of contact of the other, it may be done as follows: Draw a diameter to pass through the given point, and parallel to its conjugate draw a chord from the known point of contact, then the other extremity of this chord will be the other point of contact.

It ought to be noticed here that we must not, as in the preceding case, conclude that because the value of the subtangent is independent of B′, and that this may therefore be equal to A′, that the ellipse may become a circle, and yet the subtangent for the same abscissa remain the same. For the conjugate diameters of the ellipse, which are here taken for axes, are not conjugate diameters of the circle described upon either of them, because, in the circle, the systems of conjugates are all rectangular (50).

(60.) By multiplying the foregoing general expression for the subtangent by x' we have $T_{\prime} x' = (A' + x')(A' - x')$, which shows that *the rectangle of the subtangent and abscissa of the point of contact is equal to the rectangle of the parts into which the diameter is divided by the ordinate.*

Thus, in the annexed diagram,

$$OM \cdot MR = A'M \cdot MB'.$$

This property furnishes a commodious and expeditious method of drawing a tangent to an ellipse from a point without the curve, when the centre of the ellipse only is given.

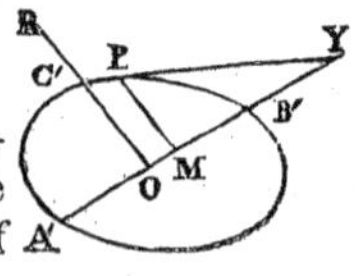

Let R be the given point, and draw RB′A′ through the centre, and let CD be the diameter conjugate to A′B′. Upon A′B′ and OR, as diameters, describe arcs intersecting in m, m', then the line joining m, m', will cut from OB′ the abscissa, OM, of the point of contact sought, so that the parallel, P, P′ to OC, through this point will intersect the curve at the points where the tangents from R must touch it.

For, as the circles upon A′B′, OR intersect in the line mMm', it follows that $mM \cdot Mm' = A'M \cdot MB' = OM \cdot MR$; hence, from the above property, the tangent from P must have the subtangent MR.

(61.) By referring to (47) we find that the rectangle of the radii vectores of any point, P, in an ellipse is $F'P \cdot FP = A^2 - e^2x^2$, and comparing this with the expression for the normal, we find, therefore, $N^2 = \dfrac{B^2 \cdot F'P \cdot FP}{A^2}$ that is, *the rectangle of the radii vectores of any point in an ellipse is to the square of the normal as the square of the major axis is to the square of the minor.*

Also, since (54) $A^2 - e^2x^2 = B'^2$, we have $A \cdot N = B \cdot B'$, that is, *the rectangle of the major axis and the normal is equivalent to the rectangle of the minor axis and the semi-diameter parallel to the tangent.*

Thus in the annexed diagram,

$F'P \cdot FP : PN^2 :: AB^2 : CD^2$,

and $AB \cdot PN = CD \cdot OC'$.

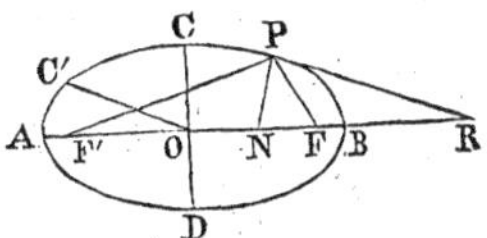

(62.) Let us now examine the equation of the tangent as exhibited in (56). If in this equation we make $x = 0$, the resulting value of y will be the ordinate, OT, at the origin; hence we obtain the property $OT \cdot y'$ or $OT \cdot PM = OC'^2$.

In like manner, by making $y = 0$, in the same equation, we shall have for the resulting value of x the abscissa OR; hence $OR \cdot x'$ or $OR \cdot OM = OB'^2$; hence, we infer, first, That *the rectangle of the ordinate of the ellipse at the point of contact, and the ordinate of the tangent at the centre, is equal to the square of that semi-diameter which is taken for axis of ordinates;* and second, that *the rectangle of the abscissa of the point of contact, and of the point where the tangent meets the axis of abscissas, is equal to the square of that semi-diameter which is taken for axis of abscissas.*

Both these properties are also derivable from the general expression for the subtangent, as will appear by adding x' thereto for the second property, and changing x' into y' for the first property.

(63.) Let us now suppose, in the same general equation of the tangent, that $x = A'$ instead of 0, then the resulting value of y will be the ordinate $B't$, which is also a tangent to the curve at B', since it is parallel to the diameter conjugate to $A'B'$ (see preceding diagram). In like manner, if instead of A' we put $-A'$ for x, the corresponding ordinate will be the tangent $A'T'$. Making then these successive substitutions for x, we obtain the values $y = B't = \frac{B'^2}{A'y'}(A' - x')$,

$$y = A'T' = \frac{B'^2}{A'y'}(A' + x'), \text{ and } \therefore A'T' \cdot B't = \frac{B'^4}{A'^2 y'^2}(A'^2 - x'^2).$$

But, from the equation of the curve, $y'^2 = \frac{B'^2}{A'^2}(A'^2 - x'^2)$; hence, by substitution, $A'T' \cdot B't = B'^2$, that is, *if, at the extremities of any diameter, lines parallel to its conjugate be drawn terminating in any tangent to the curve, their rectangle will be equal to the square of the semi-conjugate to which they are parallel.*

Still confining ourselves to the same general equation of the tangent, the conjugates $A'B'$, $C'D'$, being the axes of reference, let the supplemental chords, $B'M$, $A'M$, be drawn, the former parallel to the tangent at the point P, or (x', y'); then we already know (51) that in the equations of these chords, viz. $y = m(x - A')$, and $Y = m'(x + A')$, the coefficients m, m', must be so related that $m m' = -\frac{B'^2}{A'^2}$ (1). But, since $B'M$ is parallel to the tangent, the coefficients of x must be the same in the equations of these lines, that is, we must have $m = -\frac{B'^2 x'}{A'^2 y'}$. It follows, therefore, from the relation (1), that $m' = y' \div x'$.

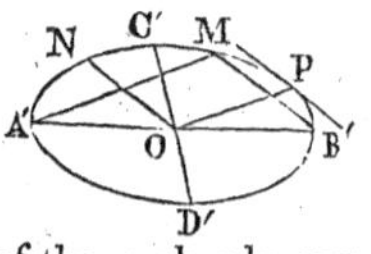

Let now OP be drawn to the point of contact, and let ON be parallel to $B'M$, then the equation of OP is $y = ax$, and for the point (x', y'), $y' = ax'$; therefore $a = y' \div x' = m'$, consequently OP is parallel to $A'M$, and OP, ON are semi-conjugates, ON being parallel to the tangent at the vertex of OP. Hence we may conclude that *diameters drawn parallel to any system of supplemental chords are conjugate*, which is an extension of the theorem at (50). As the principal diameters are the only conjugates, which contain a right angle, it follows that any system of supplemental chords which include a right angle must be parallel to the principal diameters. These latter, therefore, can always be found when we know the centre of the ellipse: it will be necessary merely to describe a semi-circle on any diameter, and to draw supplemental chords from the point where it cuts the curve, the diameters parallel to these will be those sought. It appears, moreover, that a semi-circle described on any diameter of an ellipse can cut it in but one point.

(64.) From what has just been shown, it follows that, when any system of conjugates are taken for axes, if the equation of any diameter referred to them be $y = mx$, the equation of its conjugate will be $y = -\frac{B'^2}{A'^2 m}x$. Bearing this in mind, suppose that at the vertex of the diameter taken for the axis of x, a tangent is drawn, it must be parallel to the axis of y; hence the part thereof intercepted by the line $y = mx$ will be the value of y, which its equation gives for $x = A'$, viz. $y = mA'$. In like manner, the part intercepted by the conjugate to this will be the value of y, which the equation of this conjugate gives for $x = A'$, viz. $y = -\frac{B'^2}{A'm}$.

Now the product of these two values is $-B'^2$, it follows, therefore, that *if a tangent be drawn at any point of an ellipse, the square of half the diameter conjugate to that from the point of contact will be equal to the rectangle of the two portions of the tangent intercepted between the point of contact, and any system of conjugates whatever.*

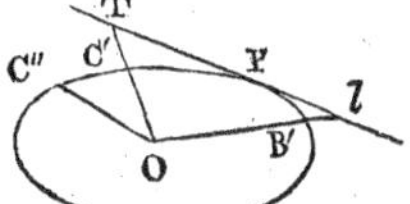

Thus, if TPt be a tangent, and OC″ a semi-diameter, parallel to it, then OB′, OC′, being any system of semi-conjugates, we shall always have, $OC''^2 = PT \cdot Pt$.

If the tangent be drawn through the vertex of one of the least conjugate diameters, and be terminated by the principal conjugates produced, it will be bisected at the point of contact. For the least conjugates bisect the principal supplemental chords, and, as one of these is parallel to the tangent, both this and the tangent must be bisected by the same conjugate; it follows, therefore, that this tangent is equal to the least conjugate diameter. It is, moreover, evident that this tangent is the shortest that can be included between the prolonged principal diameters, for every other tangent included between them must exceed its parallel diameter, since the rectangle of two unequal parts of it is equal to the square of half that diameter. It follows also that, if a tangent be drawn through the vertex of one of the least conjugates, the portion intercepted by the prolonged principal diameters will be less than the portion intercepted by any other system of conjugates.

(65.) If a tangent pass through the extremity of the latus rectum, and be referred to the principal diameters, the ordinate at the point of contact will be $y' = \frac{1}{2}p = B^2 \div A$, so that the equation of this tangent will be, $y - \frac{B^2}{A} = -\frac{x'}{A}(x - x')$, and $\therefore y = \frac{B^2 + x'^2 - x'x}{A}$, or, since $x' = c$, $y = \frac{A^2 - cx}{A} = A - ex$.

But (47,) $A - ex$ expresses the distance of the focus from that point in the ellipse whose abscissa is x; hence *the length of any ordinate to a tangent through the extremity of the latus rectum is equal to the distance of the focus from the point where this ordinate intersects the curve,* therefore, *the tangent through the vertex of the latus rectum cuts from the tangent through the vertex of the major diameter, a part equal to the distance of the focus from the vertex.*

Thus, in the annexed diagram, $MN = FP$, also $AT = AF$ and $BT' = BF$.

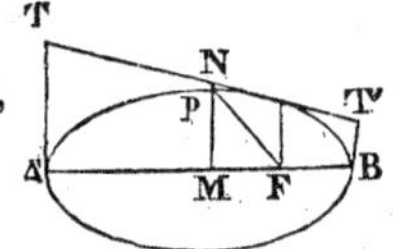

(66.) If in the equation of the normal we put $y = 0$, the resulting value of x will express the distance, ON, (*see the diagram at p.* 132,) that is, we shall have $ON = \frac{A^2 - B^2}{A^2} x' = e^2 x'$; hence, adding c or Ae to this, there results $F'N = e(A + ex')$. Now, in the triangle $PF'F$, the sum of the sides $PF' + PF$ is to the base $F'F$, so is one of the sides, PF', to the distance intercepted between F' and the line bisecting the vertical angle, P, (*Geom. p.* 90;) but $2A : 2Ae :: A + ex' : e(A + ex') = F'N$. It follows, therefore, that *the normal at any point of the ellipse bisects the angle formed by the radii vectores of that point,* and consequently the radii vectores are equally inclined to the tangent, also the angle included by one radius vector and the prolongation of the other is bisected by the tangent.

We can arrive at the same conclusion without the aid of the geometrical property here employed; thus: Since the equation of FP, passing through the points $(c, 0)$ and (x', y'), is $y = \frac{y'}{x' - c}(x - c)$, the coefficient $\frac{y'}{x' - c}$ must express the tangent of the angle PFR, and we already know, from the equation of PR, that $-\frac{B^2 x'}{A^2 y'}$ is the tangent of the angle PRX. Hence, in order to obtain the tangent of the angle FPR, we shall merely have to substitute the preceding expressions, for a and a', in the formula (11), $v = \frac{a' - a}{1 + aa'}$ which then becomes $v = \frac{B^2 cx' - (A^2 y'^2 + B^2 x'^2)}{(A^2 - B^2) x'y' - A^2 cy'}$. In this expression, if we substitute for $A^2 y'^2 + B^2 x'^2$ its equal $A^2 B^2$, and for $A^2 - B^2$ its equal c^2, we shall finally obtain $v = \frac{B^2}{cy'}$. If the other radius vector, F'P, had been employed, c would have been negative, and therefore the tangent of its inclination to PR would differ from that here deduced only in its sign; thus showing that the angles F′PR, FPR are supplements of each other, and therefore, that the angles FPR, F′PR′, are equal.

The same property admits of a simple geometrical proof; thus:

Let one radius vector, F′P, be produced, till PG is equal to the other, FP, then the line PR, bisecting the angle GPF will be a tangent to the curve at the point P. For join FG, then, since PR bisects the vertical angle of the isosceles triangle, PFG, it also bisects the base, FG, at right angles; therefore from whatever point, N, in PR, lines be drawn to F and G, we shall always have $NF = NG$ (*Geom. p.* 20); consequently, if there existed any point, N, in PR, besides P which was common to the curve, we should have $F'N + FN = 2A$ and therefore $F'N + NG = F'G$, which is impossible, so that PR, which bisects the angle FPG, is a tangent to the curve, and therefore, conversely, the tangent at P must bisect the angle FPG.

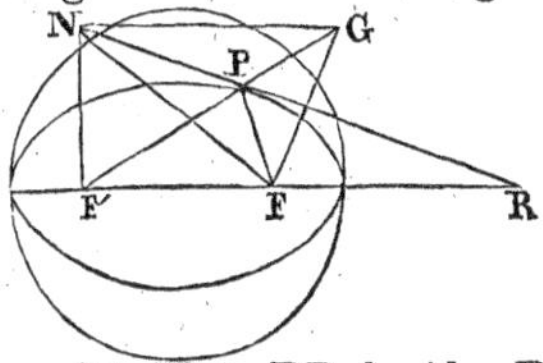

If OM be drawn from the centre to the middle of FG, it will be parallel to F′G, because O is the middle of F′F, therefore $FO : FF' :: OM : F'G$; hence $OM = \frac{1}{2}F'G$, but $F'G = 2A$ by construction, consequently $OM = A$, therefore, if from either focus a perpendicular to any tangent be drawn, its intersection therewith will be always at the same distance from the centre, viz. at the distance, OB, in other words, *the locus of these intersections is the circumference of the circle described on the major axis as a diameter.*

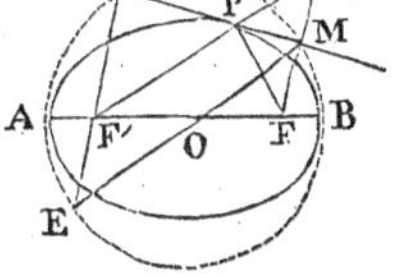

If MO be produced to meet this circumference in E, and F′E be drawn, the triangles F′OE, FOM, having two sides, and the included angle in each equal are themselves equal, therefore F′E is both equal and parallel to FM, and is therefore the continuation of the perpendicular, F′M′, from the us, F, to the tangent; hence, from the property of the circle $\cdot F'E = M'F' \cdot MF = AF' \cdot F'B$, but $AF' \cdot F'B = (\ -c)(A + c) = A^2 - c^2 = B^2$; hence $M'F' \cdot MF = B^2$, that is *the rectangle of the perpendiculars from the foci upon any tangent to the curve is equal to the square of half the miner axis.*

Since the triangles F′PM′, FPM, are similar, we have the equation $\frac{M'F'}{MF} = \frac{F'P}{FP}$, which, multipled by that just deduced, gives

$M'F'^2 = B^2 \cdot \frac{F'P}{FP} = B^2 \cdot \frac{A + ex'}{A - ex'}$ and, multiplying its reciprocal by the same, we have $MF^2 = B^2 \frac{FP}{F'P} = B^2 \cdot \frac{A - ex'}{A + ex'}$; x' being the abscissa of the point P.

(67.) The property in (66) furnishes a simple method of drawing a tangent to the ellipse from a point either in the curve or without it, when the foci are known. Thus, if the point be in the curve, as at P, then it will be necessary merely to draw the lines FP, F′PG, and to bisect the angle FPG.

If the point be without the curve, as at N, then, from the focus, F′, as a centre, with a radius equal to the principal diameter, describe an arc, and from the given point, N, as a centre, with its distance from the other focus, describe another arc, intersecting the former in the point G, then the line F′G will cut the curve in the point through which the required tangent must pass. For, let NPR be drawn through this point and draw PF, NG, NF; then, since, by construction, $F'G = F'P + PF$, $PF = PG$, also $NF = NG$, consequently the triangles NFP, NGP, are equal, therefore the angles NPF, NPG are equal, as also their supplements FPR, GPR, hence NPR is a tangent to the ellipse. As the distance, F′N, of the centres from which the arcs intersecting in G are described is less than the sum, and greater than the difference, of the radii, (*Geom. p.* 19,) it follows (28) that these arcs intersect also in another point, and thus two tangents may be drawn from N.

We shall terminate the present chapter with the following problem.

PROBLEM V.

(68.) Pairs of tangents to an ellipse being always supposed to intersect at right angles, to find the locus of the points of intersection.

Let MT, NT, be any pair of tangents intersecting at right angles in and, parallel thereto, draw FP, F′P′, from the foci, then the points P, P′ will be in the circumference of a circle described upon the major diameter, AB. Produce PF to meet this circumference again in M′, draw M′T′, FM, each perpendicular to the tangent TT′, and lastly, draw the chords PP′, M, M′, the points M, M′ being

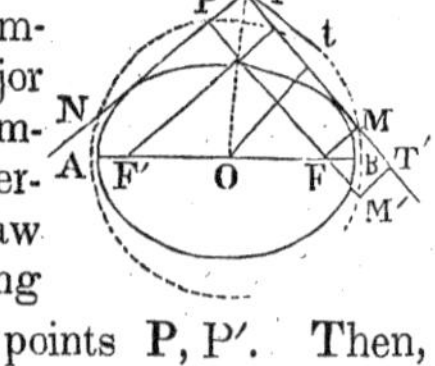

obviously on the same circumference as the points P, P′. Then, since TM′ is a rectangle, $PT = M'T'$, and on account of the parallels PM′, P′M, the arcs PP′, MM′ are equal, their chords are therefore equal; hence the rectangles PP′, MM′ are equal, and $TP' = T'M = FM'$, consequently $TM \cdot TP' = PF \cdot FM'$, but, by the property of the circle, $PF \cdot FM' = AF \cdot FB$ and $\cdot$ TM TP′ $= Tt^2$; Tt being a tangent to the circle, therefore $Tt^2 = AF \cdot FB$, a constant quantity and, as the radius, Ot, is also constant, the distance, OT, must be constant, therefore *the locus of T is a circle* of which the radius OT is $\sqrt{(A^2+A^2-c^2)} = \sqrt{(A^2+B^2)}$.*

CHAPTER III.

ON THE HYPERBOLA.

Its equation and Properties.

(69.) An hyperbola is a curve from any point, P, in which, if two straight lines be drawn to two fixed points, F, F′, their difference shall always be the same.

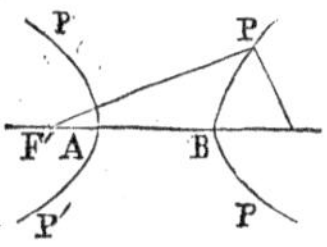

The given points, F, F′, are called the foci of the hyperbola, and the lines, FP, F′P, drawn therefrom to any point, P, in the curve, are called the radii vectores, or focal distances of that point.

This curve may be described, by means of points, thus: From one of the foci, F, as a centre, with any assumed radius, describe an arc, and from the other focus, F′, with any other radius exceeding the former, describe a second arc, intersecting the first in two points, P, p. Let this operation be repeated with two new radii, taking care that the second of these shall exceed the first by the same difference as before, and two new points will be determined; and this determination of points in the curve may thus be continued till its tract becomes obvious. That the locus of these points will be an hyperbola is plain from the definition, since the distance of any one of them from F′ always exceeds its distance from F by the same constant difference. If of each pair of intersecting arcs employed to determine the several points, we had supposed the greater to have been described from F, and the less from F′, the same constant difference being preserved, we should obviously have obtained a series of points, P′, p', &c. equally belonging to the hyperbola, although none would be situated in the locus of the former series. It appears, therefore, that the hyperbola consists of two separate branches, PBp, P′Ap'.

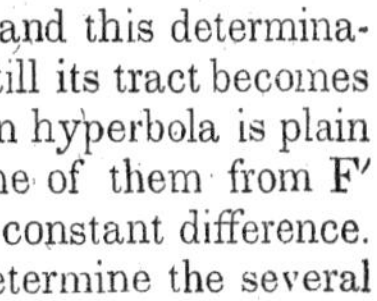

Any portion of this curve may be described by continuous motion, by employing a ruler and a cord. Thus, let a ruler, FR, be fixed to F, so that it may be turned round this point, in the plane whereon the curve is to be described; then having assumed any other point, F′, in this plane, connect it by means of a cord shorter than the ruler to the extremity, R; then a pencil, P, keeping this cord always stretched, and at the same time pressing against the edge of the ruler, will, as the ruler

* The analytical investigation of this problem will be given hereafter.

revolves round F, describe an arc of an hyperbola, of which F, F′, are the foci, for the difference of the distances of the describing point, P, from the fixed points F, F′, will be always the same.

(70.) Let us now seek the equation of this curve, by means of its characteristic property.

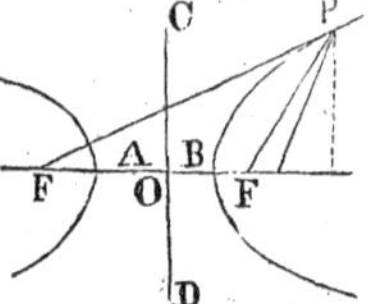

Draw F′F, and let O, the middle point of this line, be taken for the origin of the rectangular axes, and let the constant difference of the radii vectores of any point in the curve be represented, as in the ellipse, by 2A. Put c for OF, or OF′, and x, y, for the coordinates of any point, P, in the curve. Then we shall have these equations, viz. $y^2 + (x - c)^2 = \text{PF}^2$, (1), $y^2 + (x + c)^2 = \text{PF}'^2$ (2) and $\text{FP}' - \text{PF} = 2\text{A}$, (3). Hence, by first adding and then subtracting equations (1) and (2) we have $2y^2 + 2x^2 + 2c^2 = \text{PF}'^2 + \text{PF}^2$ (4), $4cx = (\text{PF}' + \text{PF})(\text{PF}' - \text{PF})$, $\therefore \text{PF}' + \text{PF} = 2cx \div \text{A}$, Combining this with equation (3), we have, $\text{PF}' = \frac{cx}{\text{A}} + \text{A}$ and $\text{PF} = \frac{cx}{\text{A}} - \text{A}$, and these values substituted in equation (4) give $y^2 + x^2 + c^2 = \frac{c^2 x^2}{\text{A}^2} + \text{A}^2$, whence $\text{A}^2 y^2 + (\text{A}^2 - c^2) x^2 = \text{A}^2 (\text{A}^2 - c^2)$, (5). This equation would be identical with that at p. 120, and would thus characterize an ellipse, were it not that here A represents half the difference, instead of half the sum of the radii vectores. With this condition, therefore, equation (5) is the equation of the hyperbola.

(71.) In order to determine the points of intersection with the axes, suppose $y = 0$, in equation (5), and there results for x the value $x = \pm \text{A}$, which intimates that the hyperbola, like the ellipse, intersects the axis of x in two points, B, A, equidistant from O, the one to the right, and the other to the left, and that A expresses this distance.

If, in the same equation, we suppose $x = 0$, we have for the corresponding value of y the expression $y = \sqrt{\{\text{A}^2 - c^2\}}$.

Now since $2c$ is the base, and 2A the difference, of the sides of a triangle, PF′F, it follows (*Geom. p.* 19) that c must exceed A, and consequently the value of y, in the foregoing expression, is impossible; the curve, therefore, can never meet the axis of ordinates. In this respect, therefore the hyperbola differs entirely from the ellipse. Let us, however, mark on the axis of ordinates two points, C, D, each at the distance of $\sqrt{\{c^2 - \text{A}^2\}}$ from O, and, calling this distance B, we shall then have the equation of the hyperbola in a form analogous to that of the ellipse, viz. $\text{A}^2 y^2 - \text{B}^2 x^2 = -\text{A}^2\text{B}^2$, or $y^2 = \frac{\text{B}^2}{\text{A}^2}(x^2 - \text{A}^2)$ 6.

The only difference between this equation and that of the ellipse is that here the sign of B^2 is negative. The general expressions, therefore, for the coordinates of any point in the curve is

$x = \pm \frac{A}{B} \sqrt{\{y^2 + B^2\}}$; $y = \pm \frac{B}{A} \sqrt{\{x^2 - A^2\}}$. From the first of these expressions it appears, that there are two values of x numerically equal, but differing in sign for every value of y. We conclude, therefore, that chords drawn parallel to AB are bisected by CD, or its production. In like manner, in the second expression, we are furnished with two values of y numerically equal, but differing in sign for every value of x. If, however, x be assumed numerically less than A, the resulting value of y will be imaginary; now x is less than A for every ordinate drawn between the points A and B, hence none of these ordinates can meet the curve; but, if $x = \pm A$, then, since $y = 0$, we infer that parallels to CD, drawn through the points A and B, are tangents to the curve. It further appears, from this expression, that so long as x is assumed numerically greater than A, there will always correspond a possible value of y, which will increase as x increases; hence the two branches of the curve are unlimited, proceeding onwards, in opposite directions, to infinity. Let us actually suppose, that x takes a succession of values from $\pm$ A to infinite, then, putting the expression for y under the form $y = \frac{Bx}{A} \sqrt{\{1 - \frac{A^2}{x^2}\}}$ we see that as x increases, the fraction $A^2 \div x^2$ diminishes, so that the values of y go on approaching to the value $\pm Bx \div A$, which value, however, is never reached till x becomes infinite, rendering the fraction $A^2 \div x^2$ nothing. It is obvious from this, that if through the origin, O, two straight lines, KL, MN, be drawn, making angles, KOX, MOX, with AB whose tangents are respectively $+ B \div A$ and $- B \div A$, these two lines will continually approach the curve, and yet can never meet it. For the equation of these lines are $y = \frac{B}{A} x$ and $y = -\frac{B}{A} x$, and it has just been seen that in the curve y can never be so great as $\pm Bx \div A$, till x becomes infinite, although throughout the course of the curve, y continually approaches to this value, that is, the differences between the ordinates of the curve and those of the lines just drawn, for the same abscissas continually diminish as the abscissas increase.

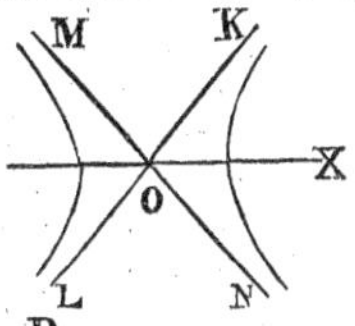

The two lines KL, MN, are called *asymptotes* to the hyperbola.

(72.) Let us now, as in the ellipse, seek the expression for the distance of any point (x, y) in the curve from the origin, O.

For this purpose, we must substitute for y^2, in the expression $D = \sqrt{\{x^2 + y^2\}}$, its value $\frac{B^2}{A^2}(x^2 - A^2)$, which reduces it to

$D = \sqrt{\{\frac{A^2 + B^2}{A^2} x^2 - B^2\}}$. This expression is obviously independent of the signs of x and y, and, therefore, it may be shown here in precisely the same way as the fact was established with regard to the ellipse, that every chord passing through O is bisected at that point, and hence O is called the centre of the curve, and chords drawn through it, diameters. Indeed, all lines drawn through the centre of the hyperbola are, for the sake of uniformity, called diameters, although an infinite number may be drawn, so as not to meet the curve, viz. all those comprehended between the asymptotes, or that inclined to the axis of x at an angle of which the tangent is not numerically less than $B \div A$.

The asymptotes may, therefore, be regarded as separating those diameters which are chords, called *transverse diameters*, from those which are not chords, called *second diameters*. Of the former, least is AB, since the above expression for D is least when x is least, that is, when $x = \pm A$; but, as there is no limit to the value of x, the other transverse diameters increase from this least value to infinity.

(73.) As in the ellipse, so here, 2A, 2B, that is, AB, CD, are called the principal diameters, or the principal axes of the curve. The distinction of these diameters into major and minor, as in the ellipse, cannot, however, be here used with propriety, for in the expression $\sqrt{\{c^2 - A^2\}}$, which B represents, it is only necessary that c exceed A (p. 142,) so that B may either be greater or less than A, or indeed equal to it. From these remarks then it appears that the equation $A^2y^2 - B^2x^2 = -A^2B^2$, or $y^2 = \frac{B^2}{A^2}(x^2 - A^2)$, (6), is *the equation of the hyperbola related to its principal diameters.*

If we suppose $B = A$, the equation is $y^2 - x^2 = -A^2$. In this case, the hyperbola is called *equilateral*, on account of the equality of its principal diameters. Thus the same modification transforms the equation of the common hyperbola into that of the equilateral hyperbola, that changes the equation of the ellipse into that of the circle. We may here remark that, in the equilateral hyperbola, since $B \div A = 1 =$ tangent of 45°, the angles which the asymptotes make with the axis of x are 45°, and 90° + 45°; hence, in this case, the asymptotes are perpendicular to each other.

(74.) By removing the origin of the axes of coordinates from the centre, O, to the vertex, A, of the transverse axis, by a transformation similar to that employed in the ellipse, the equation becomes $A^2y^2 - B^2x^2 + 2AB^2x = 0$, or $y^2 = \frac{B^2}{A^2}(x^2 - 2Ax)$ (7), *the equation of the hyperbola when the origin is at the vertex of the transverse axis*

(75.) If we wish for the equation in terms of the eccentricity, c, we may obtain it from equation (6), by substituting for B^2 its equal $c^2 - A^2$, which gives $y^2 = \frac{c^2 - A^2}{A^2}(x^2 - A^2)$ or putting e for $\frac{c}{A}$

$y^2 = (e^2 - 1)(x^2 - A^2)$ *the equation of the hyperbola as a function of the eccentricity.*

From the intimate analogy which subsists between the equations of the ellipse and those of the hyperbola, it may easily be conceived that the principal properties of the former curve belong also to the latter. This is in fact the case; and hence most authors exhibit the theory of these two curves in conjunction. With a view to simplicity, it has, however, been here thought preferable to consider these curves separately; but, as we propose to develope the properties of the hyperbola by imitating the steps which led us to those of the ellipse, we shall frequently have occasion to refer to the preceding chapter for details, which need not be repeated in this.

Properties of the Hyperbola related to its principal diameters.

(76.) From equation (6), $\dfrac{y^2}{(x + A)(x - A)} = \dfrac{B^2}{A^2}$ hence $y^2 : (A + x)(x - A) :: B^2 : A^2$. Now $x + A$ and $x - A$ are the distances of the ordinate, y, from the vertices of the transverse axis, hence the square of any ordinate is to the product of its distances from the vertices of the transverse axis as the square of the conjugate axis is to the square of the transverse; consequently *the squares of the ordinates are as the products of the parts into which they divide the transverse axis.*

If the hyperbola is equilateral, that is, if $B = A$, then $y^2 = (x + A)(x - A)$; so that, in the equilateral hyperbola, the square of any ordinate is equal to the products of the parts into which it divides the transverse axis, a property analogous to that of the circle.

(77.) From equation (8) the *parameter* or double ordinate through the focus is easily determined, for, putting c for x, in that equation, and extracting the square root of each side, there results $Ay = c^2 - A^2 = B^2$, in which equation y is the semi-parameter; therefore, calling the parameter p, we have $p = 4B^2 \div 2A$ that is, *the parameter is a third proportional to the transverse and second axes.*

Hence the equation of the hyperbola, as a function of the parameter, is obtained by substituting, in equation (6), $p \div 2A$ for its equal, $\dfrac{B^2}{A^2}$ so that $y^2 = \dfrac{p}{2A}(x^2 - A^2)$, is *the equation of the curve, in terms of the parameter.*

In the equilateral hyperbola, since $4B^2 \div 2A = 2A$, we have $p = 2A$, that is, the parameter is equal to the transverse axis.

PROBLEM I.

(78.) To find the expression for the angle contained by supplemental chords drawn from the extremities of the transverse axis.

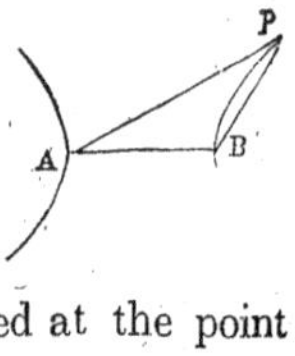

Referring to the corresponding problem on the ellipse, we find that the tangents of the angles which the supplemental chords, meeting in the point (x, y), make with the axis of x are

$a = \frac{y}{x + A}$ and $a' = \frac{y}{x - A}$ (1), and that, consequently, the general expression for the angle P, formed at the point (x, y), is $\tan. P = \frac{2Ay}{y^2 + x^2 - A^2}$ (2). Now the point (x, y) being in the hyperbola, we must have, from the equation of the curve, $x^2 - A^2 = A^2y^2 \div B^2$ hence, by substitution, equation (2) becomes $\tan. P = \frac{2AB^2}{(A^2 + B^2)y}$ .... (3), the expression sought.

As this result is independent of the sign of x, it follows that the angle is the same, whether the chords be drawn to (x, y) or to $(-x, y)$.

Since the preceding expression for the tangent is always positive for any point in the curve above the axis of x, it follows that *all the angles subtended by the transverse axis are acute.* These angles diminish as y increases, till they become 0, when y is infinite.

Multiplying the expressions (1) together, we have

$aa' = \frac{y^2}{x^2 - A^2} = \frac{B^2}{A^2}$ (4), therefore *the product of the trigonometrical tangents of the two angles formed by the transverse axis produced to the right, and its supplemental chords, is constant and equal to* $B^2 \div A^2$.

As this product is positive, we conclude that the two angles must be either both acute or both obtuse.

If the hyperbola be equilateral, then $\tan. P = \frac{A}{y} = \frac{A}{\sqrt{(x^2 - A^2)}}$

and $aa' = 1$ or $a = \frac{1}{a'}$, but, since $\tan. = \frac{1}{\cot.}$, it follows that a, a' must represent the tangent and cotangent of the same angle, consequently, *in the equilateral hyperbola, the angles which the supplemental chords make with the axis of x are together equal to a right angle.*

PROBLEM II.

(79.) To find the expression for the radius vector of any point in the curve.

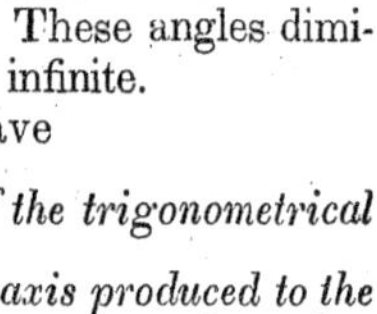

Let x, y be the coordinates of any point, P, in the curve, then $FP^2 = (x - c)^2 + y^2 = x^2 - 2cx + c^2 + y^2$. Now equation (8) p. 145 gives $y^2 = (e^2 - 1)(x^2 - A^2)$; hence by substitution, $FP^2 = x^2 - 2ex + c^2 + (e^2 - 1)(x^2 - A^2) = e^2x^2 - 2Aex + A^2$ $\therefore$ $FP = ex - A$, and since $F'P - FP = 2A$, it follows that $F'P = ex + A$; hence the radius vector of any point is always a rational function of the abscissa.

For the sum of the radii vectores we have $FP + F'P = 2ex$, and for their product, $FP \cdot F'P = e^2x^2 - A^2$.

Properties of the Hyperbola, when related to its conjugate diameters.

(80.) By transforming the equation of the hyperbola from rectangular axes to oblique, the origin still remaining at the centre of the curve, we shall have the equation

$$\left.\begin{array}{l} A^2 \sin.^2 \alpha' \\ -B^2 \cos.^2 \alpha' \end{array}\right| y^2 + \left.\begin{array}{l} 2A^2 \sin. \alpha \sin. \alpha' \\ -2B^2 \cos. \alpha \cos. \alpha' \end{array}\right| xy + \left.\begin{array}{l} A^2 \sin.^2 \alpha \\ -B^2 \cos.^2 \alpha \end{array}\right| x^2 = -A^2 B^2$$

where α, α' denote the angles which the oblique coordinates x, y make with the primitive axis of x.

Let us now, as in the ellipse, determine the relation which must subsist between the angles α, α', in order that the term containing xy may disappear from this equation. That such may be the case, the coefficient of xy must obviously be assumed equal to 0, or dividing this coefficient by the expression $2 \cos. \alpha \cos. \alpha'$, we must have the equation $A^2 \tan. a \tan. \alpha' - B^2 = 0$. Hence the relation between the angles α, α' will be thus expressed, viz. $\tan. \alpha' = \dfrac{B^2}{A^2 \tan. \alpha}$ (1), so that one of the angles being chosen at pleasure, the other will be determined by this equation, and thus an infinite number of oblique axes exist that will render the transformed equation of the form $(A^2 \sin.^2 \alpha' - B^2 \cos.^2 \alpha') y^2 + (A^2 \sin.^2 \alpha - B^2 \cos.^2 \alpha) x^2 = -A^2 B^2 \ldots$ (2). This equation will be simplified by putting

A'^2 for, $\dfrac{-A^2 B^2}{A^2 \sin.^2 \alpha' - B^2 \cos.^2 \alpha}$, and B'^2 for $\dfrac{A^2 B^2}{A^2 \sin.^2 \alpha' - B^2 \cos.^2 \alpha'}$

for then equation (2) may be put under the form

$\dfrac{1}{B'^2} y^2 - \dfrac{1}{A'^2} x^2 = -1$, as will readily appear, by making the proposed substitutions. Hence finally, $A'^2 y^2 - B'^2 x^2 = -A'^2 B'^2$ (3). This, therefore, is the equation of the hyperbola, when the oblique axes of reference originate at the centre and form angles α, α' with the primitive axes of x, related as in equation (1).

If in this equation we put $x = 0$; $y = \sqrt{(-B'^2)} = \pm B' \sqrt{(-1)}$, and $y = 0$; $x = \pm A'$. Since the value of y for $x = 0$ is impossible, it follows that the curve does not meet the axis of y, but $\pm A'$ being the value of x for $y = 0$, it follows that the curve meets the axis of x at the extremities of the diameter, $2A'$.

We must here remark that the imaginary expression $\pm B \sqrt{(-1)}$, denoting the ordinate of the origin, merely indicates that such ordinate does not belong to the curve, or, in other words, that the point determined by its extremity has no existence therein; we are not, therefore, to infer that the ordinate itself has no existence, for its absolute value considered independently of the curve, is $= B'$, since the absolute value of its square is B'^2. Now, as lines through the centre are diameters, whether they meet the curve or not, it is plain that, if we assume $2B'$ for the length of the diameter, which coincides with the axis of y just in the same way as we before assumed, $2B$ for the length of the principal second diameter, equation (3) will be analogous to equation (6), p. 144, for the semi-diameters, A', B', enter into the former equation in the same manner that the semi-diameters, A, B, enter into the latter.

(81.) By reason of this analogy in the forms of these two equations, analogous properties of the curve are deducible from each. The following are obvious.

1. *Each diameter*, $2A'$, $2B'$, *bisects the chords drawn parallel to the other*, as was shown of the principal diameters, $2A$, $2B$, (*art.* 71.) Such are called *conjugate diameters*, and the equation $A'^2y^2 - B'^2x^2 = -A'^2 B'^2$, is *the equation of the hyperbola related to conjugate diameters.*

Hence (72) of every system of conjugate diameters, one is a *transverse* and the other a *second* diameter.

2. *Straight lines drawn at the extremities of a transverse diameter parallel to its conjugate are tangents to the curve*, (71).

3. *A transverse diameter is divided by an ordinate parallel to its conjugate into two parts, such that their rectangle is to the square of the ordinate as the square of the transverse is to the square of the conjugate*, (76).

4. *The line which bisects parallel chords is a straight line.*

For the conjugate to that diameter which is parallel to the chords must bisect them. Hence the method of finding the centre of an hyperbola, and of determining the conjugate to any given diameter are analogous to those already given for the ellipse (*art.* 49).

From equation (1) it is obvious that the principal diameters are the only system of conjugates which are rectangular, for in that equation there can never be $\tan. \alpha' = -\dfrac{1}{\tan. \alpha}$, except when $\alpha = 0$, in which case we must have $\alpha' = 90°$.

It is further evident, from the same equation, that both the tangents $\tan. \alpha$, $\tan. \alpha'$, must have the same sign, so that accordingly as any semi-transverse is above or below the principal semi-transverse, so will the semi-conjugate to the former be to the right or the left of the principal semi-conjugate, the one pair being always included between the other.

(82.) Since, from equation (1), $\tan. \alpha \cdot \tan. \alpha' = B^2 \div A^2$, and from (4), p. 146, $aa' = B^2 \div A^2$, a, a' being the tangents of the two angles formed with the diameter, $2A$, by supplemental chords from its extremities, it follows that *diameters drawn parallel to a system of supplemental chords from the principal transverse diameter are conjugate and conversely.*

From what has been shown of the supplemental chords (78), it is obvious that the angle included by a system of conjugate diameters may be any magnitude from 0 to 180°.

(83.) Referring again to equation (1), we find that, if the equation $y = ax$ represent any diameter of an hyperbola, when referred to its principal axes, then will $y = B^2 \div A^2 a$ be the equation of its conjugate. If the problem at (78) be solved with regard to the transverse diameter, $2A'$, we shall have $mm' = B'^2 \div A'^2$ where, as in the ellipse, m, m', are the coefficients of x, in the equations of two supplemental chords drawn from the transverse, $2A'$.

(84.) Having established these properties, let us now, as in the ellipse, return from oblique to the original rectangular conjugates, by substituting, in the equation $A'^2 y^2 - B'^2 x^2 = - A'^2 B'^2$ the values for x and y already employed at (52), and for the transformed equation we shall have

$$\left.\begin{matrix} A'^2 \cos.^2 \alpha \\ -B'^2 \cos.^2 \alpha' \end{matrix}\right| y^2 \left.\begin{matrix} -2A'^2 \sin.\alpha \cos.\alpha \\ +2B'^2 \sin.\alpha' \cos.\alpha' \end{matrix}\right| xy \left.\begin{matrix} +A'^2 \sin.^2 \alpha \\ -B'^2 \sin.^2 \alpha' \end{matrix}\right| x^2 = -A'^2 B'^2 \sin^2 [A', B']$$

Therefore, this being identical with $A^2 y^2 - B^2 x^2 = - A^2 B^2$, we have the following equations, viz. $A'^2 \cos.^2 \alpha - B'^2 \cos.^2 \alpha' = A^2$ (1), $A'^2 \sin.^2 \alpha - B'^2 \sin.^2 \alpha' = -B^2$ (2), $-A'^2 B'^2 \sin.^2 [A', B'] = -A^2 B^2$ (3).

By adding (1) and (2), $A'^2 - B'^2 = A^2 - B^2$ (4), that is *the difference of the squares of any system of conjugate diameters is equal to the difference of the squares of the principal diameters.*

From equation (3) there results $4A' B' \sin. [A' B'] = 4AB$. Hence, as in the ellipse, *the parallelogram constructed on any system of conjugate diameters is equivalent to the rectangle on the axes of the curve.*

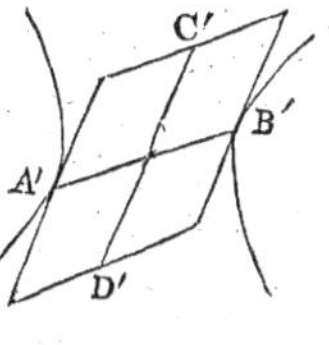

These parallelograms are said to be *inscribed* in the hyperbola, as they are included between the two branches of the curve, the sides parallel to that diameter which does not meet the curve being tangents at the extremities of the conjugate thereto (81). The property (4) shows that if $A' = B'$ then $A = B$, and conversely, so that there cannot be a system of equal conjugates, except in the equilateral hyperbola, and in this, each system consists of equal conjugates. Hence, in the equilateral hyperbola, all the inscribed parallelograms are rhombusses, but in the common hyperbola none are.

(85.) It has been shown (71) that the equation of the asymptotes, when referred to the principal diameters, is $y = \pm Bx \div A$; and in precisely the same manner may it now be shown that when any system of conjugates, $2A'$, $2B'$, are substituted for the principal diameters, the equation of the asymptotes will be $y = \pm B'x \div A'$.

If, in this equation, we give to x the value A′, the resulting expression for y will be the length of the tangent drawn at the extremity of the diameter 2A′, and terminated by the asymptote. This length is therefore $\pm$ B′, and thus we obtain a correct notion of the absolute length of any imaginary diameter of the hyperbola.

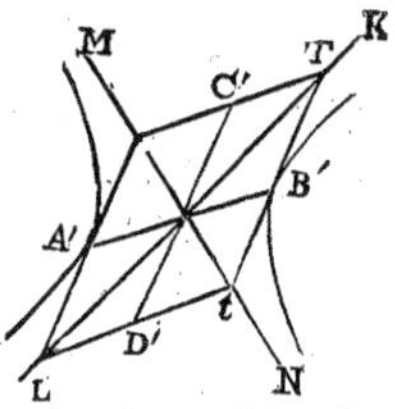

Thus, if A′B′ be any transverse diameter, and KL, MN the asymptotes of an hyperbola, then will the tangent, Tt, be the length of the diameter, C′D′, conjugate to the former.

It moreover follows, that all the inscribed parallelograms having their sides parallel to a system of conjugate diameters, have their vertices in the asymptotes, and in the equilateral hyperbola the angle included by any system of conjugates is bisected by the asymptote.

We shall terminate this division of the present chapter with the two following problems, corresponding to those already solved for the ellipse.

PROBLEM I.

(86.) The axis and vertex of any diameter being given to find the length of that diameter and of its conjugate.

Let $(x', y',)$ represent the vertex of the diameter, 2A′, the length of which we are to determine. Then for the distance of this point from the centre, we have $A'^2 = x'^2 + y'^2$. But by this equation of the curve $y'^2 = \frac{B^2}{A^2}x'^2 - B^2$; therefore by substitution $A'^2 =$ $\frac{A^2+B^2}{A^2}x'^2 - B^2 = \frac{c^2}{A^2}x'^2 - B^2 = e^2x'^2 - B^2$. Also since $A'^2 - B'^2 = A^2 - B^2$, we have $B'^2 = e^2x'^2 - A^2$. $\therefore A = \sqrt{\{e^2x'^2 - B^2\}}$ and $B' = \sqrt{\{e^2x'^2 - A^2\}}$.

Now the above expression for B'^2 is the same as that for $FP \cdot F'P$ in (79), hence as in the ellipse *the product of the radii vectores of any point is equal to the square of the semi-diameter conjugate to that passing through the point.*

PROBLEM II.

(87.) The axes of the curve and the inclination of a system of conjugate diameters being given to determine them in length and direction. Let the semi-conjugates be A′, B′.

Then we have from equations (3) and (4), p. 149,

$$B'^2 = \frac{A^2B^2}{A'^2 \sin.^2[A',B']} \text{ and } A'^2 - B'^2 = A^2 - B^2.$$

By substituting for B'^2 in the second equation its value in the first we have after reducing $A'^4 - (A^2 - B^2) A'^2 = \frac{A^2 B^2}{\sin.^2 [A', B']}$

$$A'^2 = \frac{A^2 - B^2}{2} + \sqrt{\left\{\frac{(A^2 - B^2)^2}{4} + \frac{A^2 B^2}{\sin.^2 [A', B']}\right\}}$$

$$\therefore A' = \sqrt{\left(\frac{A^2 - B^2}{2} + \sqrt{\left\{\frac{(A^2 - B^2)^2}{4} + \frac{A^2 B^2}{\sin.^2 [A', B']}\right\}}\right)}$$

Having thus found an expression for A', the value of B' may be obtained from the first of the above equations.

We have now to ascertain at what angles A' and B' are inclined to A. Let us denote the tangent of the angle at which A' is inclined by a, then the tangent of the angle at which B' is inclined will be $\frac{B^2}{A^2 a} = a'$. Hence, denoting the tangent of the angle $[A', B']$ by v, we have $v = \frac{a' - a}{1 + aa'} = \frac{B^2 - A^2 a^2}{(A^2 + B^2) a}$, and from this we obtain by reduction, the quadratic $a^2 + (1 + \frac{B^2}{A^2}) va = \frac{B^2}{A^2}$ which, solved, gives

$$a = -\frac{1}{2A^2}\left((A^2 + B^2) v \pm \sqrt{\{(A^2 + B^2) v^2 + 4A^2 B^2\}}\right)$$

and this value of a is always possible, whatever be the value of v.

The geometrical construction of this problem is analogous to the corresponding problem in the ellipse, it being only necessary to describe on the transverse axis a segment of a circle capable of containing the proposed angle, for then the diameters parallel to the supplemental chords from the points where the circular and hyperbolic arcs intersect, will furnish two systems of conjugates, each including the proposed angle.

Properties of the Tangent to the Hyperbola.

(88.) It appears from (9) that the equation of a straight line passing through two points (x', y') and (x'', y'') is $y - y' = \frac{y' - y''}{x' - x''}(x - x')$ (1.)

If these two points belong to an hyperbola, we must have, from the equation of the curve $A'^2 y'^2 - B'^2 x'^2 = -A'^2 B'^2$ (2), and $A'^2 y''^2 - B'^2 x''^2 = -A'^2 B'^2$ (3). Subtracting (3) from (2) there remains

$$A'^2 (y' + y'') (y' - y'') = B'^2 (x' + x'') (x' - x'')$$

$\therefore \frac{y' - y''}{x' - x''} = \frac{B'^2}{A'^2} \cdot \frac{x' + x''}{y' + y''}$. Hence, by substitution, equation (1) becomes $y - y' = \frac{B'^2}{A'^2} \cdot \frac{x' + x''}{y' + y''} (x - x')$.

Let us now suppose that the points (x', y''), $x'', y'')$ are identical, then the secant will become a tangent to the curve at the point (x', y') and its equation will be $y - y' = \frac{B'^2}{A'^2} \cdot \frac{x'}{y'} (x - x')$, or $A'^2 yy' - B'^2 xx'$

$= - A'^2 B'^2$, so that the equation of the tangent is obtained from that of the curve, as in the ellipse, by substituting yy' for y^2 and xx' for x^2.

When the rectangular conjugates are taken for axes, the equation is $y - y' = \frac{B^2 x'}{A^2 y'}(x - x')$, in which the coefficient $\frac{B^2 x'}{A^2 y'}$ expresses the trigonometrical tangent of the angle PRX, formed by the tangent to the curve and the axis of x. This angle is obviously acute or obtuse, according as the abscissa of the point of contact is positive or negative.

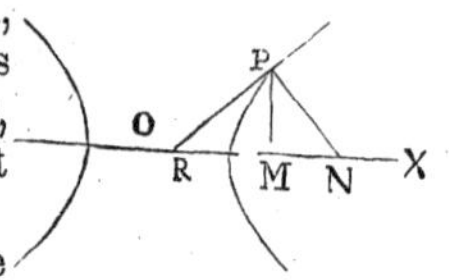

For the normal, or perpendicular, PN, to the tangent, from the point of contact, P, the equation is

$$y - y' = -\frac{A^2 y'}{B^2 x'}(x - x').$$

To determine the length of the subtangent, MR, put $y = 0$, in the equation of the tangent, and we shall have the corresponding value of x, $x = OR = \frac{A^2}{x'}$ which, deducted from $x' = OM$, leaves

$$MR = \frac{x'^2 - A^2}{x'}.$$

From the equation of the normal we readily find the length of the subnormal, MN. For, putting in that equation, 0 for y, we have ON for the resulting value of x, and taking OM, or x', from this, we have $x - x' = MN = \frac{B^2}{A^2}x'$. From the two last expressions we obtain those for the tangent and normal. Thus, since $PR = \sqrt{\{MR^2 + PM^2\}}$, we have, by substitution, $PR = \sqrt{\left\{\frac{(x'^2 - A^2)^2}{x'^2} + y'^2\right\}}$ or, since $y'^2 = \frac{B^2}{A^2}(x'^2 - A^2)$, $PR = \sqrt{\left\{\frac{(x'^2 - A^2)^2}{x'^2} + \frac{B^2}{A^2}(x'^2 - A^2)\right\}}$

In like manner, because $PN = \sqrt{\{MN^2 + PM^2\}}$, we have, by substitution, $PN = \frac{B}{A}\sqrt{\left\{\left(1 + \frac{B^2}{A^2}\right)x'^2 - A^2\right\}} = \frac{B}{A}\sqrt{(e^2 x'^2 - A^2)}$.

(89.) Collecting these several formulas together, we have the following expressions.

The equation of the tangent is $x - y' = \frac{B^2 x'}{A^2 y'}(x - x')$.

The equation of the normal is $y - y' = -\frac{A^2 y'}{B^2 x'}(x - x')$.

The length of the tangent is $T = \sqrt{\left\{\frac{(x'^2 - A^2)}{x'^2} + \frac{B^2}{A^2}(x'^2 - A^2)\right\}}$.

The length of the normal is $N = \frac{B}{A}(e^2 x'^2 - A^2)$.

The length of the subtangent is $T_{,} = \dfrac{x'^2 - A^2}{x'}$.

The length of the subnormal is $N_{,} = \dfrac{B^2}{A^2}x'$.

(90.) Since the expression for the subtangent remains the same for all values of B we infer, as in the ellipse, that in every hyperbola described upon the same axis, 2A, the subtangent is the same for the same abscissa.

Had the tangent been referred to oblique conjugates, instead of rectangular, the expression for the subtangent would have differed from that above only in this, that A′ would have occupied the place of A, that is, we should have had $T_{,} = \dfrac{x'^2 - A'^2}{x'}$, and as this is independent of the sign y', we conclude that from any point without an hyperbola two tangents may be drawn to the curve, viz. one to the point (x', y'), and the other to the point $(x', -y')$.

That more than two tangents cannot be drawn from the same point is obvious, for the subtangent must vary if x' does.

From the general expression for the subtangent just given, it follows that $T_{,}x' = (x' + A')(x' - A')$, that is, as in the ellipse, *the rectangle of the subtangent and abscissa of the point of contact is equal to the rectangle of the sum and difference of the same abscissa and semi-transverse axis*

Thus $OM \cdot MR = A'M \cdot MB'$.

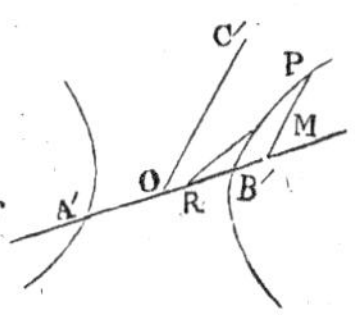

(91.) If we compare the expression for the normal with that for the product of the radii vectores, we find, as in the ellipse that $N^2 = \dfrac{B^2 \cdot F'P \cdot FP}{A^2}$, that is, *the rectangle of the radii vectores of any point is to the square of the normal as the square of the transverse axis is to the square of the conjugate.*

If the hyperbola be equilateral, then $N^2 = F'P \cdot FP$; consequently (84) and (86) *the normal at any point of an equilateral hyperbola is equal to the distance of that point from the centre.*

If B′ represent the semi-diameter conjugate to that passing through P, we shall have (86) $F'P \cdot FP = B'^2$, hence, by substitution, in the expression for N^2, above $A \cdot N = B \cdot B'$; so that, as in the ellipse, *the rectangle of the transverse axis and the normal is equivalent to the rectangle of the second axis and the semi-diameter parallel to the tangent.*

If, in the annexed diagram, OC′ be the semi-diameter parallel the tangent at P, then from these theorems we must have

$F'P \cdot FP : PN^2 :: AB^2 : CD^2$,

and $AB \cdot PN = CD \cdot OC'$.

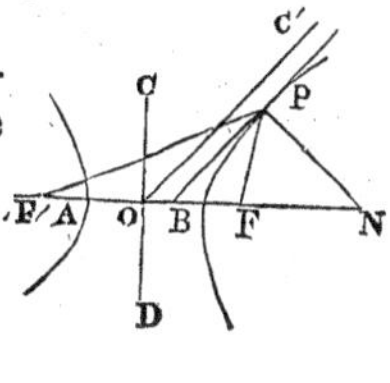

(92.) In the general equation of the tangent (88) if we put $x = 0$, we shall have for y the ordinate, OT, at the origin, and the equation will become $OR \cdot y' = OT \cdot PM = -B'^2 = OC'^2$; and if, in the same general equation, we put $y = 0$, we shall have for x the abscissa OR, and the equation then gives $OR \cdot x' = OR \cdot OM = A'^2 = OB'^2$.

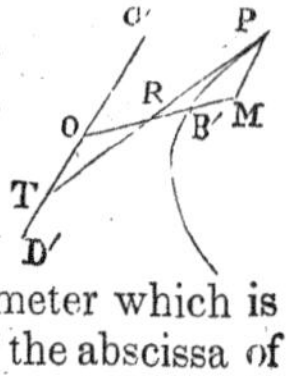

Hence, as in the ellipse, the rectangle of the ordinate at the point of contact and ordinate of the tangent at the centre is equivalent to the square of that semi-diameter which is taken for the axis of ordinates; also, the rectangle of the abscissa of the point of contact and of the point where the tangent meets the axis of abscissas is equivalent to the square of that semi-diameter, which is taken for the axis of abscissas.

If we put $x = A'$, in the same general equation of the tangent, the resulting value of y will express the ordinate $B't$, which being parallel to the diameter conjugate to that through B' will be also a tangent to the curve. If instead of A', $-A$ be substituted for x, we shall in like manner have the value of the tangent $A't'$ for the corresponding ordinate. By making these substitutions, and proceeding as in the ellipse (63), we shall arrive at the same property in the hyperbola, viz. $A't' \cdot B't = B'^2 = OC'^2$.

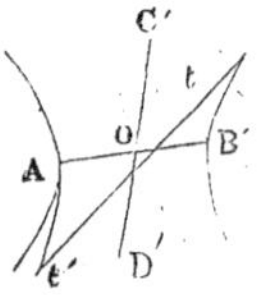

(93.) By thus imitating the investigations already given at length for the ellipse, we shall find that the properties of that curve established at pages 138, 140, equally belong to the hyperbola. With regard to the property at (66), however, when it is shown that the normal at any point bisects the angle formed by the radii vectores, it will be found that in the hyperbola the *exterior* angle is to be understood, and the reference will here be to *Geom.* p. 208, instead of p. 90, as for the ellipse. We shall now leave the student to investigate for himself these remaining properties of the hyperbola, which, after what has already been shown of the ellipse, he will find an easy, and, at the same time, an instructive exercise. In each case the diagram should be neatly sketched, and the disposition of the several lines employed compared with the corresponding lines in the ellipse.

On the Asymptotes of the Hyperbola.

(94.) It has already been seen (71), that the asymptotes, KOL, MON, of the hyperbola, make angles, KOB, MOB, with the principal transverse, of which the tangents are respectively $\frac{B}{A}$ and $-\frac{B}{A}$; so that these angles being supplements of each other, the angles, KOB, NOB, are equal. Let it now be required to transform the equation of the hyperbola, from the rectangular axes, OB, OC, to the oblique axes ON, OK, the new axis of x being *below* the primitive. For this pur-

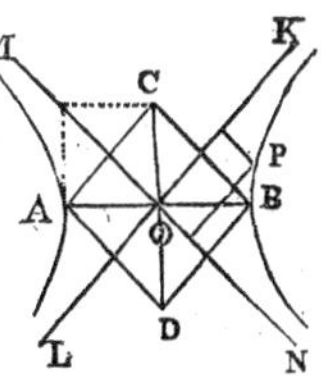

pose we must employ the formula (2′), at p. 117, which, since $\alpha = \alpha'$, becomes for this case

$x = (x + y) \cos. \alpha,\ y = (y - x) \sin. \alpha.$

Substituting, therefore, these values for x and y, in equation (6), p. 144, we have $A^2 (y - x)^2 \sin.^2 \alpha - B^2 (x + y)^2 \cos.^2 \alpha = - A^2 B^2$. But, since $\tan. \alpha = \frac{B}{A}$ $\therefore \sin.^2 \alpha = \frac{B^2}{A^2 + B^2}$ and $\cos.^2 \alpha = \frac{A^2}{A^2 + B^2}$, therefore this equation becomes, by substitution, $\frac{A^2 B^2}{A^2 + B^2} \times (y - x)^2 - \frac{A^2 B^2}{A^2 + B^2} (x + y)^2 = - A^2 B^2$ that is $\frac{4 A^2 B^2}{A^2 + B^2} xy = A^2 B^2$ $\therefore xy = \frac{1}{4}(A^2 + B^2)$ the equation of the hyperbola when referred to its asymptotes. The asymptotes are the only system of axes that can render the equation of the hyperbola of this form, in which both x^2 and y^2 are absent; this will be readily proved by taking the general transformed equation at (80), and determining the values of α and α', so that the terms involving x^2 and y^2 may disappear, for we shall then find that these angles can be no other than those formed by the asymptotes with the axis of x.

Calling the angle KON, included by the asymptotes φ, we have from the foregoing equation, $xy \sin. \varphi = \frac{1}{4}(A^2 + B^2) \sin. \varphi$.

The first member of this equation expresses the parallelogram, PO, contained by the coordinates of any point, P, in the curve. The second member is constant, and expresses a fourth of the parallelogram, ACBD, formed by joining the vertices of the principal diameters for $A^2 + B^2 = AC^2 = AC \cdot AD$ $\therefore \frac{1}{4}(A^2 + B^2) \sin. \varphi = \frac{1}{4}$ parallelogram AB; and this is obviously equivalent to half the rectangle of the two principal semi-diameters. Hence we infer this remarkable property, viz. If from any number of points in the hyperbola lines be drawn parallel to, and terminating in, the asymptotes, the parallelograms so formed will all be equal to each other, and to half the rectangle of the principal semi-diameters.

In the equilateral hyperbola, these parallelograms are all rectangles, and the rectangle of the principal semi-diameters is a square

PROBLEM I.

(95.) To find the equation of the tangent to an hyperbola when referred to the asymptotes as axes.

First, let us consider a secant or line passing through two points of the curve (x', y'), (x'', y''), then there must exist the following equations, viz. $y - y' = \frac{y' - y''}{x' - x''}(x - x')$, $x' y' = x'' y''$. Adding to the second equation the identity $- x' y'' + x' y'' = 0$, it may be put under the form $x'(y'-y'')+y''(x'-x'')=0$, $\therefore \frac{y'-y''}{x'-x''}=-\frac{y''}{x'}$. Hence, by substitution, the equation of the secant becomes $y-y'=-\frac{y''}{x'}(x-x')$. If we suppose $x'=x''$ and $y'=y''$, the secant will become a tangent, and the equation will then be $y-y'=-\frac{y'}{x'}(x-x')$ (1.)

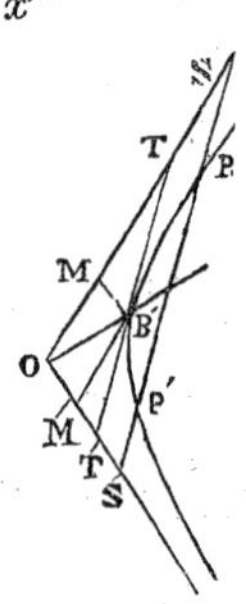

Putting, for brevity, m^2 for $x'y'$, or $\frac{1}{4}(A^2+B^2)$, and multiplying each side of the equation just deduced by x', the equation of the tangent will appear under the more simple form $yx'+xy'=2m^2$. . . . . (2).

In this equation let $x=0$, then $y=OT=\frac{2m^2}{x'}=2y'$ (3).

Let $y=0$, then $x=OT'=\frac{2m^2}{y'}=2x'$ . . . . (4).

$\therefore OT\cdot OT'=\frac{4m^4}{x'y'}=4m^2$, $\therefore \frac{OT\cdot OT'}{2}$. sin. $\varphi=2m^2$ sin. φ.

This proves that the area of the triangle formed by a tangent and the portions of the asymptotes intercepted between it and the centre is constant, and equal to the rectangle of the principal semi-diameters.

Equations (3) and (4) show that $OT=2OM$ and $OT'=2OM'$, from either of which equations it follows that the tangent, TT′, between the asymptotes is bisected at the point of contact, as we already knew from (85). Since OB′ bisects TT′ it must also bisect every line, SS′, parallel to TT′.* The same line also bisects the chord PP′; hence SP = S′ P′, that is, if any line cut the asymptotes and curve, the parts included between the asymptotes and curve are equal.

(96.) The analytical value of either of these intercepts will be expressed by subtracting the ordinate of the curve from the ordinate of the assymptote, for any proposed abscissa. Thus, if x denote the abscissa common to the two points S, P, then for the intercept SP we have $SP=\frac{B'}{A'}x-\frac{B'}{A'}\sqrt{\{x^2-A'^2\}}$, and for PS′ the expression will be $PS'=\frac{B'}{A'}x+\frac{B'}{A'}\sqrt{\{x^2-A'^2\}}$. Multiplying these two values together, we have $SP\cdot PS'=B'^2$.

Hence, *if a straight line be drawn through any point*, P, *in an hyperbola, the rectangle of the parts thereof intercepted between that point and the asymptotes will be equal to the square of the parallel semi-diameter.* Consequently, *if any number of parallels be drawn, and*

* This might also have been inferred from the equations of the asymptotes, viz. $y=B'x\div A'$ and $y=-B'x\div A'$, for they show that $y=-y$.

are terminated by the asymptotes, the rectangles of the parts into which they are cut by the curve will be equal.

We shall terminate the present chapter with the following problem;

PROBLEM II.

(97.) Knowing the asymptotes and a point in the curve to construct the hyperbola, and to determine the lengths and directions of the principal diameters.

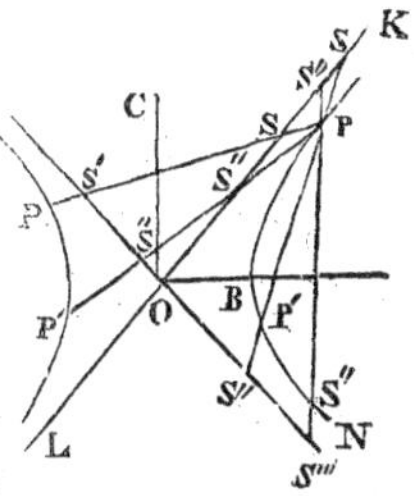

Let KL, MN be the asymptotes, and P the given point. Through the point P draw any number of secants, SS′, S″S‴, *s*, *s*′, &c. to the asymptotes, and so many points of the hyperbola will then be determinable. For, if we make the distances S′P′, S‴P″, &c. respectively equal to the distances SP′, S″P, &c. the points P, P′, P″, &c. will belong to the hyperbola, of which KL, MN are the asymptotes, and by thus determining a sufficient number of points the curve may be traced.

To determine the directions of the principal diameters of the curve thus constructed, we shall have merely to bisect the angle φ, or KON, and its supplement, MOK. As to the magnitudes of these diameters we have, by denoting the coordinates of the given point, P, by x', y', $x'\,y' = \frac{1}{4}(A^2 + B^2)$, or $4x'\,y' = A^2 + B^2 = A^2\{1 + (B^2 \div A^2)\}$.

But, (71), $\tan.\,\frac{1}{2}\varphi = \pm\,(B \div A)$ hence, by substitution, $4x'\,y' =$ $A^2(1 + \tan.^2\frac{1}{2}\varphi)$, whence $A^2 = \dfrac{4x'\,y'}{1 + \tan.^2\frac{1}{2}\varphi} = 4x'\,y'.\cos.^2\frac{1}{2}\varphi$,

$\therefore A = 2\cos.\frac{1}{2}\varphi\,\sqrt{\{x'\,y'\}}$, and $B = A\tan.\frac{1}{2}\varphi =$ $2\cos.\frac{1}{2}\varphi\tan.\frac{1}{2}\varphi\,\sqrt{\{x'\,y'\}} = 2\sin.\frac{1}{2}\varphi\,\sqrt{\{x'y'\}}$.

By means of this problem we can construct the hyperbola, when any system of conjugate diameters are given. For if at P, the extremity of one of the diameters, we draw a line parallel to the other, and mark on this line two points, one on each side of P, at distances from it equal to the parallel semi-diameter, these points will be on the asymptotes; these, therefore, may be drawn, and we shall then have the asymptotes and a point, P, in the curve to construct the hyperbola.

CHAPTER IV.

ON THE PARABOLA.

Its Equation and Properties.

(98.) A parabola is a curve in which any point P, is equally distant from a fixed point, F, and a straight line, KK′, given in position. Thus PF is always equal to the perpendicular, PD

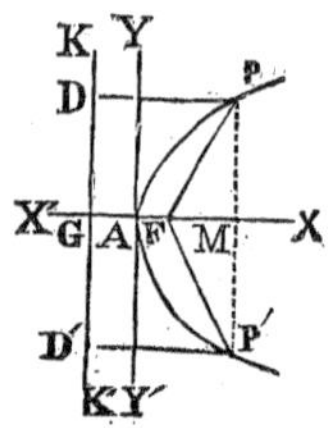

The point F is called the focus of the parabola, and the line KK′ the directrix. The distance of the focus from any point in the curve is called the focal distance, or radius vector of that point.

This curve is described by points, as follows: Through the focus, F, draw the straight line, GFX, perpendicular to the directrix. Bisect FG in A, then A will be a point in the curve for AF = AG. Take now any point, M, in AX, from which draw a perpendicular to AX, and from the focus as a centre, with the distance MG, for radius; describe an arc cutting this perpendicular in P, P′, then P, P′, will be points in the curve, for, by construction, PF = PD and P′F = P′D′. A like construction will furnish two new points, and this determination of points may be continued till the path of the curve becomes obvious.

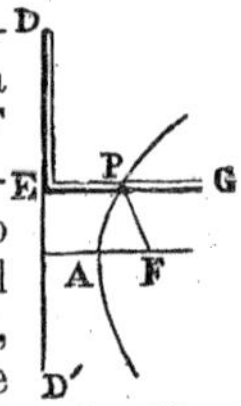

Any portion of this curve may be described by continuous motion. Thus, suppose we had to describe a parabolic arc, having the point A for its vertex, and F for the focus; then, having drawn the directrix, DD′, apply to it one of the sides, DE, of a square, DEG, and to the points G and F fasten the extremities of a cord, equal in length to EG; then with a pencil, P, stretch this cord, so that the part PG may always coincide with the side EG of the square, while the other side, DE, moves along the directrix. The point P will thus describe the required curve, for we shall always have FP + PG = PE + PG, and therefore always FP = PE.

(99.) In order to determine the equation of the parabola, (see diagram, art. (98) above) let us take the rectangular axes, AX, AY, originating at the point A on the curve. Put m for AF, and x, y for the coordinates of any point, P, in the curve, then $FP^2 = PD^2 = y^2 + (x - m)^2$, But $PD^2 = (AG + AM)^2 = (m + x)^2$, $\therefore y^2 + (x - m)^2 = (x + m)^2 \;\therefore y^2 = 4mx$ (1), the equation of the parabola required.

From this equation we get $x = y^2 \div 4m$; $y = \sqrt{\{4mx\}}$.

The first of these equations shows that the curve is entirely to the right of the axis YAY′, since the abscissas are all positive; it shows, moreover, that this axis is a tangent to the curve at A, since x can be 0 only when y is. The second equation shows that for the same abscissa there are two equal ordinates situated on opposite sides of AX, that is, AX bisects all the chords drawn parallel to AY. The same equation shows that y will always be possible so long as x is positive, so that the curve extends indefinitely to the right of AY, both above and below AX. The point A is called the *vertex*, the line AX the *principal axis* of the parabola, and AY the *principal second axis*. These axes are unlimited, but, as the former bisects all the chords drawn parallel to the latter, it is called a diameter.

(100.). The equation of the parabola, in terms of the principal *parameter*, or double ordinate, through the focus, is at once obtained from equation (1), for, since the distance of this double ordinate from the origin is m, we have, by substituting this value for x, and putting p for the parameter, $y^2 = 4m^2$, or $y = 2m = \frac{1}{2}p$ (2), $\therefore$ $y^2 = px$ (3), *the equation of the parabola, in terms of the parameter.*

From equation (2) it appears that the semi-parameter is equal to the distance of the focus from the directrix.

Since, from equation (3), the square of any ordinate is equal to the abscissa multiplied by a constant quantity, it follows that for every point in the curve *the abscissas are as the squares of the ordinates.*

Properties of the Parabola, when related to its conjugate Axes.

(101. Let us now transform the equation of the parabola from rectangular coordinates to oblique, in order to determine, as in the preceding curves, those systems in reference to which the equation of the parabola preserves the same form. For this purpose, let us substitute for x and y, in equation (1), the values $x = a + x \cos. \alpha + y \cos. \alpha'$, and $y = b + x \sin. \alpha + y \sin. \alpha'$, and this equation will then become, after transposing,

$$\left.\begin{array}{l} y^2 \sin.^2 \alpha' + 2xy \sin. \alpha \sin. \alpha' + x^2 \sin.^2 \alpha + b^2 - ap \\ + (2b \sin. \alpha' - p \cos. \alpha')\, y + (2b \sin. \alpha - p \cos. \alpha)\, x \end{array}\right\} = 0.$$

Now, in order that this equation may be of the form $y^2 = kx$, we must have the following conditions, viz. $\sin. \alpha \sin. \alpha' = 0$, $\sin.^2 \alpha = 0$, $2b \sin. \alpha' - p \cos. \alpha' = 0$, and $b^2 - ap = 0$ (1), and, when these subsist, the equation reduces to $y^2 = \dfrac{p}{\sin.^2 \alpha'} \cdot x \ldots$ (2).

In the foregoing general transformation we have supposed the origin, as well as the inclination of the axes, altered, because had the transformation been confined to the direction of the axes, in which case a and b would have been 0, we could never have obtained the form (2) for any oblique system of axes, for the conditions (1) would have led us back to the original rectangular system, since we should then have had $\alpha = 0$, $\alpha' = 90°$. The second of the conditions (1), which necessarily establishes the first, shows that the angle included by the old and new axes of x must be 0, that is, these axes are parallel, α' therefore will express the inclination of the new axes, and this inclination, as determined by the third condition is

$\tan. \alpha' = \dfrac{p}{2b}$ (3); and for the value of b we have, by the fourth condition, $b^2 = pa$ (4), in which equation a may be any value whatever; hence b, and consequently α', may take an indefinite number of values, so that the systems of coordinates that will render the equation of the parabola of the form (2) are unlimited in number, and from equation (4) it is obvious that in each of these systems the origin must always be on the curve, for that equation represents a point in the parabola, of which the coordinates are a and b.

(102.) The form of the equation (2) shows that for every positive value of x there are two values of y equal numerically, but opposite in sign, therefore the axis of x bisects all the chords drawn parallel to the axis of y; hence the axis of x is always a diameter. From this and the preceding article, it follows that *all the diameters are parallel to each other.* The same equation shows that no part of the curve can be to the left of the axis of y, for y is impossible for all negative values of x; moreover, since y must be 0 when x is, it follows that the axis of y is a tangent to the curve. These two axes are called *conjugate axes.*

(103.) Since (3) $\tan.^2 \alpha' = \frac{p^2}{4b^2}$ $\therefore \sec.^2 \alpha' = 1 + \frac{p^2}{4b^2}$

$$\therefore \sin.^2 \alpha' = \frac{\tan.^2 \alpha'}{\sec.^2 \alpha'} = \frac{p^2}{4b^2 + p^2} = \frac{p}{4a+p}$$

$$\therefore \frac{p}{\sin.^2 \alpha'} = 4a + p = 4\left(a + \frac{p}{4}\right) = 4(a+m)$$

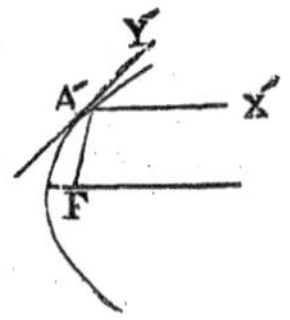

Now $a + m$ expresses the distance of a point, A′, in the parabola of which the abscissa is a, from the focus. This distance, A′F, is the *radius vector* of the point A′, therefore representing it by r, equation (2) becomes, by substitution, $y^2 = 4rx$ . . . . (5).

This, therefore, is the equation of the curve, related to any system of conjugate axes. It obviously includes equation (1), p. 158, in which the value of r, the radius vector of the origin, is m.

(104.) In article (100), it was found that the principal parameter is equal to $4m$, the coefficient of x, in the equation of the curve, when referred to rectangular conjugates. By analogy, the coefficient $4r$, when any system of conjugates are employed, is called the parameter of that diameter, which is taken for the axis of x, so that generally the parameter of any diameter is equal to four times the distance of its vertex from the focus. By equation (5) the parameter is always equal to the double ordinate, corresponding to the abscissa $x = r$. We shall hereafter see that this double ordinate always passes through the focus. The property in (100), viz. that *the abscissas are as the squares of the ordinates,* is obviously true for every system of conjugate axes.

Since, in the condition (4), a and consequently b, may take any value from 0 to infinite, therefore $\tan. \alpha'$, in the condition (3), may take any value between 0 and infinite, so that conjugate axes may exist inclined to each other, at any angle not exceeding a right angle.

Properties of the Tangent to the Parabola.

(105.) The equation of a straight line passing through two points (x', y') and (x'', y'') is $y - y' = \frac{y' - y''}{x' - x''}(x - x') \ldots . (1)$.
If the two points are on a parabola, then we must have $y'^2 = 4rx$ (2) and $y''^2 = 4rx''$ (3) $\therefore (y' + y'')(y' - y'') = 4r(x' - x'')$ $\because \frac{y' - y''}{x' - x''} = \frac{4r}{y' + y''}$ hence, by substitution, equation (1) becomes $y - y' = \frac{4r}{y' + y''}(x - x')$, which when the points (x', y'') and (x'', y'') coincide, reduces to $y - y' = \frac{2r}{y'}(x - x')$ (4), the equation of the tangent.

This equation may be simplified; for, multiplying by y', and transposing $yy' = 2rx - 2rx' + y'^2$; or, substituting for y'^2, its value in (2), $yy' = 2rx - 2rx' + 4rx'$, $\therefore yy' = 2r(x + x')$ (5).

This equation differs from that of the curve (103) only in this, that y^2 and $2x$ are here replaced by yy' and $x + x'$.

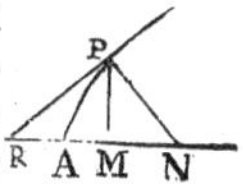

From the equation of the tangent we readily find that of the *normal*, or perpendicular, PN, to the tangent passing through the point of contact, x', y'. This equation, in reference to the rectangular conjugates, or when $x = m$, is $y - y' = -\frac{y'}{2m}(x - x')$.

From the equation of the tangent we may also determine the length of the subtangent, MR, being the distance intercepted between the ordinate of the point of contact and the intersection of the tangent with the axis of x. For, putting, in equation (5), $y = 0$, we have for the resulting value of x, whatever system of conjugates be employed, $x = -x'$ $\therefore$ MR $= 2x'$.

To find the expression for the length of the subnormal, MN, put, in the equation of the normal, $y = 0$, then for x we have the length, AN, and for $x - x'$, the length required; hence $x - x' =$ MN $= 2m$.
For the length, PR, of the tangent we have PR $= \sqrt{\{MP^2 + MR^2\}}$ $= \sqrt{\{y'^2 + 4x'^2\}} = 2\sqrt{\{mx' + x'^2\}}$. And for the length of the normal, PN $= \sqrt{\{MP^2 + MN^2\}} = \sqrt{\{y'^2 + 4m^2\}} = 2\sqrt{\{mx' + m^2\}}$.

(106.) Collecting these expressions into one point of view, for more convenient reference,

The equation of the tangent is $y - y' = \frac{2r}{y'}(x - x')$.

The equation of the normal is $y - y' = -\frac{y'}{2m}(x - x')$.

The length of the tangent is $T = 2\sqrt{\{mx' + x'^2\}}$.

The length of the subtangent is $T_{,} = 2x'$.
The length of the normal is $N = 2\sqrt{\{mx' + m^2\}}$.
The length of the subnormal is $N_{,} = 2m$.

These expressions lead immediately to several properties of the curve. Thus the expression for the subtangent at once shows that *the subtangent is always double the abscissa* for every system of conjugate axes; in other words, the subtangent is always bisected by the curve. As the same expression is independent of m, it follows that for every parabola having the same principal axis the subtangent measured on that axis is the same for the same abscissa.

The first-mentioned property above suggests an easy and obvious method of drawing a tangent to a parabola, when the abscissa of the point of contact is given. From the last of the above expressions it appears that, for every point in the curve, *the subnormal is constant, and equal to the distance of the focus from the directrix.*

(107.) Comparing together the expressions for the tangent and normal, the subtangent and subnormal, we find that, when $x' = m$ that is, when the tangent passes through the extremity of the latus rectum, the tangent and normal are equal, as also the subtangent and subnormal. The equation of the same tangent, in reference to the principal axes is $y - 2m = \frac{2m}{2m}(x - m)$, $\therefore y = x + m$.

But $x + m$ expresses the distance of any point (x, y) in the curve from the focus; therefore *any ordinate to the focal tangent is equal to the radius vector of the point where this ordinate cuts the curve,* so that *the focal tangent cuts from the tangent, through the vertex of the principal axis, a part equal to the distance of this vertex from the focus.*

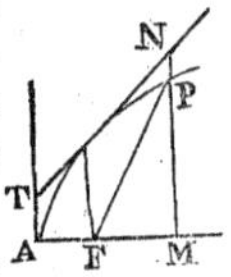

Thus, in the annexed diagram, MN=FP, & AT=AF.

(108.) If to the subnormal, $MN_{,} = 2m$, we add $FM = x' - m$, we have, $FN = x' + m = FP$; therefore the angle FPN = angle FNP=NPX′, that is, *the radius vector and the diameter at the point of contact are equally inclined to the tangent.*

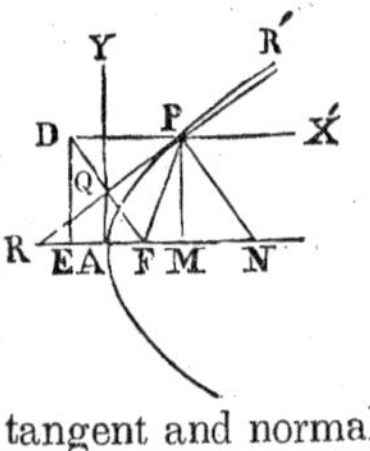

The same property might have been derived from (106); for, since $AR = AM = x'$, $\therefore FR = x' + m = FP$ $\therefore$ EPR = FRP = R′P′X′.

From this it appears that the points where the tangent and normal intersect, the axis are at the same distance from the focus as the point of contact, and therefore either may be easily drawn. If X′P be produced to cut the directrix in D, and FD be drawn, then, since DP = PF, and DPR = FPR, $\therefore$ PQ bisects FD at right angles, and Q is always on the axis, AY; for this line, bisecting FE, must bisect every other line, FD, drawn to ED from F; it follows, therefore, that *a tangent and a perpendicular to it from the focus always intersect on the principal second axis;* therefore the square of the perpendicular is equal to the product of the distances of the point of contact and vertex from the focus, that is, $FQ^2 = FP \cdot FA = FR \cdot FA$.

From this property we may derive an easy method of drawing a tangent to a parabola from a point either within or without the curve. Thus, let **P** be a point either within or without the curve, through which it is required to draw a tangent. Draw **PF**, upon which describe a semi-circle, and through the point **Q**, where it meets the second axis, **AY**, draw **PQ**, which will be the tangent required.

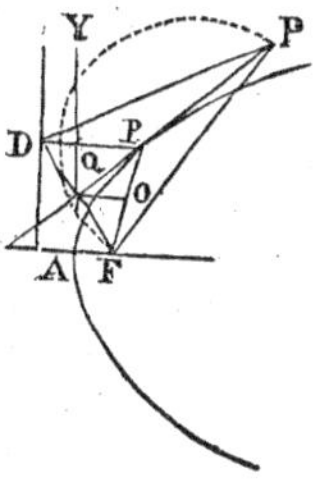

If the point be in the curve, the semi-circle will meet the axis in but one point; but, if it be without the curve, there will be two points of intersection. For, in the former case, since **FP** = **PD**, a parallel, **OQ**, through the middle of **FP** must be equal to PO, and at the same time less than any other line drawn from O to the axis AY. In the latter case, the circle on **PF** must cut AY in another point beside **Q**, for, if AY were a tangent to both circles at this point, the circles would touch there, which is impossible, since they meet also in F; hence from any point without a parabola two tangents may be drawn to the curve, a fact which might obviously have been inferred from the expression for the subtangent, which, being independent of the sign of y, shows that for the same abscissa there are two tangents, the one above and the other below the diameter, passing through the proposed point. When the directrix is given, another means of drawing the tangent is suggested from (108). Thus from the given point, P, as a centre, with a radius equal to **PF**, describe an arc, cutting the directrix in D, from which point draw a parallel to the diameter, and it will cut the curve in the point of tangence.

We shall conclude this chapter with the following problem:

(109.) Pairs of tangents to a parabola being always supposed to intersect at right angles, to find the locus of the points of intersection.

Let PD, P′D be any pair of tangents intersecting at right angles in D, and parallel thereto draw **FQ′**, **FQ** from the focus, then (108) the points **Q**, **Q′** will be on the second axis, AY, which will divide the rectangle, FD, into two equal triangles, **DQQ′**, **FQQ′**, therefore Dd, the altitude of the former, is equal to FA, the altitude of the latter; hence *the locus of* D *is the directrix.* Since D′ is a right angle, and the angle DPF = DPD′, and **PF** = **PD′**, ∴ DFP is a right angle. In like manner, DFP′ is a right angle; hence, first, *the part of the tangent intercepted between the point of contact and the directrix, subtends a right angle at the focus;* second, *the line joining the points of contact of perpendicular tangents always passes through the focus.*

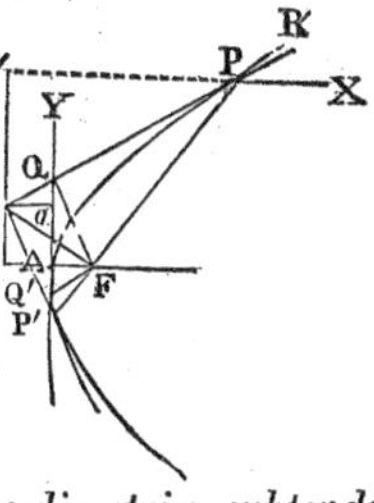

CHAPTER V.

ON POLAR COORDINATES.

(110.) In order to determine the analytical representation of a curve line, our object has hitherto been to obtain an equation between the *rectilineal* coordinates of any point in it. But besides this method there is another, and which consists in first assuming a point upon a fixed straight line, and then determining the position of any point in the curve, by means of an equation between its distance from the assumed point and the angle formed by this distance and the fixed line. The assumed point is called the *pole;* its distance from any point in the curve the *radius vector;* and the radius vector, together with its angle of inclination to the fixed line, are called the *polar coordinates* of the point.

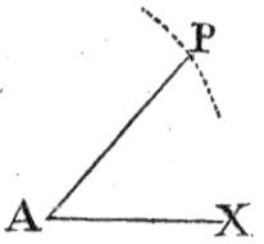

Thus assuming the point A on the fixed line, AX, as pole, then the polar coordinates of any point, P will be the radius vector AP, and the angle PAX.

To deduce the polar equation of a curve from the rectilineal is an easy operation; Thus:

Let P be a point in a curve related to the rectangular axes AX, AY, then AM, MP will be the coordinates of this point; but, if the same curve be related to polar coordinates, A′ being the pole, and A′X′ the fixed line, then will the coordinates of the same point be A′P′ and PA′X′. Draw A′X″ parallel to AX, and denote the radius vector, A′P, by r, the angle PA′X′ by ω, and the angle X′AX″ by α, then the angle PA′X″ will be $\omega + \alpha$. Let, also, a, b represent the rectangular coordinates of the pole. Then, since, AM = AB + A′Q, PM = A′B + PQ, and A′Q = A′P · cos. PA′Q; PQ = A′P · sin. PA′Q, we have, by substitution, $x = a + r \cdot \cos.(\omega + \alpha)$; $y = b + r \cdot \sin.(\omega + \alpha)$ (1); therefore, by substituting these values for x and y, in the primitive equation of the curve, we shall obtain the transformation desired.

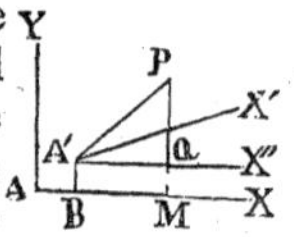

If A′X′ is parallel to AX, then $\alpha = 0$, and the preceding formulas become $x = a + r \cdot \cos.\omega$; $y = b + r \cdot \sin.\omega$ . . . . (2).

If the pole coincide with the origin of the primitive axes, and the original axis of x be taken for the fixed line, then, in the last formulas, a and b are 0, so that, in this case, $x = r \cos.\omega$; $y = r \sin.\omega$ (3).

Let the curve proposed be a circle, of which A, the origin of the rectangular axes, is the centre, then the equation, in reference to these axes, is $x^2 + y^2 = R^2$, and the polar equation, in its most general form, is found, by means of formulas (1), to be

$$r^2 + 2r(b \sin.[\omega + \alpha] + a \cos.[\omega + \alpha]) + a^2 + b^2 = R^2.$$

We shall now proceed to determine in succession the polar equations of the three other curves.

PROBLEM I.

(111.) To find the polar equation of the ellipse, either focus being the pole.

Let F be the pole, and AB the fixed axis; then the coordinates of the pole are, in this case, $a = c = \text{A}e$, $b = 0$; therefore the formulas (2) become $x = \text{A}e + r \cos. \omega$; $y = r \sin. \omega$; these substituted for x and y, in the equation of the curve, $y^2 = (1 - e^2)(\text{A}^2 - x^2)$, transforms it to

$$r^2 \sin.^2 \omega = (1 - e^2)\left[\text{A}^2 - (\text{A}e + r \cos. \omega)^2\right]$$
$$= (1 - e^2)\left[\text{A}(1 + e) + r \cos. \omega\right]\left[\text{A}(1 - e) - r \cos. \omega\right]$$
$$= \text{A}^2(1 - e^2)^2 - 2\,\text{A}e(1 - e^2)\,r \cos. \omega - (1 - e^2)\,r^2 \cos.^2 \omega,$$

or, since $r^2 \sin.^2 \omega = r^2 - r^2 \cos.^2 \omega$, we have, by substitution, transposition, &c. $r^2 + \dfrac{2\text{A}e(1 - e^2)\cos. \omega}{1 - e^2 \cos.^2 \omega}\, r = \dfrac{\text{A}^2(1 - e^2)^2}{1 - e^2 \cos.^2 \omega}$.

This quadratic, solved, gives $r = \dfrac{\pm \text{A}(1 - e^2) - \text{A}e(1 - e^2)\cos. \omega}{1 - e^2 \cos.^2 \omega}$

$$= \text{A}(1 - e^2)\frac{\pm 1 - e \cos. \omega}{(1 + e \cos. \omega)(1 - e \cos. \omega)}$$
$$= \text{A}\frac{1 - e^2}{1 + e \cos. \omega} \text{ or } - \text{A}\frac{1 - e^2}{1 - e \cos. \omega} \Bigg\} \ldots . (1).$$

It hence appears that there are two values of r measured in opposite directions, for the first value is essentially positive, whether ω be acute or obtuse, since both e and cos. ω are always less than 1, and for like reasons the second value is essentially negative. These two values therefore represent the two portions into which the focus divides the focal chord, inclined at the angle ω to the major diameter.

If we disregard the signs of these portions, and consider only their absolute lengths, we shall have no occasion for the second expression, since it will then be included in the first, provided we conceive the variable angle ω to go through all the degrees of magnitude from 0° to 360°, while the radius vector revolves round the pole.* With this condition, therefore, $r = \text{A}\dfrac{1 - e^2}{1 + c \cos. \omega}\Bigg\}$ (2), is *the polar equation of the ellipse, when the focus is the pole.*

The equation would have been similar if the other focus had been taken for the pole.

The polar equation would have been obtained more expeditiously by employing the expression already found for r at (47), and substituting therein $\text{A}e + r \cos. \omega$ for x, as upon trial the student will find.

---

* Thus the angle BF′P′ will be a re-entrant angle equal to $180° + \omega$.

PROBLEM II.

(112.) To find the polar equation of the ellipse, when the centre is the pole.

The solution of this problem is left for the student to perform. If the radius vector be denoted by r', and the variable angle by ω', the equation will be $r' = A\sqrt{\{(1-e^2) \div (1-e^2 \cos.^2 \omega')\}}$.

PROBLEM III.

(113.) To find the polar equation of the hyperbola, either focus being the pole. See the diagram at p. 141.

We shall solve this problem by the method indicated at (111).

Taking F′ for the pole, put ω for the angle PF′X, then (79) we have $r = ex + A$; hence, substituting for x the value given in formula (1), where $a = -c = -Ae$, and $b = 0$, there results $r = er \cos. \omega - Ae^2 + A \therefore r(1 - e \cos. \omega) = -A(e^2 - 1)$. $\therefore r = -A \frac{(e^2-1)}{1 - e \cos. \omega}$ } (1).

If the other focus had been taken for the pole, Ae would have been positive and A negative (79) so that, in the value of r A would have been positive, that is, we should have had $r = A \frac{e^2 - 1}{1 - e \cos. \omega}$ } . . (2).

In this case, however, it is customary to employ not the angle PFX, but its supplement, PFB $= 180° - \omega$, the cosine of which is $-\cos \omega$; hence, putting ω' for the angle PFB, the last equation becomes $r = A \frac{e^2 - 1}{1 + e \cos. \omega'}$ } (3). If, in these equations, ω, and ω' be negative, the value of r will remain unaltered, because the cosine of a negative arc is the same as the cosine of a positive arc; hence the whole of the branch PBp is represented by the equation (1) or (3), according as the pole is at F′ or F. If we transpose the poles, and then consider the angles ω, and ω' measured in the opposite directions to those above, it is obvious, from the symmetry of the two branches, that the same equations will characterize the other branch of the curve.

PROBLEM IV.

(114.) To find the polar equation of the hyperbola, when the centre is the pole.

By substituting $r' \cos. \omega$ for x, and $r' \sin. \omega$ for y, in the equation of the curve, the expression for the radius vector will be found to be $r' = A \sqrt{\{(e^2 - 1) \div (e^2 \cos.^2 \omega - 1)\}}$.

PROBLEM V.

(115.) To find the polar equation of the parabola, the focus being the pole.

Put ω for the angle PFA, then cos. PFM $= -\cos. \omega$. Hence we shall have to substitute in the equation $y^2 = 4mx$ of the curve $m -$

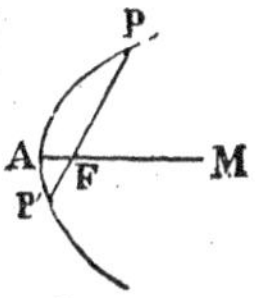

r cos. ω for x, and r sin. ω for y, which, after substituting $1-\cos^2\omega$ for $\sin.^2\omega$, transforms it to $(1-\cos.^2\omega)\,r^2+4m\cos.\omega\,r=4m^2$

$$\therefore r=\frac{-2m\cos.\omega\pm\sqrt{\{4m^2\cos.^2\omega+4m^2(1-\cos^2\omega)\}}}{1-\cos.^2\omega}$$

$$=\frac{2m(-\cos.\omega\pm1)}{1-\cos.^2\omega},\ \therefore r=\left.\frac{2m}{1+\cos.\omega}\text{ or }-\frac{2m}{1-\cos.\omega}\right\}\ldots(1).$$

The first of these expressions is always positive, and the second always negative. The first denotes FP, and the second FP′. The second expression is included in the first, abstracting from the sign, provided we conceive the angle ω to pass through all stages of magnitude from 0 to 360°, while the radius vector revolves round the pole, since then the sign of cos. ω will change as soon as the radius vector descends below FM. We may here remark that, by adding together the two values of r, in equation (1), we obtain for the absolute length of any focal chord $PP'=\frac{2m}{1+\cos.\omega}+\frac{2m}{1-\cos.\omega}=\frac{4m}{\sin.^2\omega}$.

This expression is identical with the coefficient of x, in equation (2,) p. 159; we may now, therefore, infer the property alluded to in art. 103, viz. that *in the parabola the parameter of any diameter is always equal to the double ordinate passing through the focus.*

(116.) If from the focus there be drawn a perpendicular to the radius vector of any point to meet a tangent to the curve at that point, the part intercepted is called the *polar subtangent* of the curve at that point. We have already seen (108) that, in the parabola, the locus of the extremities of the polar subtangents is the directrix. Let us now investigate the locus when the curve is the ellipse or hyperbola.

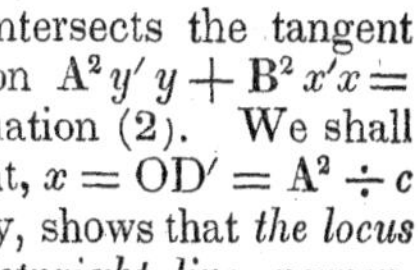

First. In the ellipse, the equation of FP, passing through the two points $(-c, 0)$ and $(-x', y')$, is $y=\frac{-y'}{c-x'}(x-c)$ (1), OX, OY, originating at the centre, being the rectangular axes. The equation of a perpendicular to this line through the point F or $(-c, 0)$ is $y=\frac{c-x'}{y'}(x-c)$ (2). Now, in order to find the abscissa of the point in which this last line intersects the tangent through P, we must equate y in the equation $A^2y'y+B^2x'x=A^2B^2$ (3) of the tangent, with its value in equation (2). We shall then find, upon reduction, for the abscissa sought, $x=OD'=A^2\div c=A\div e$ (4). This, being a constant quantity, shows that *the locus of the extremities of the polar subtangents is a straight line, perpendicular to the major diameter, and at the distance* $A\div e$ *from the centre.*

By analogy to the parabola, this line is called the *directrix* of the ellipse. There are obviously two directrices to the ellipse at equal distances from the centre, for if the semi-ellipse ADC be conceived to revolve round the axis, CD, till it coincide with the other semi-ellipse, taking the directrix along with it, this latter line will necessarily preserve its same distance from the centre.

If the radius vector, PF, be produced to P′, then a tangent through P′ must intersect that through P in D″, because the polar subtangents of the points P, P′ are both measured on the same line, viz. FD″.

From the expression (4) we may infer that *the distance of any point in an ellipse from the focus has a constant ratio to the distance of the same point from the directrix.* For the distance of any point in the curve from the directrix is $\frac{A}{e} - x = \frac{A - ex}{e}$ and its distance from the focus is (47) $A - ex$, and these expressions are to each other as 1 is to e. Because e is less than 1, the distance of any point in the ellipse from the focus is less than its distance from the directrix.

Second. In the hyperbola, the equation of FD, perpendicular to the radius vector FP, is, as before, $y = \frac{c - x'}{y'}(x - c)$, and, if we substitute this value of y in the equation $A^2 y' y - B^2 x' x = -A^2 B^2$ of the tangent we shall obtain for x the value $x = \frac{A^2}{c} = \frac{A}{e}$

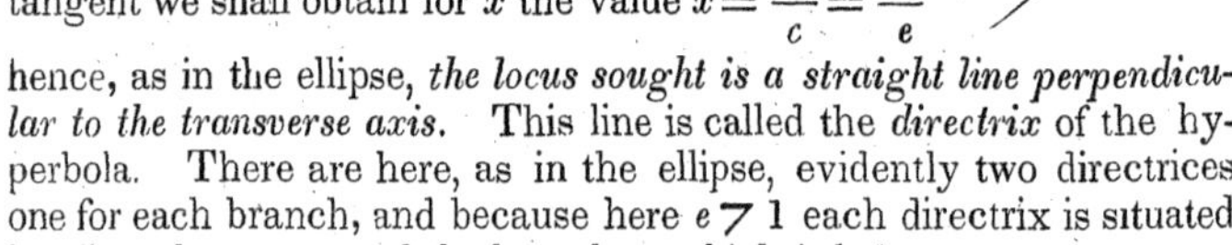

hence, as in the ellipse, *the locus sought is a straight line perpendicular to the transverse axis.* This line is called the *directrix* of the hyperbola. There are here, as in the ellipse, evidently two directrices one for each branch, and because here $e > 1$ each directrix is situated between the centre and the branch to which it belongs.

If FP be produced to P′, the tangent at P′ must meet that through P on the directrix, as in the ellipse, and for a like reason. Comparing the expression for the distance of any point in the curve from the directrix, viz. $x - \frac{A}{e} = \frac{ex - A}{e}$ with $ex - A$, the distance of the same point from the focus, we find that *their ratio is constant,* viz. as 1 to e; e being > 1, the distance of any point from the focus is greater than its distance from the directrix.

From what has now been said, we may define the directrix of either of the three curves to be *the locus of the extremities of the polar subtangents,* or else as *the locus of the intersection of pairs of tangents at the extremities of the focal chords.* In the parabola, the distance of any point from the focus is *equal* to its distance from the directrix, but in the ellipse it is *less,* and in the hyperbola *greater.*

(117.) By referring to the value of p, the principal parameter in each of the three curves, we find that in the ellipse $\frac{1}{2}p = A(1 - e^2)$, in the hyperbola $\frac{1}{2}p = A(e^2 - 1)$, and in the parabola $\frac{1}{2}p = 2m$; hence, by substituting $\frac{1}{2}p$ for these values in the polar equations of the respective curves, they will then all become comprehended in a single equation, viz. $r = \frac{\frac{1}{2}p}{1 + e \cos. \omega}$. If we suppose r produced to meet the curve again *below* the principal axis, we shall have the value of r', the radius vector of this second point, by changing ω into $180° + \omega$, that is $r' = \frac{\frac{1}{2}p}{1 - e \cos. \omega}$. Hence, for any focal chord, we have $r + r' = \frac{p}{1 - e^2 \cos.^2 \omega}$ also $rr' = \frac{\frac{1}{4}p^2}{1 - e^2 \cos.^2 \omega}$ consequently $(r + r')\, p = 4rr' \therefore \frac{rr'}{r + r'} = \frac{1}{4}p$; hence focal chords are to each other as the rectangles of the parts into which they are divided by the focus, likewise half the principal parameter is always an harmonical mean between the parts into which any focal chord is divided by the focus.

The three sections now completed comprise the first principal division of this work. It embraces a pretty comprehensive treatise on the *conic sections*, as the curves we have been discussing are usually called, for reasons to be hereafter given. In a future chapter will be thrown together several interesting and very general properties of these curves, which we could not conveniently revert to in the preceding articles.

SECTION IV.

CHAPTER I.

ON THE LOCI OF INDETERMINATE EQUATIONS OF THE SECOND DEGREE.

(118.) We have already seen that each of the curves whose properties have now been investigated is analytically represented by an indeterminate equation of the second degree, and we now propose to show, conversely, that every indeterminate equation of the second degree is geometrically represented by one or other of these curves. We shall do this by first showing that any equation whose form agrees with that characterizing one of the preceding curves, must have that curve for its locus, we shall afterwards prove that any indeterminate equation of the second degree, of whatever form, may be transformed to one or other of those particular forms already shown to characterize lines of the second order.

(119.) First, let us seek the locus of the equation $My^2 + Nx^2 = P$ (1), agreeing in form with the equation of the ellipse, M, N, and P being positive.

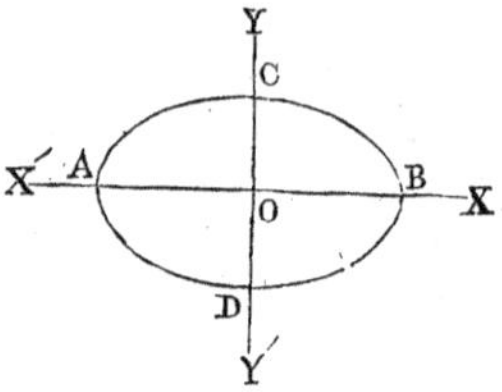

Assume a system of rectangular axes, XX′, YY′, in reference to which the locus of (1) is to be constructed. Then, since for, $y = 0$ we have $x = \pm\sqrt{\frac{P}{N}}$, and for $x = 0$ we have $y = \pm\sqrt{\frac{P}{M}}$, If we make OB, OA each equal to $\sqrt{(P \div N)}$, and OC, OD each equal to $\sqrt{(P \div M)}$, the points, A, B, C, D, thus determined will be those in which the locus cuts the axes. Let us now represent $\sqrt{(P \div N)}$, that is, the abscissa OB by A and $\sqrt{(P \div M)}$, or the ordinate OC by B, then we shall have $N = \frac{P}{A^2}$, and $M = \frac{P}{B^2}$, and equation (1) will become, by substitution, $\frac{P}{B^2}y^2 + \frac{P}{A^2}x^2 = P$, or $A^2y^2 + B^2x^2 = A^2B^2$ (2.)

Let us now suppose that upon the lines AB, CD, as principal diameters, an ellipse is constructed; we know that this ellipse is analytically represented by equation (2); in other words, that it is the locus of equation (2). But equations (1) and (2) are identical; hence the same curve is the locus of equation (1). We have here proceeded upon the supposition that $\sqrt{\frac{P}{N}} > \sqrt{\frac{P}{M}}$ or $M > N$; if, however, this be not the case, but $N > M$, then, putting X^2 for y^2, and Y^2 for x^2, in equation (1), it would have become $NY^2 + MX^2 = P$, which equation we should have shown, as above, to be the analytical representation of the ellipse constructed on the axes $2\sqrt{\frac{P}{M}}$ and $2\sqrt{\frac{P}{N}}$.

Secondly. When $M = N$, equation (1) reduces to $y^2 + x^2 = (P \div N)$ which represents a circle whose radius $\sqrt{\frac{P}{N}}$.

To express the distance, c, of the centre of the ellipse, which is the locus of equation (1), from the focus, we have $A^2 = \frac{P}{N}$, $B^2 = \frac{P}{M}$ $\therefore A^2 - B^2 = c^2 = \frac{P(M-N)}{MN}$. Let us now suppose that the equation proposed is $My^2 - Nx^2 = -P$ (3) agreeing in form with the equation of hyperbola. In this case for $y = 0$ we have $x = \pm\sqrt{\frac{P}{N}}$ and for $x = 0$ we have $y = \pm\sqrt{-\frac{P}{M}}$ hence the locus of (3) cuts the axis of

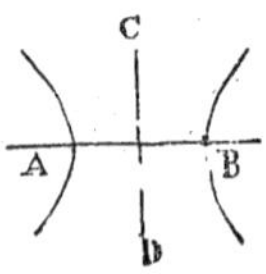

x in two points, equally distant from the origin, but it does not meet the axis of y.

Let $\sqrt{\frac{P}{N^2}}$ be represented by A, and $\sqrt{-\frac{P}{M}}$ by $\sqrt{-B^2}$, then $N = \frac{P}{A^2}$ and $M = \frac{P}{B^2}$; and equation (3) becomes, by substitution, &c.

$$\frac{P}{B^2}y^2 - \frac{P}{A^2}x = -P \text{ or } A^2y^2 - B^2x^2 = -A^2B^2 \ldots . (4).$$

If, now, upon the principal diameters, $AB = 2\sqrt{\frac{P}{N}}$; or 2A and $CD = 2\sqrt{\frac{P}{M}}$; or 2B, an hyperbola be supposed to be constructed, it will be analytically represented by the equation (4), that is, this curve will be the locus of equation (4); therefore, since equations (4) and (3) are identical, the hyperbola is also the locus of equation (3).

If, in equation (3), P had been positive, instead of negative, the locus would still have been an hyperbola, for then by putting X^2 for y^2, and Y^2 for x^2, the equation becomes, by changing the sides, $-P = NY^2 - MX^2$, the locus of which has just been shown to be an hyperbola. The distance, c, between the centre and focus of the hyperbola, which is the locus of equation (3), is found, as in the preceding case, to be $c = \pm\sqrt{\frac{P(M+N)}{MN}}$. Thirdly. Let the equation to be constructed be $y^2 = Qx$ (5), corresponding in form to the equation of the parabola. Then from the origin O of the given rectangular axes take two distances, OF, OG, each equal to $\frac{1}{4}Q$; then, having drawn the perpendicular, GD, if on the proposed axes a parabola be described, having F for its focus, and GD for its directrix, its equation will obviously be identical with equation (5), the locus of this equation is therefore a parabola.

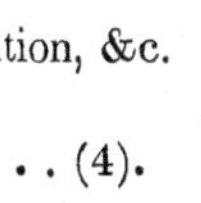

(120.) It must here be remarked that the generality of the foregoing conclusions is not in the least diminished, because the axes to which the several loci are referred have been supposed rectangular. For, if they had been in each case oblique, we might, by employing the formulas at (38), have obtained the equation of the same locus for rectangular axes, after which we could, as in articles (48,) (80), &c. have so determined the angles α and α' as to have preserved the form of the equation unaltered.

It must be further observed that if, in equation (1), P be supposed negative, the locus will not be an ellipse, but an imaginary curve, since the general value for any ordinate is in this the imaginary expression, $y = \sqrt{(-\frac{N}{M}x^2 - \frac{P}{M})}$. This imaginary curve, as also a circle, and a point, all arising from equation (1), under different modifications, are called *varieties of the ellipse*.

By supposing N as well as P negative, in equation (1), this equation becomes identical with (3), characterizing an hyperbola. It will not however represent an hyperbola, if $P=0$, but a system of two straight lines intersecting at the origin, for any ordinate of the locus will then be, $y= \pm \sqrt{\frac{N}{M}}x$; so that the *varieties of the hyperbola* are an equilateral hyperbola, or a system of two intersecting straight lines. The equation (5) of the parabola furnishes no variety; the change in the sign of Q merely changes the position of the curve from the right to the left of the axis of y.

(121.) It has been shown (44) that the equation of the ellipse, in terms of the parameter, the origin being at the vertex, is $y^2=px-\frac{p}{2A}x^2$. If in this equation, we suppose A to increase indefinitely, while p, or the value of $2B^2 \div A$, remains constant, it is plain that the coefficient $\frac{p}{2A}$ will continually diminish, and will at length vanish, when A becomes infinite, that is, the equation will then become $y^2=px$, agreeing in form with equation (5), above, which belongs to a parabola. It follows, therefore, that the parabola may be considered as a species of the ellipse, since it is the form the ellipse takes, when the major diameter becomes infinite. Considering the parabola in this light, several properties of it established in sect. iii. chap iv. might have been deduced from the properties of the ellipse. Thus, for instance, since it was shown in (66) that the locus of the intersection of tangents to the ellipse, with the perpendiculars drawn to them from the focus, was the circumference of a circle described on the major diameter, we might have inferred that, when the centre of this circle became infinitely distant from the vertex, A, any infinite portion of the circumference might be considered as a straight line; and have thence concluded that the locus becomes a straight line, when the ellipse becomes a parabola. As, however, this mode of deduction is both objectionable and unnecessary, we have in no instance thought proper to resort to it. Many of the properties of the parabola demonstrated in that chapter have been established, by independent processes, in a manner much more simple than the corresponding properties of the ellipse; on this account, therefore, it would have been wrong to have made them depend upon these latter. The property here referred to is an illustration of this remark.

(122.) We now proceed to show that the locus of every indeterminate equation of the second degree, containing two variables, can be no other than one of the curves already considered. The proof of this will be established, provided we can show that the general equation $Ay^2+Bxy+Cx^2+Dy+Ex+F=0$ (1), may always be transformed into another, either of the form $My^2+Nx^2=P$, or $y^2=Qx$ by merely altering the axes to which the locus of (1) is referred.

It is obvious that we are at liberty to consider (1) as the equation of the locus, in reference to rectangular axes, since, if the axes were oblique, we should, by employing the formulas (4), p. 117, be able to transform the equation into another, referring the curve to rectangular axes; and, as this transformed equation would have the same form as the primitive, we may therefore consider equation (1) as the result.

1. *To remove the Term containing the Product of the Variables.*

For x and y in equation (1) substitute the values

$$\left.\begin{aligned} x &= x\cos.\alpha - y\sin.\alpha \\ y &= x\sin.\alpha + y\cos.\alpha \end{aligned}\right\} \ldots (2)$$

by means of which values we pass from one system of rectangular coordinates to another, having the same origin. The result of this substitution is

$$\left.\begin{array}{l|l|l|l} \text{A}\cos.^2\alpha & y^2 + 2\text{A}\sin.\alpha\cos.\alpha & xy + \text{A}\sin.^2\alpha & x^2 \\ -\text{B}\sin.\alpha\cos.\alpha & \quad\text{B}\cos.^2\alpha & \quad\text{B}\sin.\alpha\cos.\alpha & \\ \text{C}\sin.^2\alpha & \quad -\text{B}\sin.^2\alpha & \quad\text{C}\cos.^2\alpha & \\ & \quad -2\text{C}\sin.\alpha\cos.\alpha & & \end{array}\right.$$

$$\left.\begin{array}{l|l|l} +\text{D}\cos.\alpha & y + \text{D}\sin.\alpha & x + \text{F} = 0 \\ -\text{E}\sin.\alpha & \quad\text{E}\cos.\alpha & \end{array}\right\} (3).$$

Now the value of α is arbitrary; we may therefore assume it of such value, that the second term of the transformed equation may vanish; this value will be determined by the equation, $2\text{A}\sin.\alpha\cos.\alpha + \text{B}\cos.^2\alpha - \text{B}\sin.^2\alpha - 2\text{C}\sin.\alpha\cos.\alpha = 0$, or $(\text{A} - \text{C})\,2\sin.\alpha\cos.\alpha + \text{B}\,(\cos.^2\alpha - \sin.^2\alpha) = 0$; which by substituting $\sin.2\alpha$ for $2\sin.\alpha\cos.\alpha$ and $\cos.2\alpha$ for $\cos.^2\alpha - \sin.^2\alpha$, becomes $(\text{A} - \text{C})\sin.2\alpha + \text{B}\cos.2\alpha = 0$, from which we obtain $\tan.2\alpha = -\dfrac{\text{B}}{\text{A}-\text{C}}\Big\} \ldots\ldots (4)$.

If, therefore, in the formulas (2), we give to the angles α a value such that the tangent of double that angle may be that number $-\dfrac{\text{B}}{\text{A}-\text{C}}$, no term containing xy can appear in the transformed equation. This term, therefore, is removed, by changing the directions of the rectangular axes, and the transformed equation then takes the form

$$\text{M}y^2 + \text{N}x^2 + \text{R}y + \text{S}x + \text{F} = 0 \ldots\ldots (5).$$

II. *To remove the terms containing the first power of the variables.*

For x and y in equation (5) substitute the values $x = a + x$, and $y = b + y$. By means of which the locus of (5) will become referred to new axes, parallel to the primitive; equation (5) will be transformed to

$$\text{M}y^2 + \text{N}x^2 + \begin{array}{l|}2\text{M}b \\ \quad\text{R}\end{array}\, y + \begin{array}{l|}2\text{N}a \\ \quad\text{S}\end{array}\, x + \text{M}b^2 + \text{N}a^2 + \text{R}b + \text{S}a + \text{F} = 0$$

in which equation, in order that the terms containing x and y may disappear, there must exist the conditions $2\text{M}b + \text{R} = 0$, and $2\text{N}a + \text{S} = 0$, which give $b = -\dfrac{\text{R}}{2\text{M}}$ and $a = -\dfrac{\text{S}}{2\text{N}}$. These values of a and b, therefore, reduce equation (5) to the form $My^2 + Nx^2 = P$, P being put for $- Mb^2 - Na^2 - Rb - Sa - F$.

(123.) We have here proceeded upon the supposition that neither of the terms My^2, Nx^2 is absent from equation (5). If, however, one of these, as Nx^2, is absent, that is, if $N = 0$, then the coefficient of x, in the transformed equation, will be simply S, and consequently the term containing the first power of x can in this case vanish only when $S = 0$, that is, when it is also absent from equation (5). If then this term be not absent from equation (5), neither can it be removed from the transformed equation when $N = 0$. We can, however, in this case, remove the term which is independent of the variables, for when $N = 0$ this term is $Mb^2 + Rb + Sa + F$, and, in order to find what value must be given to the arbitrary quantity, a, that this expression may be 0, we must determine a from the condition $Mb^2 + Rb + Sa + F = 0$, which gives $a = -\frac{Mb^2 + Rb + F}{S}$, with this value of a, therefore, and the value $-\frac{R}{2M}$ for b, equation (5), in the case proposed, is reduced to the form $My^2 + Sx = 0$, or $y^2 = Qx$, Q being put for $- (S \div M.)$

If we had supposed $M = 0$, instead of $N = 0$, the resulting equation would have been $Nx^2 + Ry = 0$, agreeing with the former in form, when the axes are transposed.

We cannot suppose that, in equation (5), both $M = 0$ and $N = 0$ at the same time, or, which is the same thing, that the three first terms vanish from equation (3). For, from inspecting the coefficients of these terms, it is obvious that the first and third cannot vanish, unless $A = -C$, and $B = 0$, and upon this supposition the second term must remain, unless we moreover suppose that A and C are both 0, when equation (1) will no longer be of the second degree, but of the first, which is contrary to the hypothesis; so that the supposition of both M and N disappearing from equation (5) is inadmissible.

If both the first and second powers of one of the variables, as Sx and Nx^2, are absent from equation (5), then the form of that equation is $My^2 + Ry + F = 0$, which is no longer an equation containing two variables, and represents not a curve, but a system of two parallel straight lines, for there are two constant values for each ordinate, viz. $y = -\frac{R}{M} \pm \frac{1}{2M} \sqrt{\{R^2 - 4FM\}}$. The two parallels characterized by this equation coincide, if $R^2 = 4MF$, and they become imaginary, if $4FM > R^2$. This *variety* of the general equation, arising from the supposition that $N = 0$ and $S = 0$, in its transformed state (5), is considered as a variety of the parabola, because, in the equation of this curve, N is always likewise 0.

(124.) We have now shown that every indeterminate equation of the second degree, containing two variables, may, by means of a double transformation of coordinates, be always reduced to one of the forms $My^2 + Nx^2 = P$, or $y^2 = Qx$, except in the particular case, where the removal of xy by the first transformation takes away also the terms containing the first and second powers of one of the variables, in which case, the locus is a system of parallel straight lines. Hence the locus of (1) can be no other curve but one of the three already discussed, that is, this locus must be either, An ellipse, having for varieties a circle, a point, or an imaginary curve. An hyperbola, having for varieties an equilateral hyperbola, or a system of two straight lines, intersecting at the origin. Or a parabola, having for varieties a system of two parallel straight lines, a single straight line, or two imaginary straight lines.

It is of importance to be able to ascertain readily when any equation of the second degree is given to which of the three curves it belongs. The following process will lead to a criterion for this purpose. By adding and subtracting the values of M and N, and substituting, in the first result, 1 for $\sin.^2\alpha + \cos.^2\alpha$, we have

$$M = A\cos.^2\alpha - B\sin.\alpha\cos.\alpha + C\sin.^2\alpha$$
$$N = A\sin.^2\alpha + B\sin.\alpha\cos.\alpha + C\cos.^2\alpha$$
$$M + N = A$$
$$M - N = (A - C)(\cos.^2\alpha - \sin.^2\alpha) - B\,2\sin.\alpha\cos.\alpha$$
$$= (A - C)\cos.2\alpha - B\sin.2\alpha.$$

If, in this last equation, we substitute for $\cos.2\alpha$, $\sin.2\alpha$, their values in terms of $\tan.2\alpha = -\frac{B}{A - C}$, which are

$$\cos.2\alpha = \frac{A - C}{\sqrt{\{(A-C)^2 + B^2\}}},\ \sin.2\alpha = \frac{-B}{\sqrt{\{(A-C)^2 + B^2\}}}$$

we have $M - N = \dfrac{(A-C)^2 + B^2}{\sqrt{\{(A-C)^2 + B^2\}}} = \sqrt{\{(A-C)^2 + B^2\}}$;

consequently, $M = \frac{1}{2}[(A + C) \pm \sqrt{\{(A - C)^2 + B^2\}}]$

$N = \frac{1}{2}[(A + C) \mp \sqrt{\{(A - C)^2 + B^2\}}]$.

By multiplying these two expressions together, we have

$$M \cdot N = \tfrac{1}{4}[(A + C)^2 - \{(A-C)^2 + B^2\}] = \tfrac{1}{4}(4A \cdot C - B^2).$$

From this equation it follows that M and N must have the same sign, so long as $4A \cdot C > B^2$, that they must have different signs, when $4A \cdot C < B^2$, and that one of these coefficients must be 0, when $4A \cdot C = B^2$. Hence the general equation of the second degree, characterizes, when

$B^2 - 4A \cdot C < 0$, the ellipse, and its varieties;

$B^2 - 4A \cdot C > 0$, the hyperbola, and its varieties;

$B^2 - 4A \cdot C = 0$, the parabola, and its varieties;

From the equation $M + N = A + C$ it follows, that, if $A = -C$, then $M + N = 0$, that is, $M = -N$; the equation, therefore, de notes an equilateral hyperbola. Since, when the general equation belongs to the parabola and its varieties, there must be $B^2 = 4A \cdot C$, the preceding expressions for M and N become, in this case,

$$M = \tfrac{1}{2}[(A + C) \pm (A + C)] = A + C, \text{ or } 0,$$
$$N = \tfrac{1}{2}[(A + C) \mp (A + C)] = 0, \text{ or } A + C.$$

The upper sign of the quantity, $\pm \sqrt{\{(A - C)^2 + B^2\}}$, which occurs in the general expressions for M and N, is to be used when B is negative, and the lower sign when this coefficient is positive. For the sine of an angle being positive, whether the angle be acute or obtuse, it follows that the above quantity which forms the denominator in the expression for sin. 2α, above, must agree in sign with the numerator.

(125.) Having thus determined the coefficients, M, N, of the equation $My^2 + Nx^2 + Ry + Sx + F = 0$, we may thence obtain the values of R and S. We shall, in order to do this, have to substitute for sin. α, cos. α, in the coefficients of x and y, in equation (3) their respective values cos. $\alpha = \sqrt{\dfrac{1 + \cos. 2\alpha}{2}}$, sin. $\alpha = \sqrt{\dfrac{1 - \cos. 2\alpha}{2}}$ and, as cos. 2α has already been expressed in terms of the coefficients of the proposed equation, we shall thus obtain known values for all the coefficients of the above equation, which may then be simplified by art. (122) or (123). When the locus of the proposed equation is not a parabola, that is, when $B^2 - 4AC \lessgtr 0$, it is plain, from the foregoing expressions for cos. 2α, that the above values of cos. α, sin. α, will be rather complicatad, much more so than when the locus is a parabola, or when $B^2 = 4AC$, since the expression for cos. 2α becomes then free from radicals. On this account it will be found more convenient, when we have actually to construct the equation in the cases $B^2 - 4AC = \lessgtr 0$, first to remove the terms containing the first power of the variables and afterwards to remove that containing their product. The two curves comprised in these two cases are called central curves, to distinguish them from parabolas, which have no centre, their diameters being infinite. We shall now proceed to determine formulas for the construction of central curves, by reversing the order of transformation before used.

Construction of Central Curves.

(126.) Resuming the general equation,

$$Ay^2 + Bxy + Cx^2 + Dy + Ex + F = 0 \ldots (1),$$

Let us remove the origin of coordinates, by means of the formulas $x = a + x$, $y = b + y$, and the transformed equation becomes

$$Ay^2+Bxy+Cx^2+\left.\begin{array}{r}2Ab\\ Ba\\ D\end{array}\right|y+\left.\begin{array}{r}2Ca\\ Bb\\ E\end{array}\right|x+Ab^2+Bab+Ca^2+Db+Ea+F=0.$$

In order that the terms containing x and y may disappear from this equation, we must have the conditions $2Ab+Ba+D=0$ (2), and $2Ca+Bb+E=0$ (3), whence $a=\frac{2AE-BD}{B^2-4AC}$, $b=\frac{2CD-BE}{B^2-4AC}$. These values of a and b therefore reduce the transformed equation to the form $Ay^2+Bxy+Cx^2+F'=0$ (4), of which the three first coefficients are the same as those in the primitive equation, and where $F'=Ab^2+Bab+Ca^2+Db+Ea+F$. This expression for F' may be simplified by means of the conditions (2) and (3), for, multiplying the first by b, and the second by a, we have for their sum $2Ab^2+2Bab+2Ca^2+Db+Ea=0$,

$$\therefore\ Ab^2+Bab+Ca^2=-\frac{Db+Ea}{2}\ \therefore\ F'=F+\frac{Db+Ea}{2}.$$

The transformation just employed brings the origin of the axes to the centre of the curve, for equation (4) will remain unaltered, if for x we substitute $-x$, provided we at the same time change y into $-y$, so that, if $(x'\,y')$ be one point in the curve, $(-x',-y')$ will always be another, and the line joining them will obviously pass through, and be bisected by, the origin; as, therefore, the origin bisects all the chords passing through it, it must be at the centre. Equation (4) is therefore the general equation of central curves, when the axes originate at the centre, and have any inclination whatever to each other. Had we known that the ellipse and hyperbola were the only curves coming under this denomination, the same thing might have been inferred from the general equations of them in (48) and (80). Although equation (4) is entirely independent of the inclination of the axes, yet, for simplicity, we shall, as in the former mode of transformation, consider the axes as rectangular. To pass from these to the axes of the curve, we shall have to remove from equation (4) the term containing xy, by a transformation which has in (122) already been effected for the general equation, with which equation (4) agrees when $D=0$, $E=0$, and $F\quad F'$; so that the transformed equation (5), at p. 173, will here be $My^2+Nx^2+F'=0$: hence, putting $P=-F'$, we have, finally, $My^2+Nx^2=P$ .... (5).

(127.) We may now, for more convenient use, collect together the formulas to be employed, in order to transform an equation from the form (1) to the form (5), in those cases were $B^2-4AC\angle 0$, or $B^2-4AC > 0$.

1. Proposed Equation of the curve.

$$Ay^2+Bxy+Cx^2+Dy+Ex+F=0.$$

1. *Formulas to be employed for removing the terms* Dy, Ex.

$$a=\frac{2\text{AE}-\text{BD}}{\text{B}^2-4\text{AC}},\ b=\frac{2\text{CD}-\text{BE}}{\text{B}^2-4\text{AC}},\ \text{F}'=\text{F}+\frac{\text{D}b+\text{E}a}{2}$$

II. Resulting Equation,

The origin of the rectangular axes being at the centre,

$$\text{A}y^2+\text{B}xy+\text{C}x^2+\text{F}'=0.$$

2. *Formulas for the removal of the term* Bxy, (*see art.* 124).

$$\tan.\ 2\alpha=-\frac{\text{B}}{\text{A}-\text{C}},\ \text{M}=\tfrac{1}{2}[(\text{A}+\text{C})\pm\sqrt{\{(\text{A}-\text{C})^2+\text{B}^2\}}],$$

$$\text{N}=\tfrac{1}{2}[(\text{A}+\text{C})\mp\sqrt{\{(\text{A}-\text{C})^2+\text{B}^2\}}],\ \text{P}=-\text{F}'.$$

III. Resulting Equation,

The axes being the principal diameters of the curve $\text{M}y^2+\text{N}x^2=\text{P}$. Since $\text{P}=-\text{F}'$, it follows from (120), that, when $\text{F}'=0$, the locus will be two intersecting straight lines, if $\text{B}^2-4\text{AC}>0$, and a point, if $\text{B}^2-4\text{AC}<0$. In the case $\text{B}^2-4\text{AC}\angle 0$, the locus will be an imaginary curve, provided F′ be positive.

EXAMPLES.

(128.) 1. Construct the locus of the equation

$$y-2xy+3x^2+2y-4x-3=0.$$

Comparing this with the general equation (1), we find $\text{A}=1$, $\text{B}=-2$, $\text{C}=3$, $\text{D}=2$, $\text{E}=-4$, $\text{F}=-3$; hence, substituting these values in the first class of formulas, above, we have $a=\frac{1}{2}$, $b=-\frac{1}{2}$, $\text{F}'=-\frac{9}{2}$; therefore the equation of the curve, when the origin is removed to the centre, is $y^2-2xy+3x^2-\frac{9}{2}=0$. By the second class of formulas we have $\tan.\ 2\alpha=-1$, $\text{M}=2+\sqrt{2}$, $\text{N}=2-\sqrt{2}$, $\text{P}=\frac{9}{2}$; hence the equation of the curve, when related to its principal diameters, is $(2+\sqrt{2})y^2+(2-\sqrt{2})x^2=\frac{9}{2}$. To determine the values of the diameters 2A, and 2B, we have, by supposing $y=0$, in this equation, $x^2=\frac{9}{2(2-\sqrt{2})}=\frac{9}{4}(2+\sqrt{2})\ \therefore \text{A}=\frac{3}{2}\sqrt{(2+\sqrt{2})}=2\cdot 7$, also for $x=0$ we have $y^2=\frac{9}{2(2+\sqrt{2})}=\frac{9}{4}(2-\sqrt{2})\ \therefore \text{B}=\frac{3}{2}\sqrt{(2-\sqrt{2})}=1\cdot 1$.

The construction of the curve, is therefore, as follows:

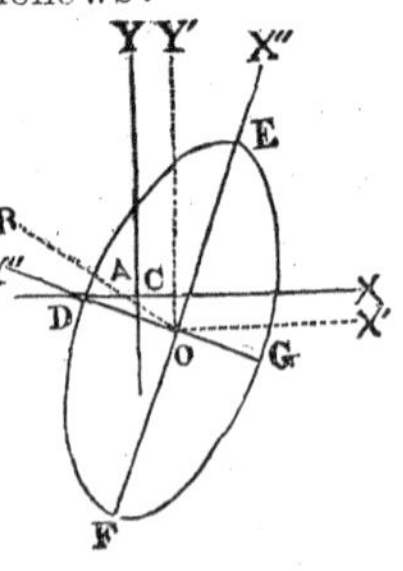

Let AX, AY be the original axes, to which the curve is referred. Make $\text{AC}=\frac{1}{2}$, $\text{CO}=-\frac{1}{2}$, then O will be the centre of the curve, and OX′, OY′, parallel to the primitive axes, will be the axes to which the first transformed equation refers the curve. From O draw the straight line OB, making with OX′ an angle of which the tangent is -1, that is, an angle of 135°. Bisect this angle by the line OX″, then OX″ and the perpendicular to it, OY″, will be the axes to which the second transformed equation refers tne curve; therefore, taking on these axes OE, OF, each equal to $2\cdot 7$, and OD, OG, each equal to $1\cdot 1$; the principal diameters of the ellipse will be determined, and thence the curve easily traced.*

2. Construct the locus of the equation $2y^2 - 2xy - x^2 + y + 4x - 10 = 0$.

Here $A = 2$, $B = -2$, $C = -1$, $D = 1$, $E = 4$, $F = -10$, and, since $B^2 - 4AC > 0$, the locus is an hyperbola; also $a = \frac{3}{2}$, $b = \frac{1}{2}$, $F' = -\frac{27}{4}$; hence, when the origin is removed to the centre, the equation is $2y^2 - 2xy - x^2 - \frac{27}{4} = 0$. Again, tan. $2\alpha = \frac{2}{3}$, $M = \frac{1}{2} + \frac{1}{2}\sqrt{13}$, $N = \frac{1}{2} - \frac{1}{2}\sqrt{13}$, $P = \frac{27}{4}$, therefore the equation of the curve, when referred to its principal diameters, is $(\frac{1}{2} + \frac{1}{2}\sqrt{13})x^2 + (\frac{1}{2} - \frac{1}{2}\sqrt{13})y^2 = \frac{27}{4}$, or $(1 + \sqrt{13})x^2 + (1 - \sqrt{13})y^2 = \frac{27}{4}$.

When $y = 0$, $x^2 = \dfrac{27}{2(1 + \sqrt{13})} = \frac{9}{8}(\sqrt{13} - 1) = 2\cdot 93 = A^2$;

when $x = 0$, $y^2 \dfrac{27}{2(1 - \sqrt{13})} = -\frac{9}{8}(\sqrt{13} + 1) = -5\cdot 18 = B^2$.

Hence, as in the preceding example, there are given the axes of the curve to construct it.

3. Determine the axes of the curve of which the equation is $2y^2 - 4xy + 5x^2 - 3x = 0$.† *Ans.* The curve is an ellipse, whose axes are $\sqrt{3}$ and $\frac{1}{2}\sqrt{2}$.

4. Required the axes of the curve which is the locus of the equation $5y^2 + 2xy + 5x^2 + 2y - 2x - \frac{3}{4} = 0$. *Ans.* The curve is an ellipse, whose axes are $2\sqrt{\frac{1}{3}}$ and $2\sqrt{\frac{1}{2}}$.

5. Required the axes of the curve which is the locus of equation

$$y^2 - 6xy + x^2 + 2y - 6x + 5 = 0.$$

Ans. The curve is an hyperbola, whose axes are $2\sqrt{2}$ and 2.

6. What is the locus of the equation $y^2 - 6xy + x^2 + 2y - 6x + 1 = 0$?
Ans. Two straight lines, characterized by the equation $y = x\sqrt{\frac{1}{2}}$.

7. What is the geometrical representation of the equation $y^2 - 4xy + 5x^2 + 2x + 1 = 0$? *Ans.* A point $(-1, -2)$.

8. What is the locus of the equation $2x^2 + 2y^2 - 3x + 4y - 1 = 0$? *Ans.* A circle, whose radius is $\frac{1}{4}\sqrt{33}$.

9. What is the locus of the equation $y^2 - 2xy + 2x^2 - 2x + 4 = 0$? *Ans.* An imaginary curve.

10. What is the locus of the equation $y^2 + 2xy - 2x^2 - 4y - x + 10 = 0$? *Ans.* An hyperbola, in which the second axis is taken for the axis of abscissa. The axes are $1\cdot 7$ and $2\cdot 2$.

11. What is the geometrical representation of the equation $2y^2 + 3x^2 - 3x - 2y + 2 = 0$? *Ans.* The equation has no geometrical representation.

12. What is the locus of the equation $3y^2 + 6x^2 - 24x + 6 = 0$,

* The expression for the distance between the centre and focus is given at (p. 170.).

† The solution is given in a Key just published.

the axes of reference being oblique? *Ans.* An ellipse in which the semi-conjugates parallel to the axes of reference are $\sqrt{3}$ and $\sqrt{6}$.

(129.) Before we proceed to the construction of parabolas, we shall remark, that if, in the equations of condition (2), (3), p. 177, we substitute for the constants a, b, the variables x, y, the equations $2Ay + Bx + D = 0$, (1), $2Cx + By + E = 0$, (2), will characterize two straight lines, passing through the centre of the locus, as is evident, since a, b, the coordinates of this centre, satisfy both equations. Were we to suppose these lines to be parallel, or the centre of the locus to be infinitely distant, as in the parabola, we could infer, from the equations (1), (2), that in the equation of the locus there must be $B^2 - 4AC = 0$. For these equations give

$$\left.y = -\frac{Bx + D}{2A} \text{ and } y = -\frac{2Cx + E}{B}\right\} (3)$$

; and since, when the lines are parallel, the difference of the ordinates corresponding to every abscissa must be constant, we have, by reducing these expressions to the same denominator,

$$B^2x + BD - 4ACx - 2AE = \text{constant}$$
$$\text{or } (B^2 - 4AC)\,x + BD - 2AE = \text{constant},$$

which can only happen when $B^2 - 4AC = 0$.

If the lines (1) and (2) coincide, we must conclude that the locus has an infinite number of centres, and all situated in the line (1) or (2). The locus therefore can be no other than a system of parallel straight lines, as GH, KL, equally distant from the line through the centres, for then every chord of the locus must be bisected by this line. We already know that this locus is a variety of the parabola. Equations (1) and (2) will also show this to be the case, and will moreover furnish an additional criterion, whereby we may readily ascertain, by inspecting the coefficients of the proposed equation, when that equation characterizes a system of parallels, and when it does not. For, since, in this case, equations (1) and (2) represent the same line, we have, equation (3),

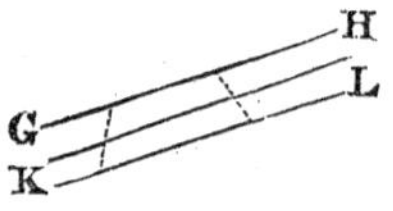

$$\frac{Bx + D}{2A} = \frac{2Cx + E}{B} \therefore B^2x + BD = 4ACx + 2AE,$$

whatever be the value of x; consequently (*Alg.* p. 156.) $B^2 = 4AC$, and $BD = 2AE$; so that, when the indeterminate equation of the second degree represents a system of parallels, there must exist among the coefficients the conditions $B^2 - 4AC = 0$, and $BD - 2AE = 0$.

These lines may be at once determined from the given equation, for, being parallel, the coefficient of x must necessarily be the same in the equation of each, that is, these equations will be of the form $y + px + q = 0$ and $y + px + r = 0$; so that the proposed equation, after having freed y^2 from its coefficient, may always, in the case we are considering, be decomposed into two factors of this form where p, the coefficient of x, in each must be equal to half the coefficients of xy, in the proposed equation, after this has been divided by A the coefficient of the first term; hence $p = \frac{B}{2A}$. With regard to q and r, it is plain that their sum must be equal to $D \div A$, the coefficient of y, in the proposed, and their product must make the last term, $F \div A$. Having thus the sum and the product of q and r we shall get their difference, by subtracting four times the product from the square of the sum, and extracting the square root, that is,

$q - r = \sqrt{(\frac{D^2}{A^2} - \frac{4F}{A})} = \frac{1}{A}\sqrt{\{D^2 - 4AF\}}$; therefore, adding the half sum to the half difference, we have, for the greater,

$q = \frac{1}{2A}(D + \sqrt{\{D^2 - 4AF\}})$; and, by subtracting the same, we get the less. Now it is plain, from this expression, that, if $D^2 = 4AF$, then $q = r$; hence the two equations (129) become, in this case, identical, and the parallels therefore coincide, and become a single straight line. If $D^2 \angle 4AF$, then the values of q and r become imaginary, so that, in this case, the locus of (1, 2,) is two imaginary lines. From what has now been said, we may conclude that the equation $Ay^2 + Bxy + Cx^2 + Dy + Ex + F = 0$, represents a system of parallel lines when $B^2 - 4AC = 0$ and $BD - 2AE = 0$, these coincide, and form but one straight line; when also $D^2 - 4AF = 0$, and they become imaginary when, instead of this, $D^2 - 4AF \angle 0$.

When the lines are real, their equations are

$y + \frac{B}{2A}x + \frac{1}{2A}(D + \sqrt{\{D^2 - 4AF\}}) = 0$, and $y + \frac{B}{2A}x + \frac{1}{2A}(D - \sqrt{\{D^2 - 4AF\}}) = 0$. When they coincide the equation is

$$y + \frac{B}{2A}x + \frac{D}{2A} = 0.$$

(130.) As, in the case we are discussing, the factors of the original equation consist of the sum and difference of the same two quantities, their product must be the difference of the squares of these quantities; hence, when the proposed equation represents two straight lines, it must consist of the difference of two squares, or at least of one square, minus a number. On the contrary, when the lines represented are imaginary, the equation must consist of the sum of two squares, or at least of one square and a number. And the equation will be a perfect square when only one straight line is represented. Hence we may frequently discover at a glance when the equation denotes a variety of the parabola, without even trying whether $BD - 2AE = 0$. Thus we see at once that the equation $y^2 - 2xy + x^2 - 1 = 0$, is the difference of two squares, viz. $(y - x)^2$, and 1, the locus of it, is therefore two parallel straight lines, the equations of which are $y - x + 1 = 0$, and $y-x-1=0$. Also the equation $y^2-4xy+4x^2+9=0$, is immediately seen to consist of the two squares $(y-2x)^2$, and 9, therefore, the locus, is imaginary.

In like manner, since the equation $y^2+4xy+4x^2+2y+4x+1=0$, is obviously a perfect square, viz. $(y+2x+1)^2$, its locus is a straight line, the equation of which is $y+2x+1=0$. Let now the equation $y^2+6xy+9x^2-2y-6x-15=0$, be proposed, which is a variety of the parabola, because $B^2-4AC=0$; and since, moreover, $BD-2AE=0$, this variety is a system of parallels, of which the equations are $y+3x+3=0$ and $y+3x-5=0$. Lastly, let the equation be $y^2-4xy+4x^2+2y-4x+4=0$, the coefficients of which furnish beside the conditions above the relation $D^2-4AF \angle 0$, therefore the locus is imaginary.

(131.) We shall now proceed to furnish formulas for the construction of parabolas, as we have already done for the central curves. In the present case, our object will be first to remove xy from the equation, and afterwards to remove the term containing the first power of one of the variables, and the absolute number. The first transformation, as we have already seen (122), brings the equation to the form $My^2+Ry+Sx+F=0$, or $Nx^2+Ry+Sx+F=0$, where, $R=D\cos.\alpha-E\sin.\alpha$, $S=D\sin.\alpha+E\cos.\alpha$. Now, since, $\cos.\alpha=\sqrt{\frac{1+\cos.2\alpha}{2}}$, and $\sin.\alpha=\sqrt{\frac{1-\cos.2\alpha}{2}}$. And since also the expression given at (124) for the $\cos.2\alpha$ becomes, when $B^2=4AC$, $\cos.2\alpha=\frac{A-C}{+(A+C)}$, or $\frac{A-C}{-(A+C)}$ accordingly, as B is negative or positive, we have, by substitution, the following expressions for R and S, viz. when B is negative,

$R=\frac{D\sqrt{A}-E\sqrt{C}}{\sqrt{(A+C)}}$, $S=\frac{D\sqrt{C}+E\sqrt{A}}{\sqrt{(A+C)}}$, and when B is positive,

$R=\frac{D\sqrt{C}-E\sqrt{A}}{\sqrt{(A+C)}}$, $S=\frac{D\sqrt{A}+E\sqrt{C}}{\sqrt{(A+C)}}$

(132.) The values of M and N have already been determined (124), as also the values of a and b, employed in the second transformation (122); hence, collecting these formulas together, we have

1. Proposed Equation of the Curve

$$Ay^2+Bxy+Cx^2+Dy+Ex+F=0.$$

Formulas to be employed for removing the Terms containing the Product of the Variables and the Square of one of them.

1. When B is *negative*, $\tan.2\alpha=-\frac{B}{A-C}$, $M=A+C$, $N=0$,

$R=\frac{D\sqrt{A}-E\sqrt{C}}{\sqrt{(A+C)}}$, $S=\frac{D\sqrt{C}+E\sqrt{A}}{\sqrt{(A+C)}}$.

II. Resulting Equation,

The rectangular axes being parallel to the axes of the curve,

$$My^2 + Ry + Sx + F = 0.$$

Formulas for the removal of the terms R*y and* F.

$$b = -\frac{R}{2M}, \; a = -\frac{Mb^2 + Rb + F}{S} = \frac{R^2 - 4MF}{4MS}.$$

III. Resulting Equation,

The axes being those of the curve, $My^2 + Sx = 0$.

2. When B is *positive*,

The first class of formulas is, $\tan. 2\alpha = -\frac{B}{A - C}$, $N = A + C$, $M = 0$; $R = \frac{D\sqrt{C} - E\sqrt{A}}{\sqrt{(A + C)}}$, $S = \frac{D\sqrt{A} + E\sqrt{C}}{\sqrt{(A + C)}}$, and the resulting equation, $Nx^2 + Ry + Sx + F = 0$. The second class of formulas is, $a = -\frac{S}{2N}$, $b = \frac{S^2 - 4NF}{4NR}$; and the final equation is $Nx^2 + Ry = 0$.

EXAMPLES.

1. Construct the curve, which is the locus of the equation

$$y^2 - 4xy + 4x^2 + 2y - 7x - 1 = 0.$$

Here $A = 1, B = -4, C = 4, D = 2, E = -7, F = -1$, and, as B is negative, we must employ the first collection of formulas, which give $\tan. 2\alpha = -\frac{4}{3}$, $M = 5$, $N = 0$, $R = \frac{16}{5}\sqrt{5}$, $S = -\frac{3}{5}\sqrt{5}$. Hence, when the axes are parallel to those of the curve, the equation becomes, $5y^2 + \frac{16}{5}\sqrt{5}y - \frac{3}{5}\sqrt{5}x - 1 = 0$. Again, formulas II give, $b = -\frac{8}{25}\sqrt{5}$, $a = -\frac{89}{75}\sqrt{5}$; these, therefore, are the ordinate and abscissa of the principal vertex of the curve, the equation to which, in reference to the axes of the curve is, $5y^2 - \frac{3}{5}\sqrt{5}\,x = 0$, or $y^2 = \frac{3}{25}\sqrt{5x}$.

The construction of this curve is therefore as follows:

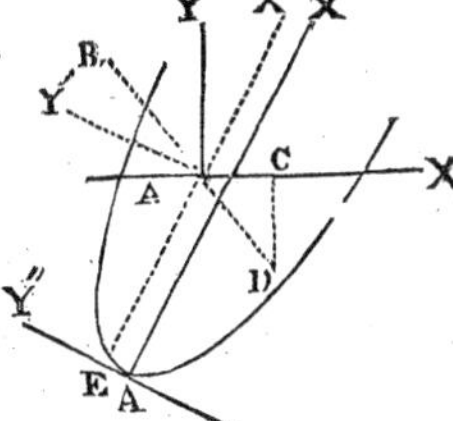

Let AX, AY be the primitive rectangular axes. On the former take AC = 1, and make the perpendicular, $CD = -\frac{4}{3}$. Draw DAB, and bisect the angle $BAX = 2\alpha$ by the line AX′; then the rectangular axes, AX′, AY′, are those to which the first transformed equation refers the curve.

Again take $AE = -\frac{89}{75}\sqrt{5}$, and the perpendicular $EA' = -\frac{8}{25}\sqrt{5}$, then the axes A′X″, A′Y″, parallel to the former, will be those of the curve. Having thus the axes and the parameter $\frac{3}{25}\sqrt{5}$, the focus and directrix are readily determined, and thence the curve constructed.

2. Construct the locus of the equation $y^2+2xy+x^2-6y+9=0$. Here $A=1$, $B=2$, $C=1$, $D=-6$, $E=0$, $F=9$, and, as B is positive, the second collection of formulas must be used, which give $\tan. 2\alpha = \frac{-2}{0} \pm M = 0$, $N = 2$, $R = -3\sqrt{2}$, $S = -3\sqrt{2}$. Hence the equation of the curve, when the axes are parallel to those of the curve, is $2x^2-3\sqrt{2}y-3\sqrt{2}x+9=0$; and when they coincide with the axes of the curve, the coordinates of whose origin are, $a=\frac{3}{4}\sqrt{2}$, $b=\frac{9}{8}\sqrt{2}$, the equation is $2x^2-3\sqrt{2}y=0$, or $x^2=\frac{3}{2}\sqrt{2}y$. Hence, as in the preceding example, the curve may be constructed. In the first transformation of axes, since tan. 2α is infinite, 2α is a right angle, so that in this transformation the new axis of x will be 45° below the primitive.

3. The equation of a parabola being $y^2-4xy+4x^2-8y+3x-2=0$, what will it become when the curve is referred to its axes?

Ans. $y^2=13\sqrt{5x}\div 25$.

4. Required the principal parameter of the parabola whose equation is $4y^2-4xy+x^2-2y-4x+10=0$. *Ans.* $p=\sqrt{\frac{4}{5}}$.

5. What is the principal parameter of the parabola represented by the equation $y^2-2xy+x^2-3y=0$? *Ans.* $p=\frac{3}{2\sqrt{2}}\div 2$

(133.) The student must bear in mind, that the various formulas given in this chapter for the construction of lines of the second order, apply only when the different equations refer the curves to rectangular axes, which, indeed, are in most cases employed. With regard, however, to the *varieties* of the three curves, the tests by which they may be discovered, and the formulas for their construction, apply generally, because in discussing these varieties we have considered the axes to have any inclination whatever, and because moreover the criteria (pp. 175, 176), by which the three classes of curves are distinguished, apply for every inclination of axes, as we are about to show in the following chapter, which has for its object the determination of the locus of the general equation when the axes are oblique.

CHAPTER II.

DISCUSSION OF THE GENERAL EQUATION.

$$Ay^2+Bxy+Cx^2+Dy+Ex+F=0,$$

By the separation of the variables.

(134.) This equation may be put under the form

$y^2+\frac{Bx+D}{A}y+\frac{C}{A}x^2+\frac{E}{A}x+\frac{F}{A}=0$, which, solved as a quad ratic, gives for y the expression $y = -\frac{Bx + D}{2A} \pm \frac{1}{2A}\sqrt{\{(B^2 - 4AC)\,x^2 + 2\,(BD - 2AE)\,x + D^2 - 4AF\}}$ (1). In like manner, we have for x, in terms of y, the expression

$$x = -\frac{By + E}{2C} \pm \frac{1}{2C}\sqrt{\{(B^2 - 4AC)\,y^2 + 2\,(BE - 2CD)\,y + E^2 - 4CF\}} \ldots (2).$$

Either of these expressions will furnish an indefinite number of points in the locus, when this is not imaginary, since the first will give the ordinates corresponding to any assumed abscissa, and the second will give the abscissa corresponding to any assumed ordinate. If we wish to determine points in the curve from the equation (2), we must first, for any assumed abscissa, x, draw an ordinate equal to $-\frac{Bx + D}{2A}$, determining some point, P; then, if this ordinate be prolonged, and the distances PM, PM′ be taken thereon, each equal to the line represented by the remaining part of the expression for y, two points of the locus will thus be determined. P therefore is the middle of the chord M, M′. The same construction for another value of x will determine another point, P′, and two new points, m, m', of the curve, which will, as before, be equally distant from P′. Hence, calling the variable ordinates of these points, P, P′, &c. Y, since we must always have

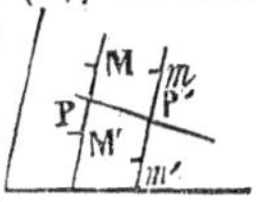

$Y = -\frac{Bx + D}{2A}$ it follows that the locus of these points is a straight line, which, because it bisects all the chords in the curve drawn parallel to the axis of y, is called a diameter of the curve. Similar reasoning applied to the expression (2) will show that the straight line represented by the equation $X = -\frac{By + E}{2C}$ is a diameter, bisecting the chords drawn parallel to the axis of x. These diameters are obviously the same as those represented at (129).

Having thus the equations of two diameters, we can always readily find the centre of any locus of the second order, to whatever axes it be referred; for, representing the coordinates of the centre by a, b, we shall have, by substituting these for the coordinates in each of the preceding equations, and solving them, as at (126), the values

$$a = \frac{2AE - BD}{B^2 - 4AC},\quad b = \frac{2CD - BE}{B^2 - 4AC}.$$

(135.) From these remarks it appears that the nature of the curve depends upon the irrational part of the expression (1), or (2), and that it cannot exist when this irrational part becomes 0, or imaginary for every value of the variable it contains. Let us examine the circumstances under which these irrational expressions can become real, imaginary, or nothing. We shall first take the expression (1), and shall suppose that the quantity under the radical is decomposed into two factors, each containing x; in other words, we shall suppose the solution of the equation $x^2 + \frac{2(BD - 2AE)}{(B^2 - 4AC)} x + \frac{D^2 - 4AF}{(B^2 - 4AC)} = 0$ (3) to be effected, and that the resulting values of x are $x = \beta$ and $x = \beta'$, then we know (*Alg.* *p.* 175-6) that the multiplication of the factors $(x - \beta)$, $(x - \beta')$ will produce this equation, and consequently the quantity under the radical (1) will be $(B^2 - 4AC)(x - \beta)(x - \beta')$ (4). If, however, $B^2 - 4AC = 0$, then the expression under the radical will have only one factor containing x, discoverable by solving the simple equation $x + \frac{D^2 - 4AF}{2(BD - 2AE)} = 0$; so that, putting δ for the value of x, in this equation, the expressions under the radical will be $2(BD - 2AE)(x - \delta)$. The form (4) therefore only exists when $B^2 - 4AC < 0$, or $B^2 - 4AC > 0$; let us examine the expression in the first case, viz.

(136.) *When* $B^2 - 4AC < 0$.

There are three circumstances to consider in this case:

1. When the roots β, β', are real and unequal.
2. When the roots are real and equal.
3, and lastly, When they are imaginary.

Suppose, first, that the roots are real and unequal, β being greater than β', then (*Alg.* *p.* 180) if in the expression $(B^2 - 4AC) \cdot (x - \beta) \cdot (x - \beta')$ any quantity greater than β, or less than β', be substituted for x, the product $(x - \beta) \cdot (x - \beta')$ will be positive, and since, by hypothesis, $B^2 - 4AC$ is negative, the whole expression will be negative, and therefore, for all such values of x, the expression for y will be imaginary. But, if we substitute for x any value between β and β', then the product $(x - \beta) \cdot (x - \beta')$ will be negative, and consequently the expression (4) will be positive; for all such values of x, therefore, there correspond real values of y.

From this discussion it follows, that, under the conditions we have supposed, the curve always exists, and that it is comprised between, or *limited*, by two parallels to the axis of ordinates drawn at the respective distances of β' and β from the origin, for between these parallels all the values of x which give possible values for y are comprehended. By applying precisely similar reasoning to the expression (2), it would result that the curve is also limited by two parallels to axis of x; as, therefore, these parallels meet the former, and form a parallelogram, circumscribing the curve, it follows that the curve must be limited in all directions, as in the annexed diagram. The curve, therefore, must necessarily be an ellipse.

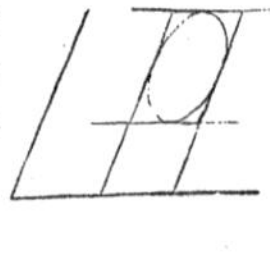

Suppose, secondly, that the roots β, β' are real and equal, then the expression (4) is $(B^2-4AC)(x-\beta)(x-\beta')$; where, since it is impossible to substitute any value for x between the roots β and β', it is, by the preceding reasoning, also impossible to render the expression for y real by any substitution for x, except in the single case $x=\beta$, which renders the irrational part of the expression 0, and reduces the value of y to $y=-\frac{B\beta+D}{2A}$ hence, when the roots β, β' are equal, the curve is reduced to a point, of which the coordinates are β and $-\frac{B\beta+D}{2A}$.

If, lastly, the roots be imaginary, then whatever value we substitute for x, in the equation containing them, the result will be positive (*Alg. p.* 173); hence every value of y will be imaginary, so that, in this case, the curve cannot exist. We may now therefore infer, that when, in the general equation, $B^2-4AC \angle 0$, whatever be the inclination of the axes, the locus is an ellipse, if the roots of the irrational part of the expression for y be real and unequal; but it is merely a point, if these roots be equal, and it is imaginary, if the roots be so.

(137.) Let us now discuss the equation upon the second hypothesis, *When* $B^2-4AC \mathrel{7} 0$.

Resuming the expression $(B^2-4AC)(x-\beta)(x-\beta')$, and reasoning as before, in reference to the roots β and β', we find that here, when these roots are real, and β greater than β', every value of x greater than β, or less than β', will, because B^2-4AC is positive, render the expression for y real; while, on the contrary, every value comprised between the limits β and β' will render the expression for y imaginary. As, therefore, without these limits x may increase indefinitely, both positively and negatively, it follows that the curve must consist of two infinite detached branches, proceeding in opposite directions, and separated from each other by the distance between two parallels to the axis of y, of which the abscissas are respectively β and β', for within these limits there exist no possible value of y. This curve therefore is the hyperbola.

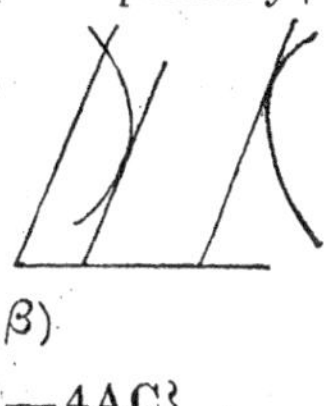

If the roots β, β' are equal, the expression above is $(B^2-4AC)(x-\beta)^2$; and hence the value of y becomes $y=-\frac{Bx+D}{2A}+\frac{\sqrt{\{B^2-4AC\}}}{2A}(x-\beta)$

or $y=\frac{-B\pm\sqrt{\{B^2-4AC\}}}{2A}x-\frac{D\pm\beta\sqrt{\{B^2-4AC\}}}{2A}$, hence the locus is *a system of two straight lines*, which intersect, since the coefficient of x is not the same in both. When the roots β, β' are imaginary, then, since every value given to x, in the equation containing them, gives a positive result, the whole expression under the radical will be positive, and, therefore, the value of y will be always real. As, therefore, x may take any value from 0 to infinity, in both directions, it follows that the curve is unlimited in both directions. It moreover consists of two distinct branches; for as each double ordinate, or chord drawn parallel to the axis of y, is bisected by the diameter whose equation is, $y = -\frac{Bx + D}{2A}$ one half of the curve must be situated entirely below this line, and the other half above it; neither can have a point in common with this diameter, because the irrational part of the value of y can never vanish; hence the curve must be an hyperbola.

($137\frac{1}{2}$.) There is a particular form of the general equation which ought here to be noticed, it is that where the squares of the variables are absent, when the equation becomes $Bxy + Dy + Ex + F = 0$, (5), which gives for y the expression, $y = -\frac{Ex + F}{Bx + D}$, and, as this value of y will always be real, whatever be the value of x, it follows that the curve extends indefinitely in opposite directions. As each value of x furnishes but one value of y each ordinate meets the curve in but one point. If the value $-D \div B$ be given to x, the corresponding value of y will be infinite, that is, if a parallel to the axes of y be drawn at the distance of $-D \div B$ from the origin, it will never meet the curve; but, as every parallel drawn on either side of this must necessarily meet the curve, because no abscissa but $x = -D \div B$ can render the ordinate infinite, it follows that the curve consists of two distinct branches, separated by the parallel whose abscissa is $-D \div B$. The curve, therefore, is an hyperbola; and the parallel, whose abscissa is $-D \div B$, is obviously one of the asymptotes, as this parallel has been seen to be the only one which does not meet the curve.

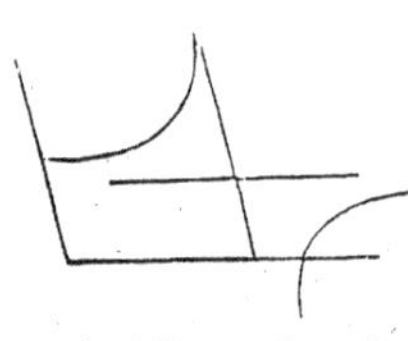

By solving the equation (5), with regard to x we have $x = -\frac{Dy + F}{By + E}$, in which expression $-\frac{E}{B}$ is the only value that can be given to y, that will render x infinite; hence we infer here, that a parallel to the axis of x, of which the ordinate is $-E \div B$, is the other asymptote of the curve. Hence equation (5) represents an hyperbola whose asymptotes are parallel to the axes of coordinates, the coordinates (x', y') of the point of intersection of the asymptotes being $x' = -\frac{D}{B}, y' = -\frac{E}{B}$. The asymptotes are therefore easily determined when the equation of the hyperbola takes the form (5).

If the term Ax^2 had appeared in the equation (5) the same reasoning with regard to the expression for y would apply; so that then also the parallel to the axis of y, of which the abscissa is $-D \div B$, is an asymptote. If Cy^2 appear in the equation, instead of Ax^2, then, reasoning as above on the expression for x, we find that a parallel to the axis of x, of which the ordinate is $-E \div B$, is also an asymptote. If both $C = 0$ and $D = 0$, the axis of y coincides with an asymptote. If both $A = 0$ and $E = 0$, the axis of x coincides with an asymptote. If both $D = 0$ and $E = 0$, the origin coincides with the intersection of the asymptotes; and when, in addition to this, the squares of the variables are absent, both axes coincide with the asymptotes, and the equation takes the form $Bxy + F = 0$.

To determine the asymptotes from the general equation, let us actually extract the root of the expression under the radical, in the general value of the ordinate (1), we shall find this root to be of the form $x \sqrt{\{B^2 - 4AC\}} + \frac{BD - 2AE}{\sqrt{(B^2 - 4AC)}} + \frac{K}{x} + \frac{K'}{x^2} + \&c.$ therefore

$$y = -\frac{Bx + D}{2A} \pm \frac{\sqrt{\{B^2 - 4AC\}}}{2A} x + \frac{BD - 2AE}{2A\sqrt{\{B^2 - 4AC\}}} + \frac{K}{2Ax}$$

$+ \frac{K'}{2Ax^2} + \&c.$ Now it is here obvious that as x increases the term $\frac{K}{2Ax}$, and all that follow will diminish, while those that precede will increase, and to these first terms the expression is finally reduced, when x becomes infinite. Hence the curve continually approaches the two straight lines denoted by,

$$Y = -\frac{Bx + D}{2A} \pm \frac{\sqrt{(B^2 - 4AC)}}{2A} \left(x + \frac{BD - 2AE}{B^2 - 4AC}\right)$$ these lines are therefore the asymptotes to the curve.

Comparing this equation with the equation at p. 188, which represents the locus when it becomes a system of two straight lines, we shall find them to be identical. For, as that equation takes place only when the roots β, β' are equal, it follows that then β must be equal to minus half the coefficient of x, in the equation (3), which contains them, that is, we must have $\beta = -\frac{BD - 2AE}{B^2 - 4AC}$, which value of β renders the equation (p. 188) identical with that above for the asymptotes. We may therefore say that, when the equation represents a system of straight lines, the hyperbola degenerates into its asymptotes.

(138.) We already know that the asymptotes intersect at the centre, this is also readily ascertained from their equation above; for since at their intersection the two values of Y coincide, we must have for x, at that point, the value, $x = \frac{2AE - BD}{B^2 - 4AC}$, which (134) is the abscissa of the centre, and the corresponding value of Y is $Y = -\frac{Bx + D}{2A} = \frac{2CD - BE}{B^2 - 4AC}$, which (134) is the ordinate of the centre. Hence, when we wish to construct the asymptotes, when the equation of the hyperbola appears under the general form, we shall have first to determine the centre from these formulas, and then to draw through this point two straight lines inclined to the axis of x, at angles α, α', whose tangents* are respectively $\tan. \alpha = \frac{-B + \sqrt{\{B^2 - 4AC\}}}{2A}$ and $\tan. \alpha' = \frac{-B - \sqrt{\{B^2 - 4AC\}}}{2A}$. The product of these two tangents is $\tan. \alpha \cdot \tan. \alpha' = \frac{4AC}{4A^2} = \frac{C}{A}$ which, when $C = -A$, becomes $\tan. \alpha \cdot \tan. \alpha' = -1$, an equation which indicates (when the axes of reference are rectangular) that the asymptotes are perpendicular to each other (11). Hence, if, in the general equation, $B^2 - 4AC > 0$, and $C = -A$, when the locus is referred to rectangular axes, we may conclude that the equation represents an equilateral hyperbola.

It must be here remarked, that, when $A = 0$, the preceding expression for $\tan. \alpha$ becomes $\frac{0}{0}$, which is not a definite result; but, by multiplying numerator and denominator by $B + \sqrt{\{B^2 - 4AC\}}$, it reduces to $\tan. \alpha = \frac{-2C}{B + \sqrt{\{B^2 - 4AC\}}} = -\frac{C}{B}$, when $A = 0$.

(139.) We shall now examine the general equation upon the third hypothesis, viz. *When* $B^2 - 4AC = 0$.

Under this condition, the general expression for any ordinate of the locus is $y = -\frac{Bx + D}{2A} \pm \frac{1}{2A} \sqrt{\{2(BD - 2AE)\,x + D^2 - 4AF\}}$.

If we put β for $-\frac{D^2 - 4AF}{2(BD - 2AE)}$, the quantity under the radical will be $2(BD - 2AE)\,(x - \beta)$, in which the factor $2(BD - 2AE)$ may be either positive, negative, or nothing.

If this factor be positive, the whole expression will be positive for every value of x greater than β, but negative for every value less than β; hence, in this case, the locus extends indefinitely to the right of a parallel to the axis of y drawn through the abscissa $x = \beta$; therefore this parallel is a tangent to the curve, to the left of which no point in the locus can be situated.

* We are here supposing the axes to be rectangular; if they are oblique, then for tan. substitute ratio of the sines.

If the factor $2(BD - 2AE)$ be negative, then, on the contrary, the locus would extend indefinitely to the left of the parallel, whose abscissa is β, and no point in the curve could be situated to the right of it.

In each of these cases, therefore, the curve will be limited to one direction, but unlimited in the opposite direction; it must therefore be a parabola.

If, lastly, $2(BD - 2AE) = 0$, then the expression for y becomes $y = -\frac{Bx + D}{2A} \pm \frac{1}{2A}\sqrt{\{D^2 - 4AF\}}$, denoting a system of parallel straight lines, which, however, coincide, when $D^2 - 4AF = 0$, and which become imaginary, when $D^2 - 4AF \angle 0$.

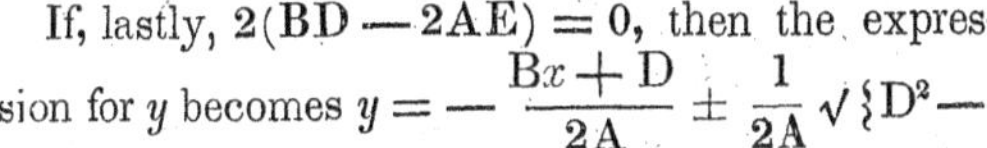

Because the condition $B^2 = 4AC$ or $B = 2\sqrt{\{AC\}}$ characterizes the parabola and its varieties, the three first terms in the general equation of this curve will always form a perfect square, viz.

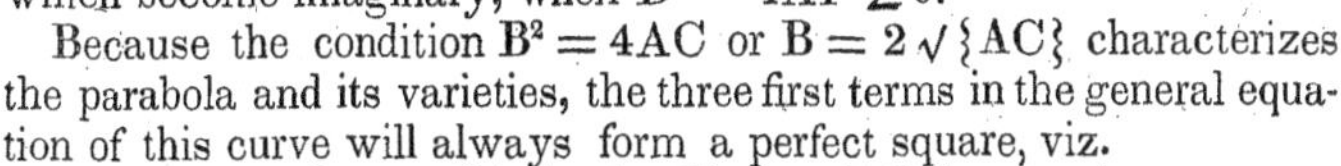

$(y\sqrt{A} + x\sqrt{C})^2 = Ay^2 + 2\sqrt{\{AC\}} \cdot xy + Cx^2$.

(140.) We might now proceed to inquire into the form of the general equation when it represents one of the varieties of the three curves, and thence derive, as in the preceding chapter, criteria by means of which these varieties may be distinguished. For the varieties of the parabola the tests of their existence which have been given in the preceding chapter are the simplest that can be employed, and may be readily applied in any case of doubt. But for the other curves, the shortest and most direct way of proceeding will generally be to solve the equation, with regard to one of the variables, and then to find the roots of that part of the resulting expression which is under the radical, the nature of these roots will make known the nature of the locus conformably to the preceding discussion. The examples we shall here give will further illustrate this.

Construction of Curves of the second order.

EXAMPLE. I.

To determine the position of the curve of which the equation is

$$y^2 - 2xy + 3x^2 + 2y - 4x - 3 = 0.$$

As, in this example, $B^2 - 4AC \angle 0$, the curve must be an ellipse; let us therefore first proceed to determine its limits. For this purpose let us put the equation under the following form, viz. $y^2 - 2(x-1)y = -3x^2 + 4x + 3$ (1), which, solved first for y and then for x, gives $y = x - 1 \pm \sqrt{\{-2x^2 + 2x + 4\}}$ (2).

$x = \frac{y+2}{3} \pm \frac{1}{3}\sqrt{\{-2y^2 - 2y + 13\}}$ (3). Equating the irrational part of (2) to 0, we have $x^2 - x - 2 = 0 \therefore x = \frac{1 \pm 3}{2}$

consequently, the roots of this equation being real and unequal, viz $\beta = 2$, and $\beta' = -1$, we know (136) that the curve exists, and that it is included between two parallels, LL′, MM′, to the axis of y, of which the abscissa, AG, of the one is equal to -1, and the abscissa, AH, of the other equal to 2.

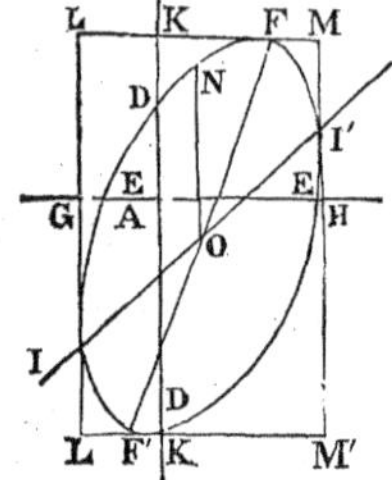

Solving, in like manner the equation $2y^2 + 2y - 13 = 0$, we obtain for the roots the values $\frac{-1+\sqrt{27}}{2}$, and $\frac{-1-\sqrt{27}}{2}$, therefore the curve is also comprehended between two parallels to the axis of x, of which the ordinate, AK, of the one is $\frac{-1+\ 27}{2}$, and the ordinate, AK′, of the other $\frac{-1-\sqrt{27}}{2}$. The curve is therefore circumscribed by the parallelogram LM′.

To find the points of contact of the parallels LL′, MM′, we must construct the diameter, $Y = x - 1$, (134); for, as the abscissas β and β' of these points render the irrational part of the equation (2) nothing, the corresponding ordinates must belong as well to the diameter as to the curve. This diameter cuts the axes in the points $x = 1$ and $y = -1$; if, therefore, through these points the line II′, be drawn, the two points of contact will be determined. Constructing also the second diameter $X = \frac{1}{3}(y + 2)$, which cuts the axes in the points $x = \frac{2}{3}$, and $y = -2$, we obtain the other two points of contact, F, F′.

To find the points where the curve intersects the axis of y, suppose $x = 0$, the equation (2), and we have for the ordinates of those points $y = 1$ and $y = -3$; hence these points, D, D′, are readily determined. In like manner, supposing $y = 0$, in equation (3), we have for the abscissas of the points E, E′, where the curve cuts the axis of x, the values of $x = \frac{2}{3} + \frac{1}{3}\sqrt{13}$ and $x = \frac{2}{3} - \frac{1}{3}\sqrt{13}$. The eight points thus determined are amply sufficient to make known the position of the curve. But there is another mode of proceeding by which an indefinite number of points in the curve may be determined. Thus:

Draw, as before, the parallels LL′, MM′, and then construct the diameter, II′, from its equation, $Y = x - 1$ The middle point, O, of this diameter is the centre of the curve, therefore the abscissa, Am, of the centre is $\frac{\beta + \beta'}{2} = \frac{2-1}{2} = \frac{1}{2}$, because β and β' are the abscissas of the *extremities* of the same diameter. Hence, drawing the ordinate mN, we shall have the direction of the diameter conjugate in II′, since this ordinate will be parallel to the tangent at the vertex of that diameter; therefore, putting for x the value $x = \frac{1}{2}$, in the expression (2), the irrational part gives for the semi-diameter, ON,

$\sqrt{\{-2x^2+2x+4\}} = \frac{3\sqrt{2}}{2}$ = ON; hence we have a system of conjugate diameters given in length and direction to construct the ellipse. This construction is as follows:

On the given diameters, AB, CD, taken as principal axes, construct an ellipse; then, if the double ordinates, ab, CD, cd, &c. of this ellipse be inclined to AB, in the given angle, while their length remains unchanged, their extremities, a', b', C′, D′, c', d', &c. will be all upon the required curve, which may therefore be drawn through them. The truth of this is obvious, for the curve thus traced will, by construction, be such, that the squares of the chords parallel to one diameter, C′ D′, are as the rectangles of the parts into which they divide the other, AB, and AB, C′D′, are the given conjugates, both as to the length and direction.

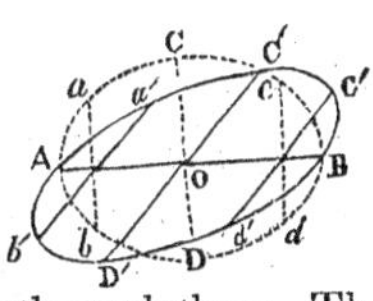

EXAMPLE II.

To construct the curve of which the equation is

$$y^2 - 2xy - 3x^2 - 2y + 7x - 1 = 0.$$

Since here $B^2 - 4AC > 0$, the curve is an hyperbola.

We shall proceed first to determine the asymptotes, because, when these are known, and a single point in the curve found, we can easily obtain as many more points in the curve as we please (97). The equation of the asymptotes is given at (137); but as it is adviseable to proceed independently of the general formulas, we shall here deduce the equation of the asymptotes from the given equation of the curve, which furnishes for y the value

$$y = x+1 \pm \sqrt{\{4x^2-5x+2\}} = x+1 \pm (2x - \tfrac{5}{4} + \frac{K}{x} + \frac{K'}{x^2} + \&c.)$$

Hence, for the two asymptotes we have the equation $Y = x + 1 \pm (2x - \frac{5}{4})$.

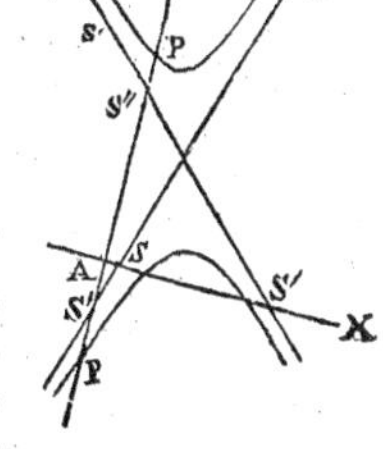

For $x = 0$ we have $Y = -\frac{1}{4}$, and $Y = 2\frac{1}{4}$; therefore, making AS $= -\frac{1}{4}$, and AS′ $= 2\frac{1}{4}$, the points S, S′ will be those in which the asymptotes cut the axis of y. In like manner, for $Y = 0$ we have $x = \frac{1}{12}$, and $x = 2\frac{1}{4}$; therefore, making A$s = \frac{1}{12}$, and A$s' = 2\frac{1}{4}$, the points s, s' will be those in which the asymptotes cut the axis of x: consequently the lines SS″, $s's''$ are the asymptotes sought. It remains now to determine a point in the curve, and for this purpose suppose $x = 0$, in the proposed equation, and their results for the ordinates of the points where the curve intersects the axis of y, the values $y = 1 \pm \sqrt{2}$; therefore, making AP $= 1 + \sqrt{2}$, and AP′ $= 1$

$-\sqrt{2}$, two points, P, P′, in the curve will be determined and thence as many more as we please (97). When the axes of reference do not meet the curve, a point must be determined, by constructing the value of y, corresponding to an assumed value of x.

EXAMPLE III.

To construct the curve of which the equation is $xy-2y+x-1=0$.

This equation represents an hyperbola, the axes of coordinates being parallel to the asymptotes (137). The expression for y is $y=-\frac{x-1}{x-2}$, which becomes infinite only when $x=2$; therefore, if AB be made equal to 2, the line LBL′, parallel to the axis of y, will be one of the asymptotes. In like manner, the expression for x, viz. $x=\frac{2y+1}{y+1}$, becomes infinite only when $y=-1$; hence, if AC $=-1$, the line HCH′, parallel to the axis of x, will be the other asymptote.

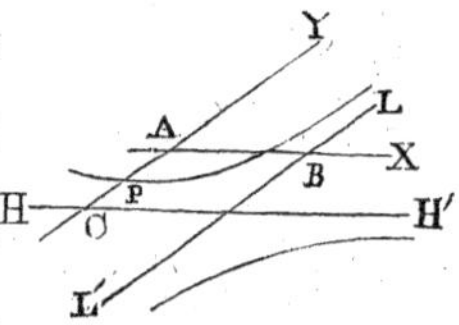

To find a point in the curve, suppose $x=0$, in the proposed equation, then $y=-\frac{1}{2}$; therefore, making AP $=-\frac{1}{2}$, the point P will be in the curve, and the construction will be effected as before.

EXAMPLE IV.

To construct the locus of the equation

$$y^2-4xy+4x^2+2y-7x-1=0.$$

This curve is a parabola, because $B^2-4AC=0$. By solving the equation, first for y, and then for x, we have $y=2x-1\pm\sqrt{\{3x+2\}}$ (1), $x=\frac{1}{2}y+\frac{7}{8}\pm\frac{1}{2}\sqrt{\{\frac{3}{2}y+4\frac{1}{16}\}}$ (2), and equating the irrational part of (1) to 0, we have $3x+2=0 \therefore x=-\frac{2}{3}$; consequently the ordinate, LL′, of which the abscissa, AG, is $-\frac{2}{3}$, will be a tangent to the curve, which will be situated to the *right* of this tangent because the coefficient 3 of x under the radical is *positive*.

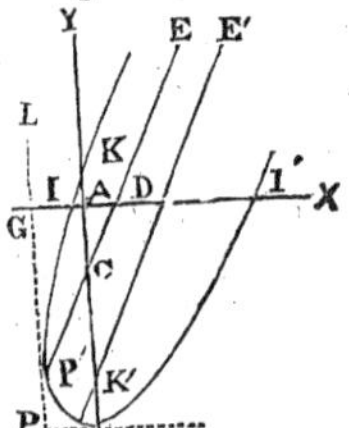

To find the point of contact we must construct the diameter $Y=2x-1$; for, as the abscissa, β, of this point renders the irrational part of (1) nothing, the corresponding ordinate must belong as well to this diameter as to the curve. Hence, supposing first $x=0$, and then $y=0$, in the equation of the diameter, we have for the points where it cuts the axes, $Y=-1$, and $x=\frac{1}{2}$; therefore, making AC $=-1$, and AD $=\frac{1}{2}$, and then, drawing the diameter, EP, we shall have the point of contact, P.

Equating, in like manner, the irrational part of the expression (2) to 0, we have $\frac{3}{2}y+4\frac{1}{16}=0 \because y=-2\frac{17}{24}$; hence a parallel to the axis of x, drawn through the point P′, of which the ordinate is $-2\frac{17}{24}$, will be also a tangent to the curve. The point of contact will be determined by constructing the diameter, E′P′, from its equation, $X = \frac{1}{2}y + \frac{7}{8}$.

To determine the points where the curve intersects the axis of x, suppose $y = 0$, in the proposed equation, and there results for the abscissas of those points the values $x = \frac{1}{8}(7 \pm \sqrt{65})$; hence these points, I, I′, are readily determined. In like manner, putting $x = 0$, we have for the ordinates of the points K, K′, where the curve cuts the axis of y, the values $y = -1 \pm \sqrt{2}$. The points thus found are sufficient to determine the track of the curve, but others, if required, may be found by assuming different values for x, in (1), and constructing the resulting values for y.

(141.) We shall terminate this chapter with a table of the conditions which must exist among the coefficients of the general equation of the second degree, in order that the locus may meet the axes of reference. The necessity of the several conditions in the various cases is obvious, from an inspection of the general values of x and y, exhibited in art. (134.) When $E^2 - 4CF > 0$, the locus has two points of intersection with the axis of x.

When $E^2 - 4CF = 0$, the locus has one point of contact with the axis of x.

When $E^2 - 4CF < 0$, the locus has no point of intersection with the axis of x.

When $D^2 - 4AF > 0$, the locus has two points of intersection with the axis of y.

When $D^2 - 4AF = 0$, the locus has one point of contact with the axis of y.

When $D^2 - 4AF < 0$, the locus has no point of intersection with the axis of y.

CHAPTER III.

PROBLEMS ON GEOMETRIC LOCI.

PROBLEM I.

(142.) Given the base of a triangle and the sum of the tangents of the angles at the base, to determine the locus of the vertex.

Let AB be the given base, through M, the middle of which, draw the perpendicular, MY; then, taking MX, MY for axes, and denoting the vertex of the triangle by (x, y), half the base by b, and the sum of the tangents by s, we have

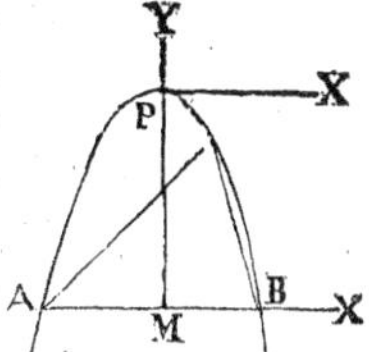

$$\tan. \angle A = \frac{y}{b + x}, \tan. \angle B = \frac{y}{b - x}$$

$\therefore s = \frac{2by}{b^2 - x^2}$ consequently, $sx + 2by - sb^2 = 0$, or $sx^2 + 2b(y - \frac{sb}{2}) = 0$ Hence the locus is a parabola. By removing the origin to a point, P, in the axis of y, of which the ordinate is $\frac{sb}{2}$, that is, by substituting $y + \frac{sb}{2}$ for y, in the equation of the locus, it becomes $sx^2 + 2by = 0$, or $x^2 = -\frac{2b}{s}y$; so that P is the vertex of the curve, and PM, PX′, its principal axes. If we substitute $-\frac{1}{2}sb$ for y, there result for x the values $x = \pm b$; hence the curve passes through the extremities of the base, A, B.

PROBLEM II.

(143.) Given the base and the difference of the tangents of the angles at the base, to determine the locus of the vertex.

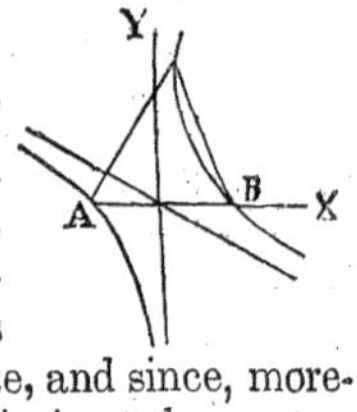

Taking the same axes as before, we have

$\tan. \angle B - \tan. \angle A = \frac{2xy}{b^2 - x^2} = d$, d being put for the difference of the tangents. Hence the equation of the locus is $2xy + dx^2 - db^2 = 0$, which belongs to an hyperbola, and, since the terms containing y^2 and y are absent from this equation, it follows (137) that the axis of y coincides with an asymptote, and since, moreover, the term containing x is also absent, the origin is at the centre. If $\pm b$ be substituted for x, in the equation of the locus, the resulting value of y is 0; hence the curve passes through the extremities of the base.

The other asymptote may be constructed by means of the expression at (138), which gives for the tangent of the angle α, which it makes with the axis of x, the value $\tan. \alpha = -\frac{1}{2}d$.

PROBLEM III.

(144.) Given the base of a triangle and the difference of the angles at the base, to determine the locus of the vertex.

Taking the same axes as before, and putting a, a', for the tangents of the angles at the base, and t for the tangent of their difference, we have $t = \frac{a - a'}{aa' + 1}$, or substituting for a, a', their respective values in terms of the coordinates of the vertex, as given in the first problem, the expression becomes

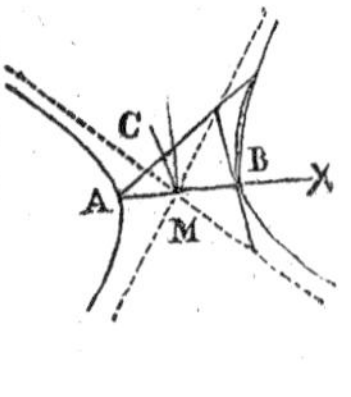

$$t = \frac{2xy}{y^2 - x^2 + b} \quad \therefore x^2 + \frac{2}{t}xy - y^2 - b^2 = 0.$$

Consequently the locus is an hyperbola, and, because the terms containing the first power of the variables is absent, the origin is at the centre. Also, since the coefficients of x^2 and y^2 are equal, and opposite in sign, the hyperbola is equilateral (138). It passes through the extremities of the base, since for $x = \pm b$, $y = 0$. When the vertex coincides with B, the angle A is 0, and the angle B is that contained by AB, and a tangent to the curve at AB; this angle therefore is equal to the given difference; consequently, if MC make an angle with AM, equal to the difference of the angles at the base of the triangle, MC, being parallel to a tangent at B, will be in the direction of the conjugate to AB, therefore the lines which bisect the angles CMA, CMB, will be the asymptotes of the curve (85).

PROBLEM IV.

(145.) It is required to find the locus of a given point in a straight line of given length, of which the extremities move along the sides of a given angle.

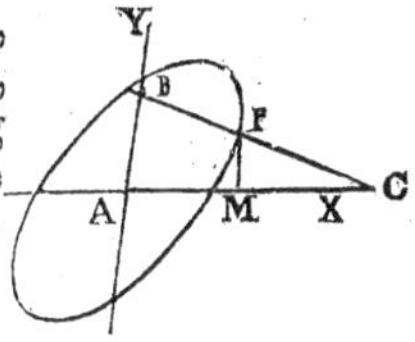

Let AX, AY, be the sides of the given angle, BC the given line, and P the given point; then, drawing the ordinate $\text{PM} = y$, and putting $\text{CP} = a$, $\text{PB} = b$, and the cosine of the angle $\text{A} = c$, we have (*Trig. p.* 54.) $a^2 = \text{MC}^2 + y^2 - 2\text{MC} \cdot cy$; but $b : a :: x : \text{MC} = \dfrac{ax}{b}$

hence, by substitution, $a^2 = \dfrac{a^2x^2}{b^2} + y^2 - \dfrac{2acxy}{b}$ $\therefore$ $a^2x^2 - 2abcxy + b^2y^2 - a^2b^2 = 0$. Hence the locus is an ellipse, of which the centre is at the origin. If the angle A is right, then $c = 0$, and the equation is $a^2x^2 + b^2y^2 = a^2b^2$; in this case, therefore, the principal diameters of the curve coincide with the sides of the given angle. Hence is suggested an easy method of tracing an ellipse; thus, having drawn the perpendicular lines, AX, AY, apply to them the extremities of a rule, BC, of which the parts BP, PC, are respectively equal to the semi-minor and semi-major axes of the proposed curve, then in every such position of BC, P will mark a point in the curve.

PROBLEM V.

(146.) Two straight lines are given in position, from any point, in one of which, a perpendicular is drawn to the other, and from a given point in this latter, with a radius equal to the perpendicular, an arc, cutting the perpendicular in P, is described. It is required to find the locus of the point P.

Let DX, DN, be the lines given in position NM, a perpendicular from the latter to the former, and A the given point; then we must

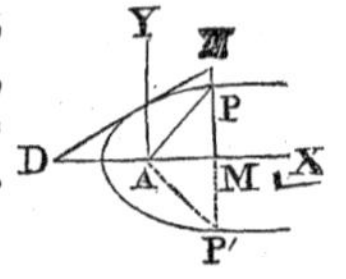

always have AP = NM. Put NM = Y, PM = y, AD = p, and the tangent of the angle D = a; then, taking the rectangular axes, AX, AY, we have for the equation of DN passing through the point $(-p, 0)$.

$$Y = a(x+p) \therefore Y^2 = a^2x^2 + 2a^2px + a^2p^2,$$

but $Y^2 = x^2 + y^2$, $\therefore x^2 + y^2 = a^2x^2 + 2a^2px + a^2p^2$; hence the equation of the locus is $y^2 + (1-a^2)x^2 - 2a^2px - a^2p^2 = 0$. If the angle D is 45°, than $a = 1$, and the equation is $y^2 = 2px + p^2 = 2p(x + \frac{p}{2})$, which characterizes a parabola, of which the abscissa of the vertex is $-\frac{1}{2}p$, that is, the vertex is at the middle of AD, and, since p is also the semi-parameter, it follows that A is the focus.

If the angle D is less than 45°, then $a \angle 1$, and the locus is an ellipse, and because y enters in the equation only in its second power, there are two equal values of y for one value of x; hence the axis of x is a principal diameter of the curve. If the equation be solved for x, the rational part of the resulting expression will be $\frac{a^2p}{1-a^2}$ this therefore, (134) is the value of the abscissa of the centre, by substituting it for x, in the equation of the locus, twice the resulting value of y, viz. $\frac{2ap}{\sqrt{\{1-a^2\}}}$, will be the length of the diameter, parallel to the axis of y. For the length of the other principal diameter, take the difference of the two values of x, which the equation gives for $y = 0$, and we obtain the expression $\frac{2ap}{1-a^2}$.

This diameter may however be found rather more easily, for since, in the equation of the locus, when the origin of the axes is removed to the centre, x^2 will preserve the same coefficient, it follows that, denoting the transformed by $y^2 + \frac{B^2}{A^2}x^2 = B^2$, we must have $\frac{B^2}{A^2} = 1 - a^2$, but we have found $4B^2 = \frac{4a^2p^2}{1-a^2}$; hence $4A^2 = \frac{4a^2p^2}{(1-a^2)^2}$.

If the angle D is greater than 45°, the locus of P is an hyperbola, of which the centre and axes may be determined as in the case of the ellipse. When the given lines are parallel the locus is obviously a circle, because then MN or AP is constant.

PROBLEM VI.

(147.) To find the locus of the vertex of a parabola which shall touch a given straight line, and have a given focus.

Let AB be the given straight line, and F, the focus; draw the perpendicular, FP, and through a vertex, V, draw FVA, join also PV. Then (108) $FA \cdot FV = FP^2$, therefore PVF must be a right angle; consequently the locus of V is a circle, of which the diameter is FP.

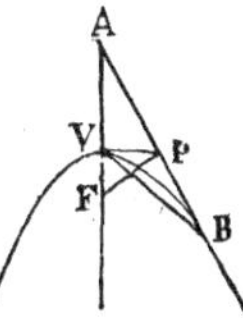

PROBLEM VII.

(148.) To find the locus of the focus of a parabola which shall touch a given straight line, and have a given vertex.

Let V be the given vertex, and AB the given tangent; then, for every position, P, of the focus, the perpendicular, FP, subtends a right angle at the vertex, V. Let VY, parallel to AB, be taken for the axis of y, and VX, perpendicular to it, for the axis of x; then, by similar triangles, VDF, PDV, we have $VD = y : DF = x :: PD = a : y \therefore y^2 = ax$. Hence the locus is a parabola, of which the axes are VX, VY, and parameter PD or VE.

PROBLEM VIII.

(149.) Given the base and altitude of a triangle to find the locus of the intersection of perpendiculars from the angles to the opposite sides.

Let c represent the base, AB, of the triangle, then the altitude, a, being constant, the locus of the vertex, C, is a parallel to AB. Hence, taking AB, AY, for the rectangular axes, and putting (x', y') for any point in the locus of C, we have always $y' = a$, and for the equation of BC, passing through the points $(c, 0)$ and (x', y'),

$y = \frac{y'}{x'-c}(x-c)$, and for the equation of a perpendicular to this,

through the origin, we have $y = \frac{c-x'}{y'}x$, at the point P, where this line intersects the perpendicular, CD, $x = x'$; therefore, substituting x for x', and a for y', in the foregoing equation, we have for the locus of P the equation $ay = cx - x^2$, which characterizes a parabola.

For $x = 0$ we have $y = 0$; therefore the curve passes through A, but does not again meet AY; so that AY is a diameter. For $y = 0$ we have not only $x = 0$, but also $x = c$; therefore, the curve passes through B; hence the principal diameter bisects AB at right angles; therefore, to find the vertex, put $x = \frac{1}{2}c$, which gives $y = \frac{c^2}{4a}$. By removing the origin of the axes (38) to the vertex, that is, to the point $(\frac{c}{2}, \frac{c^2}{4a})$, the equation becomes $ay = -x^2$; therefore the parameter is $a = \text{CD}$.

PROBLEM IX.

(150.) Given the base and the sum of the sides of a triangle, to find the locus of the point of intersection of lines from the angles bisecting the opposite sides.

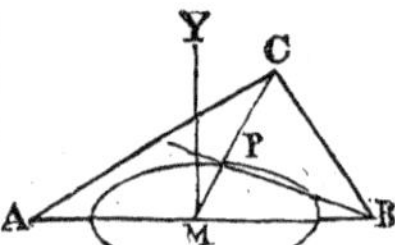

Let M be the middle of the base, AB, and take MB, MY, for rectangular axes. Put (x', y') for the vertex, C, of the triangle, and (x, y) for P, one of the points in the locus, then (19) $y = \frac{y'}{3}$ $\therefore x = \frac{x'}{3} \therefore (x', y') = (3x, 3y)$.

Now the locus of (x', y') is an ellipse, of which the principal diameter, 2A, is equal to the sum of the given sides of the triangle, and the foci A and B. The equation of the locus of $(x' \; y',)$ is therefore $A^2y'^2 + B^2x'^2 = A^2B^2$; hence, by substitution, we have for the locus of $(x, y,)$ the equation $A^2y^2 + B^2x^2 = \frac{A^2B^2}{9}$ or $(\frac{A}{3})^2 y^2 + (\frac{B}{3})^2 x^2 = (\frac{A}{3})^2 \cdot (\frac{B}{3})^2$, an ellipse of which the principal diameters are one third those of the former.

If, instead of the sum, the difference of the sides had been given, the locus would have been an hyperbola, since, in that case, the locus of $(x', y',)$ would have been an hyperbola.

But, if the sum of the tangents of the angles at the base had been constant, then the locus would have been a parabola, (*Prob.* 1.)

PROBLEM X.

(151.) Given the base and sum of the sides of a triangle, to find the locus of the centre of the inscribed circle.

Let AB be the given base, and P the centre of one of the circles, of which p is the point of contact with the base; then it is known that the distance of p from M, the middle of the base, is always equal to half the difference of the sides. This is easily proved; for, since two tangents drawn to a circle from any point are equal, we have $\text{AC} - \text{CB} = \text{A}p' - \text{B}p'' = \text{A}p - \text{B}p = 2\text{M}p$. This being premised, take MB, MY, for rectangular axes, put $\text{MB} = c$, $\text{C} = (x', y',)$ and $\text{P} = (x, y)$, then the area of the triangle, $\text{ABC} = y'c$, or putting A for the given sum, AC + BC, the area of the same triangle is $y(\text{A} + c)$; consequently, $y'c = y\,(\text{A} + c)$

$\therefore\ y' = \frac{y(A + e)}{c}$. Now, since the locus of C is an ellipse, of which A, B, are the foci, and 2A, the major diameter, we have (47) $\frac{1}{2}(AC - CB) = ex' = x \therefore x' = x \div e = Ax \div c$. Hence, substituting these values of x' and y', in the equation of the locus of (x', y'), viz. in $A^2 y'^2 + B^2 x'^2 = A^2 B^2$, we have for the locus of P, the equation $(A + c)^2 y^2 + B^2 x^2 = B^2 c^2$, which characterizes an ellipse, of which the axes coincide with the former. For $x = 0$ we have $y = \frac{Bc}{A + c}$, and for $y = 0$ we have $x = c$; these values of x and y are those of the principal semi-diameters of the locus.

If, instead of the sum, the difference of the sides had been given, then, since half this difference, that is, x, would have been constant, the locus of P would have been a straight line through p, perpendicular to the base.

PROBLEM XI.

(152.) Given the base and the sum of the sides of a triangle to find the locus of the centre of the circle touching the base, and the prolongation of the other two sides.

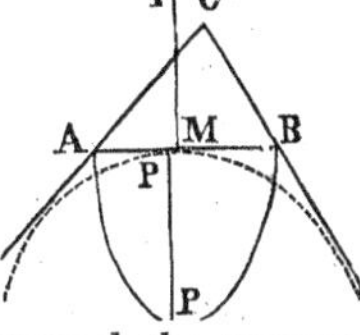

Taking the same axes as in the last problem, let P be the centre of one of the circles, and p its point of contact with the base; then, as before, $Mp = -x = \frac{1}{2}(AC - CB) = ex' \therefore x' = -Ax \div c$. Now, as the centre of the circle must always be on the line bisecting the angle C, that is, on the normal, through the point (x', y'), we have by substituting this value of x', in the equation of the normal, the expression $y = \frac{B^2 y' - Acy' - A^2 y'}{B^2} = -\frac{(A + c)cy'}{A^2 - c^2} = \frac{-c}{A - c} y'$ to determine y', which is $y' = -\frac{A - c}{c} y$. These values of x' and y', substituted in the locus of (x', y'), give for the locus sought the equation. $(A - c)^2 y^2 + B^2 x^2 = B^2 c^2$, which is that of an ellipse, of which the minor diameter is $2c = AB$, and major diameter $\frac{Bc}{A - c} = \frac{\sqrt{\{A^2 - c^2\}} \cdot c}{A - c} = \sqrt{\frac{A + c}{A - c}} \cdot c$. If, instead of the sum, the difference of the sides had been given, then, since x would have been constant, the locus of P would have been a straight line through p, perpendicular to the base.

PROBLEM XII.

(153.) Given the base and the difference of the sides of a triangle, to find the locus of the centre of the circle touching one side, and the prolongation of the base and of the other side.

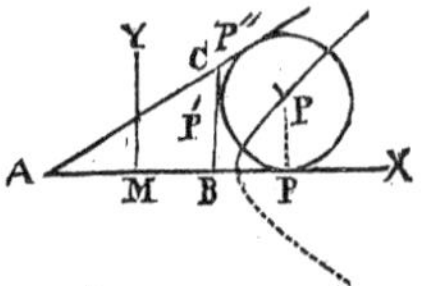

Let P be the centre of one of the circles, and p, p', p'' the several points of contact; then $Ap'' = Ap$, or $AC + Cp' = AB + Bp \therefore Ap = \frac{1}{2}(AB + AC + BC)$, and, taking AM or $\frac{1}{2}AB$ from each side, we have $Mp = \frac{1}{2}(AC + BC) = x$.

Now, since the locus of C is an hyperbola, of which AC, BC, are radii vectores, we have (79) $\frac{1}{2}(AC + BC) = ex' = x \therefore x' = Ax \div c$. Also, since P must always be on the line bisecting the angle BCP'', that is to say, on the normal through the point (x', y'), we shall have to determine the value of y' from the equation of the normal, when x' is replaced by $Ax \div c$. This equation is

$$y = \frac{B^2 y' - Acy' + A^2 y'}{B^2} = \frac{(A-c)c}{A^2 - c^2} y' = \frac{c}{A+c} y', \quad \therefore y' = \frac{A+c}{c} y.$$

These values of x' and y', substituted in the locus of (x', y'), give for the locus sought the equation $(A+c)^2 y^2 - B^2 x = -B^2 c^2$, which characterizes an hyperbola, whose principal axes are $2c$ and $\frac{2Bc}{A+c}\sqrt{-1}$. If, instead of the difference, the sum of the sides had been given, then, since half this sum, or x, is constant, the locus would have been a straight line perpendicular to, and through the extremity of, the major diameter of the ellipse, which is the locus of C.

PROBLEM XIII.

(154.) Two straight lines are perpendicular to each other, and through two given points in one, straight lines are drawn, forming, with the other, angles, the product of whose tangents is constant: what is the locus of their intersection?

Let the perpendicular lines be taken for axes, and the equations of any pair of the intersecting lines be $y = ax + b$, $y = \alpha x + \beta$. Then, by the conditions of the problem, the quantities $a\alpha$, b, and β, are constant; hence, multiplying the two equations together, and reducing, we have for the equation of the locus $y^2 - a\alpha x^2 - (b+\beta)y + b\beta = 0$, which characterizes an hyperbola, if $a\alpha$ is positive, and an ellipse if $a\alpha$ is negative. Because x enters into this equation only in its second power, there are for every value of y two equal values of x; therefore the axis of y is a principal diameter of the curve. If we put $x = 0$, the resulting values of y are obviously b and β, the difference of these is the value of the diameter, which coincides with the axis of y, that is $A = \frac{1}{2}(b - \beta)$. Also the sum of the same values gives twice the ordinate of the centre, therefore $Y = \frac{1}{2}(b + \beta)$.

To find the other principal diameter, put this value of Y for y, in the equation of the locus, and there results $x = B = \frac{b - \beta}{2\sqrt{(-a\alpha)}}$.

f $a\alpha = +1$, the locus is an equilateral hyperbola, and, if $a\alpha = -1$, t is a circle. In every case, the part of the axis intercepted by the given points is a principal diameter of the curve, as the foregoing value of A proves.

PROBLEM XIV.

(155.) From two given points two straight lines are drawn so as to intercept a given portion of a straight line given in position: what is the locus of the intersection of those lines?

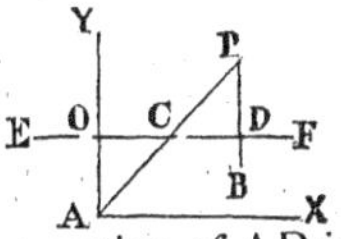

Let EF be the line given in position, and A, B, the given points; let also AP, BP, be two lines intercepting the given portion, $CD = m$, then P is a point in the locus. Draw the axes, AX, AY, the one parallel and the other perpendicular to EF, put $AO = p$ and (x', y') for the point B, then the equation of AP is $y = ax$ $\therefore$ when $y = p$, $OC = p \div a$, also the equation of BP is

$$y - y' = a'(x - x'), \because \text{ when } y = p,\ OD = \frac{p - y' + a'x'}{a'},$$

$\therefore OD - OC = CD = \dfrac{p - y' + a'x'}{a'} - \dfrac{p}{a} = m$, that is, substituting $\dfrac{y}{x}$ for a, and $\dfrac{y - y'}{x - x'}$, for a', $\dfrac{(p - y')(x - x')}{y - y'} + x' - \dfrac{px}{y} = m$, this equation becomes, after reduction, $(x' - m)y^2 - y'xy + (my' - px')y + py'x = 0$. Hence the locus is an hyperbola.

As, in this equation, the square of one of the variables, viz. x^2, is absent, the axis of x is paralleled to an asymptote (137), the ordinate of which is $py' \div y' = p$; hence the line EF is that asymptote.

To determine the centre, we may solve the equation of the locus with regard to y, and, by omitting the irrational part, in the resulting expression for y, we shall have the equation of a diameter, in which, by putting p for y, we shall obtain for x the abscissa of the centre, which is therefore thus determined. Having found the centre, we may construct the other asymptote; thus, assume any value for x, and construct the two values of y corresponding; two points of the curve will be thus determined, either of which is at the same distance from the known asymptote that the other is from the asymptote sought, the distances being measured along the line passing through the two points, and in opposite directions; hence the centre, and a point in the asymptote, being found, the line may be drawn. Or, without first finding the centre, we may determine in this way two points in the required asymptote, which will determine its position.

PROBLEM XV.

(156.) Tangents to a parabola form a given angle with each other what is the locus of their point of intersection?

Let t represent the tangent of the given angle, and $(x, y,)$ any point of intersection; then, if the equation of the parabola be $y^2 = 2mx$, the equations of tangents through any points (x', y'), (x'', y'') of the curve will be (105) $yy' = m(x + x')$, $yy'' = m(x + x'')$. As the tangents of the angles which these tangents form with the axis of x are respectively $m \div y'$, $m \div y''$, we have for t the tangent of their difference the expression $t = \frac{m(y'' - y')}{y' y'' + m^2}$; it remains therefore to determine y', y'', in terms of x, y. For this purpose, substitute for $2mx'$, $2mx''$, in the equations of the tangents, their equals y'^2, y''^2, and we have the two equations $y'^2 - 2yy' + 2mx = 0$, $y''^2 - 2yy'' + 2mx = 0$, in which the roots of the one are the same as those of the other; therefore, by the theory of equations, $y'y'' = 2mx$ and $y' + y'' = 2y$ $\therefore (2y)^2 - 8mx = (y'' - y')^2$; hence, substituting, in the square of the expression, for t, we have $t^2 = \frac{m^2(4y^2 - 8mx)}{(2mx + m^2)^2} = \frac{4y^2 - 8mx}{(2x + m)^2}$, $\therefore y^2 - t^2 x^2 - (2 + t^2)mx - \frac{1}{4}t^2m^2 = 0$; consequently the locus is an hyperbola, of which a principal diameter coincides with the axis of x, since there are two equal values of y for $x = 0$. If we put $y = 0$, in the equation, the difference of the roots will be the length of this diameter, and half their sum, the abscissa of the centre; this abscissa therefore is $-(1 \div t^2 + \frac{1}{2})m = -(\cot.^2 P + \frac{1}{2})m = AO$ By substituting it for x, in the equation of the locus, we get for the square of the semidiameter parallel to the axis of y the expression

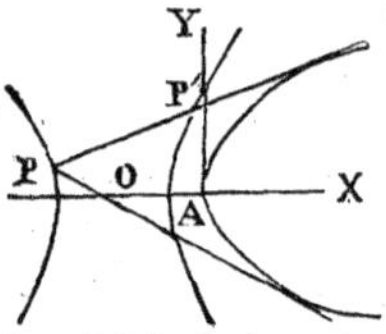

$y^2 = -m^2\{(1 \div t^2) + 1\} = -m^2(\cot^2. P + 1) = -m^2 \operatorname{cosec}.^2 P$.

Now, instead of determining the other diameter by taking the difference of the roots of the equation, as above suggested, we shall obtain it more readily from these considerations. We know that if we had removed the origin of the axes to the centre of the curve, that is, if, in the equation of the locus, we had substituted $x - (\cot.^2 P + \frac{1}{2})m$ for x, the equation would have been transformed to the form of $y^2 - (B^2 x^2 \div A^2) = -B^2$. But, in this transformed equation, the coefficient of x^2 will be the same as in the primitive, viz. $-t^2$; consequently $\frac{B^2}{A^2} = \tan.^2 P$, and $B^2 = m^2 \operatorname{cosec}.^2 P$, $\therefore A^2 = \frac{m^2 \operatorname{cosec}.^2 P}{\tan.^2 P}$,

$\therefore A = \frac{m \operatorname{cosec}. P}{\tan. P}$. If $t = 1$, that is, if the given angle be 45°, the locus will be an equilateral hyperbola. If $t =$ infinite, that is, if the given angle be 90°, then the denominator, in the expression for t^2, must be 0; that is, $2x + m = 0$, or $x = -\frac{1}{2}m$; in this case, therefore, the locus is the directrix of the proposed parabola.

If t is negative, that is, if the given angle is obtuse, the equation of the locus will remain unaltered, since t enters only in its second power.

Hence we infer that, if any pair of tangents intersect at an angle, P, and any other pair intersect at an angle, P′, supplementary to the former, the locus of P will be one branch of the hyperbola, and the locus of P′ the other branch.

PROBLEM XVI.

(157.) Tangents to a parabola form angles with the principal diameter, the product of whose tangents is given; what is the locus of the points of intersection?

Let one of the points of intersection be (x, y), and one of the points of contact (x', y'); then from the equation of the curve $y'^2 = 2mx'$ $\therefore\ x' = \left.\frac{y'^2}{2m}\right\}$ (1), and from the equation of the tangent, $yy' = mx + mx' = mx + \left.\frac{y'^2}{2}\right\}$ (2). Also, for the value of a, the tangent of the angle which the tangent through (x', y') makes with the principal diameter, we have $a = \frac{m}{y'}$, $\therefore\ y' = \frac{m}{a}$.

Substituting this value of y', in equation (2), and reducing, we have $a^2 - \frac{y}{x}a + \frac{m}{2x} = 0$. The two values of a contained in this equation belong to the two tangents drawn from the point (x, y), and, as their product, p, is given, we have, by the theory of equations $p = \frac{m}{2x}$,

$\therefore\ 2px = m$, or $x = \frac{m}{2p}$.

Hence the locus sought is a straight line, perpendicular to the principal diameter, and at the distance $\frac{m}{2p}$ from the vertex.

PROBLEM XVII.

(158.) To find the locus of the intersections of pairs of tangents to any line of the second order when they make angles with the principal diameter, such that the product of their tangents may be a given quantity.

This problem has just been solved for the parabola, and may, by employing a similar process, be extended to the other two curves. Thus, representing a point of intersection by (x, y), and a point of contact by (x', y'), we should have, from the equation of the curve, $A^2y'^2 + B^2x'^2 = A^2B^2$ (1), and from the equation of the tangent, $A^2yy' + B^2xx' = A^2B^2$ (2). Moreover, the expression for a, the trigonometrical tangent of the angle, this line makes with the principal diameter is $1 \div a = \left.-\frac{A^2y'}{B^2x'}\right\} \ldots$ (3).

Hence, by determining x' and y', from equations (1) and (2), and substituting their values in (3), we shall ultimately obtain, as in last problem, an equation between x, y, and a, and the locus will then be determined, as in the case referred to. As, however, this process will be encumbered with some very complicated expressions, we shall employ the following more simple and elegant method of investigation, including in it the preceding problem.

The general equation of a line of the second order, when referred to the principal diameter, and tangent through its vertex, is $y^2 = mx + nx^2$; therefore, from the equation of the curve, we have $y'^2 = mx' + nx'^2$; (1), and for the tangent, a, of the angle, formed by the axis of x, and a straight line through the points (x, y) and (x', y'), we have the expression $a = \frac{y - y'}{x - x'}$, $\therefore y' = y - ax + ax'$ (2). Substituting this value of y' in (1), and arranging the result according to the power of x', we have the equation

$$(a^2 - n)\, x'^2 + (2ay - 2a^2 x - m)\, x' + y^2 - 2axy + a^2 x^2 = 0 \quad (3)$$

This equation gives two values for x; but, since, by the conditions of the problem, the line through the points (x, y), (x', y') must have only one point, viz. (x', y'), in common with the curve, the two values of x', in (3), must be equal, in other words, the equation must be a complete square. Hence, by the theory of equations, $4(a^2 - n)(y^2 - 2axy + a^2x^2) = (2ay - 2a^2 x - m)^2$. Reducing this equation, and arranging the result according to the powers of a, we obtain finally

$$a^2 - \frac{my + 2nxy}{mx + nx^2}\, a + \frac{m^2 + 4ny^2}{4(mx + nx^2)} = 0.$$

The two values of a, contained in this equation, belong to the two tangents drawn from the point (x, y); and, since their product, p, is given, we have, by the theory of equations,

$$p = \frac{m^2 + 4ny^2}{4(mx + nx^2)} \text{ hence } ny^2 - pnx^2 - pmx + \frac{m^2}{4} = 0 \ldots . (4),$$

the equation of the locus required, which is, therefore, an hyperbola, or an ellipse, according as p is positive or negative. When, however, $n = 0$, that is, when the proposed curve is a parabola, this locus becomes a straight line in which $x = \frac{m}{4p}$ showing that it is perpendicular to the axis, and that it coincides with the directrix when $p = -1$, or when the intersecting tangents include a right angle. Since equation (4) gives two equal values of y for $x = 0$, it follows that a principal diameter of the locus coincides with the axis of x. If we put $y = 0$, in the equation, half the sum of the roots of the resulting equation in x will be the abscissa of the centre; this abscissa is therefore $-\frac{m}{2n}$. Substituting this abscissa for x, in (4), we have for the square

of the semi-diameter parallel to the axis of y,
$y^2 = B^2 = -\frac{m^2(p+n)}{4n^2}$. But (*see prob.* 15) $\frac{B^2}{A^2} = -\frac{pn}{n} = -p$, consequently, dividing the expression for B^2 by $-p$, we have
$A^2 = +\frac{m^2(p+n)}{4n^2p}$ therefore, if n be positive, and p negative, but numerically less than n, the locus will be impossible, for, in this case, the foregoing expressions for the squares of the semi-axes will both be negative.

If $p = -1$, that is, if the intersecting tangents to the ellipse or hyperbola form a right angle, the locus will be a circle, of which the radius is $A = B = \frac{\sqrt{\{m^2(1-n)\}}}{2n}$. But, if $p = +1$, the locus will be an equilateral hyperbola, of which the principal semi-transverse is given by the same expression.

If we suppose the intersecting tangents to be parallel to conjugate diameters, then $p = n$ (82) $= \mp\frac{b^2}{a^2}$; putting a and b for the principal semi-conjugates of the proposed ellipse, or hyperbola, also $m = \frac{2b^2}{a}$; therefore, by these substitutions, in equation (4), the locus becomes
$y^2 \pm \frac{b^2}{a^2}x^2 - \frac{2b^2x}{a} \pm b^2 = 0$, which characterizes an ellipse, when the proposed curve is an ellipse, that is, when p is negative, but when the original curve is an hyperbola, then this equation being the same as
$y^2 - \left(\frac{b}{a}x + b\right)^2 = 0 \therefore y = \pm\left(\frac{b}{a}x + b\right)$. This equation characterizes the asymptotes, which are, therefore, the loci of the intersections.

PROBLEM XVIII.

(159.) What is the locus of the centres of all the circles which pass through a given point and touch a given straight line?

Ans. A parabola.

PROBLEM XIX.

What is the locus of the centres of all the circles which may touch two given circles? *Ans.* An hyperbola.

PROBLEM XX.

The directrix and a point in a parabola being given to determine the locus of the vertex? *Ans.* An ellipse.

PROBLEM XXI.

Given the base and vertical angle of a plane triangle to determine the locus of the centre of the inscribed circle? *Ans.* A circle.

PROBLEM XXII.

Upon a given base triangles are constructed having always one

angle at the base double the other; what is the locus of their vertices?
Ans. An hyperbola.

PROBLEM XXIII.

From any point in a given straight line two straight lines are drawn, the one perpendicular to the given line, and the other to a given point; if the perpendicular be made equal to the other, what will be the locus of its extremity? *Ans.* An equilateral hyperbola.

Notes containing solutions to all the problems proposed by the Author p. 287.

CHAPTER IV.

MISCELLANEOUS PROPOSITIONS ON THE THREE CURVES.

THEOREM.

(160.) If to any line of the second order two secants, parallel to the sides of a given angle, be drawn, then the two rectangles contained by the parts intercepted between their point of intersection and the curve will have a constant ratio, wherever that point of intersection may be.

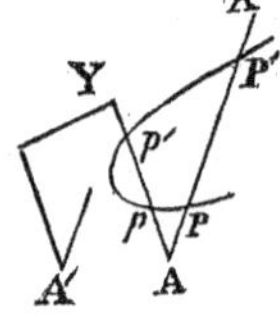

Let AX, AY be parallel to the sides of a given angle, A′, and intersect any line of the second order, in the points P, P′ and p, p'. The equation of the curve referred to these lines as axes is $Ay^2+Bxy+Cx^2+Dy+Ex+F=0$, in which, if we put $y=0$, the resulting values of x will be the abscissas AP, AP′, that is, these lines will be given by the roots of the equation $Cx^2+Ex+F=0$, or $x^2+\frac{E}{C}x+\frac{F}{C}=0$,

$\therefore AP \cdot AP' = F \div C$. In like manner, putting $x=0$, in the equation of the curve, the resulting values of y will determine the ordinates Ap, Ap', that is, these lines will be determined by the roots of the equation, $y^2+\frac{D}{A}y+\frac{F}{A}=0$, $\therefore Ap \cdot Ap' = \frac{F}{A}$

consequently $AP \cdot AP' : Ap : Ap' :: \frac{F}{C} : \frac{F}{A} :: A : C$.

Now no change can take place in the coefficients A and C of y^2 and x^2, by removing the origin, A, without altering the inclination of the axes, that is to say, these coefficients will remain the same, although we put $y+b$ for y, and $x+a$ for x; consequently, so long as the two secants remain parallel to the two given lines, the ratio of the rectangles $AP \cdot AP'$, $Ap \cdot Ap'$ will be the same, wherever A may be.

PROBLEM.

(161.) To determine the general equation of the tangent to a line of the second order.

Let (x', y'), (x'', y'') be two points on the curve, then, for the secant passing through them, we have the equation $y - y' = \frac{y'-y''}{x'-x''}(x-x')$ (1),

and from the equation of the curve,

$$\left.\begin{array}{l} Ay'^2 + Bx'y' + Cx'^2 + Dy' + Ex' + F = 0 \\ Ay''^2 + Bx''y'' + Cx''^2 + Dy'' + Ex'' + F = 0 \end{array}\right\} \dots (2)$$

Taking the difference, $A(y'^2 - y''^2) + B(x'y' - x''y'') + C(x'^2 - x''^2) + D(y' - y'') + E(x' - x'') = 0$.

Substituting, in this equation, $(y' + y'')(y' - y'')$ for $y'^2 - y''^2$,
$(x' + x'')(x' - x'')$ for $x'^2 - x''^2$,
and $x'(y' - y'') + y''(x' - x'')$ for $x'y' - x''y''$,

it becomes $(y' - y'')[A(y' + y'') + Bx' + D] + (x' - x'')[C(x' + x'') + By'' + E] = 0$,

from which we obtain $\frac{y' - y''}{x' - x''} = -\frac{C(x' + x'') + By'' + E}{A(y' + y'') + Bx' + D}$.

By substituting this value of $\frac{y' - y''}{x' - x''}$, in (1), and then, supposing the points (x', y'), (x'', y'') to coincide, we have for the equation of the tangent passing through (x', y'), $y - y' = -\frac{2Cx' + By' + E}{2Ay' + Bx' + D}(x - x')$ (3).

If the tangent is to pass through a given point (α, β), without the curve, then, in this equation, we must substitute for the general symbols x and y, the particular values α and β, and the unknown point of contact, (x', y'), may be determined, analytically, by means of equations (2) and (3), or geometrically, by constructing the loci of these equations.

*M. Puissant** has given a simple and elegant method of arriving at the general equation of the tangent. He refers the curve to *polar* coordinates, assuming the pole on the curve, and then inquires what angle the revolving line must make with the fixed axis, when the radius vector becomes 0, that is, when the line becomes a tangent.

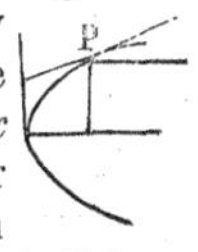

Thus if it be required to draw a tangent through any point, P, in a line of the second order, represented by the equation $y^2 = mx + nx^2$ (4), let there be substituted for x and y the values $x = x' + r\cos.\omega$, $y = y' + r\sin.\omega$, or rather, for simplicity, $x = x' + rp$, $y = y' + rq$ (5), in which x' and y' are the coordinates of the point P, and we shall have the transformed equation $(y' + rq)^2 = m(x' + rp) + n(x' + rp)^2$ (6), characterizing the proposed curve, when related to polar coordinates, of which the origin is P, and the fixed line parallel to the primitive axis of x. If this equation be developed, and the terms arranged according to the powers of r, the result will evidently be of the form $Mr^2 + Br + C = 0$.

Now it is easy to perceive that the term C, which is independent of

* Recueil de diverses propositions de Géométrie.

r, must represent $y'^2 - mx' - nx'^2$, and this being equal to 0, by equation (4), the transformed equation will be simply $Mr + B = 0$.

When, therefore, $r = 0$, that is, when the radius vector becomes a tangent at the point P, there must exist the condition $B = 0$; so that, by equating the coefficient of the first power of r, in the developement of (6), with 0, we have $2y'q - mp - 2npx' = 0$,

whence $\frac{q}{p} = \frac{\sin. \omega}{\cos. \omega} = \tan. \omega = \frac{m + 2nx'}{2y'}$. This then is the expression for the trigonometrical tangent of the angle formed by the tangent to the curve at P, and the axis of x, and consequently the equation of the tangent is $y - \beta = \frac{2nx' + m}{2y'}(x - \alpha)$, in which $(\alpha, \beta,)$ denotes any given point in the tangent, and (x', y') the point of contact. If this latter be the given point, the equation is

$y - y' = \frac{2nx' + m}{2y'}(x - x')$ (7). The same reasoning, applied to the more general equation, $Ay^2 + Bxy + Cx^2 + Dy + Ex + F = 0$, shows that, after substituting the values (5) for x and y, we need attend in the result only to the coefficient of the first power of r, which being equated to 0, will furnish the value of $\frac{q}{p}$, or $\tan. \omega$. This equation will be $2Ay'q + B(qx' + py') + 2Cpx' + Dq + Ep = 0$,

whence $\frac{q}{p} = \tan. \omega = -\frac{2Cx' + By' + E}{2Ay' + Bx' + D}$ and therefore the equation of the tangent through the point (x', y') is

$y - y' = -\frac{2Cx' + By' + E}{2Ay' + Bx' + D}(x - x')$, the same as was determined by the first method (3). By reduction, this equation becomes $(2Ay' + Bx' + D)y + (2Cx' + By' + E)x + Dy' + Ex' + 2F = 0$, in which may be observed this analogy to the equation of the curve, viz. that, if the accents be effaced, each term in the equation of the curve will appear twice in that of the tangent, from which circumstance the following rule has been contrived for arriving at the equation of the tangent, with the proper accents, by means of the equation of the curve. Substitute, in the equation of the curve, $x'x$ for x^2, $y'y$ for y^2, $x'y$ for xy, and x', y' for x, y.

Repeat the equation thus written, taking care, however, to change x' into x, and x into x', as also y' into y, and y into y' The sum of these two equations will be the equation of the tangent.

Thus, let the parabola of which the equation is $y^2 = px$ be proposed, then the two equations to be added will be $y'y = px'$ and $yy' = px$; therefore $2y'y = p(x' + x)$ is the equation sought. If we take the general equation $y^2 = mx + nx^2$, we shall have to add the equations $y'y = mx' + nx'x$, $yy' = mx + nxx'$; therefore $2'yy =$

$(m + 2nx')\,x + mx'$ is the equation of the tangent, and to this form equation (7) may be reduced.

THEOREM.

(162.) If through any given point, chords are drawn to a line of the second order, and tangents be applied to their extremities, these tangents will intersect on a straight line.

Let the coordinates of the given point be a, b, and let (x', y'), (x'', y''), represent the extremities of either of the chords passing through it, let also (α, β), be one of the points of intersection. Then for the equations of tangents passing through these points we have, by employing the expression at the conclusion of last problem,

$2y'\beta = (m + 2nx')\,\alpha + mx'$ (1), $2y''\beta = (m + 2nx'')\,\alpha + mx''$ (2).

Now the equation of the line joining the points of contact of these tangents must be $2y\beta = (m + 2nx)\,\alpha + mx$, for this equation must represent some straight line, being of the first degree; and it passes through the two points (x', y'), and (x'', y''), because equations (1) and (2) subsist; therefore it can represent no other than the *chord of contact*. As (a, b), is always a point on this chord, we have $2b\beta = (m + 2na)\,\alpha + ma$, this equation being of the first degree in α and β shows that the point (α, β) is always on a straight line.

THEOREM.

(163.) If a curve of the second order be referred to a system of conjugate axes, and a point in its plane be found, such that its distance from any point whatever in the curve be a rational function of the abscissa of that point, then the point thus found can be no other than a focus of the curve.

First, for the ellipse.

Let (x', y') denote the fixed point found, then for its distance from any point (x, y) in the curve, we have (14), $D^2 = (x - x')^2 + (y - y')^2 + 2\,(x - x')\,(y - y')\cos.\,A$, where A is the inclination of the axes.

Or, developing this expression, and putting for y its equal $\sqrt{\left\{B^2 - \frac{B^2 x^2}{A^2}\right\}}$, we have $D^2 = \frac{A^2 - B^2}{A^2}x^2 - 2x'x + x'^2 + y'^2 + B^2 - 2y'(x - x')\cos.\,A + \{2\,(x - x')\cos.\,A - 2y'\}\sqrt{\left\{B^2 - \frac{B^2 x^2}{A^2}\right\}}$

Now it is obviously impossible that D cannot be a rational function of x while the irrational function $\sqrt{\left\{B^2 - \frac{B^2 x^2}{A^2}\right\}}$ remains in this expression; hence the term in which it enters must disappear that is, the point (x', y') must be such that $2\,(x - x')\cos.\,A - 2y' = 0$, for every value of x. This condition gives the equation $x - x' = \frac{y'}{\cos. A}$.

But the first side of this equation is indeterminate, inasmuch as x is; the second side therefore must also be indeterminate, although, by hypothesis, y', and cos. A, have certain fixed values, these values therefore can be no other than $y' = 0$, and cos. A $= 0$, or A $= 90°$, in which case alone $x - x' = \frac{0}{0} =$ *an indeterminate quantity.*

It follows, therefore, that the conjugate axes must be the *principal* axes of the curve, and that (since $y' = 0$) the fixed point must be on one of them. With this condition the expression for D^2 becomes $D^2 = \frac{A^2 - B^2}{A^2} x^2 - 2x'x + (x'^2 + B^2)$, and we have now to inquire in what circumstances the root of this square can be a rational function of x. In order to this, assume $D = bx + a, \therefore b^2x^2 + 2bax + a^2 = \frac{A^2 - B^2}{A^2} x^2 - 2x'x + (x'^2 + B^2)$, then, comparing the coefficients of the like powers of x, we have for the determination of x' the conditions $b^2 = \frac{A^2 - B^2}{A^2}$, $2ba = -2x'$, $a^2 = x'^2 + B^2$. From the first two we get $a^2 = \frac{A^2x'^2}{A^2 - B^2}$, and this value of a^2, substituted in the third gives $\frac{A^2x'^2}{A^2 - B^2} = B^2 + x'^2 \therefore x' = \pm \sqrt{\{A^2 - B^2\}} = \pm\, e$. Hence x' has two values that will satisfy the proposed condition, viz. $x' = + c$ and $x' = - c$, showing that the points sought are no other than the foci.

The same process applies to the hyperbola when the sign B^2 is changed.

Second, for the Parabola.

Since, in the parabola, $y = \sqrt{\{2px\}}$, therefore $D^2 = x^2 - (2x' - 2p)\ x + x'^2 + y'^2 - 2y'\ (x - x')$ cos. A $+ \{2\,(x - x')$ cos. A $- 2y'\}\ \sqrt{2px}$; hence we must conclude, as before, that $2\,(x - x')$ cos. A $- 2y' = 0$, $\therefore x - x' = \frac{y'}{\cos. A} = \frac{0}{0}$, as before, $\therefore D^2 = x^2 + 2\,(p - x')\ x + x'^2$. Put $D^2 = (x + q)^2 = x^2 + 2qx + q^2 = x^2 + 2\,(p - x')\ x + x'^2$, then, comparing the coefficients, $q = p - x'$ and $q^2 = x'^2$, $\therefore q = x'$, $\therefore p = 2x'$, $\therefore x'\ \frac{1}{2}p$, the abscissa of the focus.

PROBLEM.

(164). To determine a cube which shall be double a given cube.

Let the side of the given cube be a, and that of the required cube x, then we are to determine x from the equation $x^3 = 2a^3$, or $x^4 = 2a^3x$.

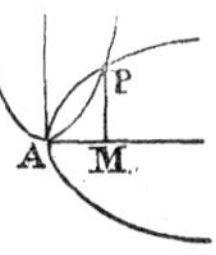

Substitute, in this equation, for x^2; $x^2 = ay$ (1), and it becomes $a^2y^2 = 2a^3x$ or $y^2 = 2ax$ (2). Now equations (1) and (2) represent parabolas referred to the same axes, the parameter of the one being a, and that of the other $2a$. At the point where these parabolas intersect the abscissa will be common to both, therefore this value of x, satisfying both the equations (1) and (2), must also satisfy the proposed, so that the side AM of the required cube may be determined by construction.

PROBLEM.

(165.) To trisect an angle.

By trigonometry, p. 43, $\cos. 3A = \frac{4 \cos. A^3 - 3R^2 \cos. A}{R^2}$,

or, putting $\cos. 3A = a$, and $\cos. A = x$, we have

$x^3 - \frac{3R^2}{4}x^2 - \frac{R^2a}{4} = 0$, from which equation we are to determine the values of x by construction.

Multiplying the terms by x, it becomes

$x^4 - \frac{3R^2}{4}x^2 - \frac{R^2a}{4}x = 0$. In this equation assume $x^2 = \frac{1}{2}Ry$ (1,)

and it becomes $y^2 - \frac{3R}{2}y - ax = 0 \ldots.$ (2).

If these two equations which denote parabolas be constructed in reference to the same axes, the abscissas of their points of intersection will be the three values of the cosine of A to radius R.

These values correspond to the three analytical values cos. A, cos. $(\frac{2}{3}\pi + A)$, cos. $(\frac{4}{3}\pi + A)$, since the given cosine, a, belongs equally to either of the three arcs 3A, $(2\pi + 3A)$, $(4\pi + 3A)$.*

The two last problems may serve to show the application of curves to the solution of equations. If the equation do not exceed the fourth degree, the roots may always be determined geometrically by the intersections of two lines of the second order. But this mode of determining the roots of equations is never employed in practice, the most accurate as well as most expeditious method being that of numerical approximation. The best method of approximating to the roots of equations is that discovered by *Mr. Horner*, and printed in the *Phi-*

* The same cosine equally belongs to the arcs $(6\pi + 3A)$, $(8\pi + 3A)$, &c. but the cosine of the third part of either of these will always be the same as one of the three cosines in the text; thus cos. $(2\pi + A = \cos. A$, cos. $(2\pi + \frac{2}{3}\pi + A) = \cos. (\frac{2}{3}\pi + A)$, &c. Also since the cosine a will remain the same, although 3A be negative, it will remain the same also for every arc in the series. — 3A, $(2\pi - 3A)$, $(4\pi - 3A)$, $(6\pi - 3A)$, &c. but for these also the same remarks apply, that is, the cosine to the third part of either will be one of the cosines in the text; thus: cos. $(\frac{2}{3}\pi - A) = \cos. [2\pi - (\frac{2}{3}\pi - A)] = \cos. (\frac{4}{3}\pi + A)$, cos. $(\frac{4}{3}\pi - A) = \cos. (\frac{2}{3}\pi + A)$, &c.

losophical Transactions for 1819, (see also my Algebra, chap. vi.) For further applications of the theory of curves to the construction of equations the student is referred to *Bourdon, Application de l'Algebre a la Geometrie, Lacroix, Trigonometrie, and Lardner's Algebraic Geometry,* in sections xx and xxi of which will be found many interesting remarks on this subject.

THEOREM.

(166.) Five points being given on a plane, of which no three are situated on the same straight line, it is possible to describe a line of the second order which shall pass through them all.

For, let the general equation $Ay^2+Bxy+Cx^2+Dy+Ex=F$ (1), be divided by A, and it will then assume the form $y^2+bxy+cx^2+dy+ex=f$ (2), so that the equation of the second degree, in its most general form, contains five coefficients, b, c, d, e, and f, the values of which may be arbitrarily assumed, they may, therefore, be so determined as to subject the curve, into whose equation they enter, to pass through the points (α, β), (α', β'), (α'', β''), (α''', β'''), (α'''', β''''), since we shall have, for this purpose, the five simple equations

$$\begin{aligned}
\beta^2 &+ b\alpha\ \beta + c\alpha^2 + d\beta + e\alpha = f \\
\beta'^2 &+ b\alpha'\ \beta' + c\alpha'^2 + d\beta' + e\alpha' = f \\
\beta''^2 &+ b\alpha''\ \beta'' + c\alpha''^2 + d\beta'' + e\alpha'' = f \\
\beta'''^2 &+ b\alpha'''\beta''' + c\alpha'''^2 + d\beta''' + e\alpha''' = f \\
\beta''''^2 &+ b\alpha''''\beta'''' + c\alpha''''^2 + d\beta'''' + e\alpha'''' = f
\end{aligned}$$

from which the values of the unknowns, b, c, d, e, and f, may obviously be determined, and these values substituted in equation (2) will render that equation the representative of the required curve. If we are not restricted in the choice of axes of coordinates, they may be so assumed as to render some of the preceding equations of condition of simpler form. Thus, by taking one of the points, as (α, β), for the origin, we shall have for the first equation merely $0=f$, and if each axis be drawn through a separate point, as (α', β'), and (α'', β''), the next two equations will be $\beta'^2+d\beta'=f$, and $c\alpha''^2+e\alpha''=f$. If the curve sought ought to be a parabola, only four arbitrary points must be assumed because, in the equation of this curve, only four of the five coefficients are arbitrary, since between the two, a and b, there must exist the relation $a^2-4b=0$; as here a has two values, a positive and a negative, for the same value of b, there may be two parabolas passing through the same four points. But no curve of the second order can intersect another in more points than four, since the coefficients determined by the five preceding equations, admit each of but one value.

PROBLEM.

(167.) To determine a curve which shall pass through any proposed number of given points.

Let us represent the given points by (α, β), (α', β'), (α'', β''), (α''', β'''), (α'''', β''''), &c. then it is obvious that several curves might be described passing through these points, some expressible by equations, and others not. But it is desirable to know which of all the possible curves is the simplest, or admits of the easiest description. Now those curves are most easily described of which any ordinate is a rational and integral function of the abscissa, because the value of the ordinate corresponding to any assumed abscissa will, in this case, never be encumbered with radicals. These curves are included in the equation $y = A + Bx + Cx^2 + Dx^3 + \&c.$ and they are called *parabolic curves*, because the parabola, of which the equation is $y = A + Bx + Cx^2$, is obviously one of them. The order of the curve depends upon the highest power of x; the common parabola is of the second order, and that in which x^3 is the highest power of x is of the third order, &c.

Hence, if we take the parabola whose order is equal to the number of proposed points, we shall have to determine the same number of coefficients, A, B, C, &c. from the simple equations

$$\left.\begin{aligned}\beta &= A + B\alpha + C\alpha^2 + D\alpha^3 + \&c.\\ \beta' &= A + B\alpha' + C\alpha'^2 + D\alpha'^3 + \&c.\\ \beta'' &= A + B\alpha'' + C\alpha''^2 + D\alpha''^3 + \&c.\\ \beta''' &= A + B\alpha''' + C\alpha'''^2 + D\alpha'''^3 + \&c.\end{aligned}\right\} \ldots\ldots (1).$$

To do this in the simplest manner, let each equation be subtracted from the next, and we shall have

$$(\beta' - \beta) = B(\alpha' - \alpha) + C(\alpha'^2 - \alpha^2) + D(\alpha'^3 - \alpha^3) + \&c.$$
$$(\beta' - \beta'') = B(\alpha'' - \alpha') + C(\alpha''^2 - \alpha'^2) + D(\alpha''^3 - \alpha'^3) + \&c.$$
$$(\beta''' - \beta'') = B(\alpha''' - \alpha'') + C(\alpha'''^2 - \alpha''^2) + D(\alpha'''^3 - \alpha''^3) + \&c$$

and by division

$$\left.\begin{aligned}\frac{\beta' - \beta}{\alpha' - \alpha} &= B + C(\alpha' + \alpha) + D(\alpha'^2 + \alpha'\alpha + \alpha^2) + \&c. = a\\ \frac{\beta'' - \beta'}{\alpha'' - \alpha'} &= B + C(\alpha'' + \alpha') + D(\alpha''^2 + \alpha''\alpha' + \alpha'^2) + \&c. = a'\\ \frac{\beta''' - \beta''}{\alpha''' - \alpha''} &= B + C(\alpha''' + \alpha'') + D(\alpha'''^2 + \alpha'''\alpha'' + \alpha''^2) + \&c. = a''\end{aligned}\right\} (2).$$

The values a, a', a'', are known, because the first member of each equation is known. Now these equations are of the same form as those originally proposed, and A is found eliminated; hence, by performing a similar process with these equations as with the first group, we shall have

$$\left.\begin{aligned}\frac{a' - a}{\alpha'' - \alpha} &= C + D(\alpha'' + \alpha' + \alpha) + \&c. = b\\ \frac{a'' - a'}{\alpha''' - \alpha'} &= C + D(\alpha''' + \alpha'' + \alpha') + \&c. = b'\end{aligned}\right\} \ldots\ldots (3).$$

In these equations, the next coefficient, B, is eliminated, and, by continuing the process, we shall eliminate the coefficients one by one, till we come to the last, the value of which may then be determined from the final simple equation, and thence all the others.

Suppose, for example, only three points (α, β), (α', β'), and (α'', β''), are proposed, then the equation of the curve is $y = A + Bx + Cx^2$, and the equations (2), (3), become

$$\frac{\beta' - \beta}{\alpha' - \alpha} = B + C(\alpha' + \alpha) = a, \quad \frac{\beta'' - \alpha'}{\alpha'' - \alpha'} = B + C(\alpha'' + \alpha') = a',$$

$\dfrac{a' - a}{\alpha'' - \alpha} = C = b$. Substituting this value of C, in equation (2), we have $B = a - b(\alpha' + \alpha)$. Also, since from the proposed equation we have $\beta = A + B\alpha + C\alpha^2$, we obtain for A, after having replaced B and C by the values just determined, the expression $A = \beta - a\alpha + b\alpha\alpha'$. Having thus determined the values of the three coefficients, we have for the equation of the required curve

$$y = \beta - a\alpha + b\alpha\alpha' + (a - b\alpha' - b\alpha)\,x + bx^2,$$
$$\text{or } y = \beta + a\,(x - \alpha) + b(x - \alpha)\,(x - \alpha').$$

Lagrange, after having given this solution from *Newton*, observes, that it may be much more simply obtained from the following considerations: (Puissant, Problèmes de Geométrié.)

Since y ought to become β, β', β'' . . . . when x becomes α, α', α'' . . . . , it is obvious that the expression for y must be of the form $y = A'\beta + B'\beta' + C'\beta'' + \ldots$, where the quantities A′, B′, C′, &c. must be functions of x, such that, when we put

$$x = \alpha, \text{ we must have } A' = 1,\ B' = 0,\ C' = 0 \ldots .$$
$$x = \alpha', \qquad A' = 0,\ B' = 1,\ C' = 0 \ldots .$$
$$x = \alpha'', \qquad A' = 0,\ B' = 0,\ C' = 1 \ldots .$$

consequently the values of A′, B′, C′, &c. must necessarily take the form

$$A' = \frac{(x - \alpha')\,(x - \alpha'')\,(x - \alpha''')\ldots}{(\alpha - \alpha')\,(\alpha - \alpha'')\,(\alpha - \alpha''')\ldots}$$

$$B' = \frac{(x - \alpha)\,(x - \alpha'')\,(x - \alpha''')\ldots}{(\alpha' - \alpha)\,(\alpha' - \alpha'')\,(\alpha' - \alpha''')\ldots}$$

$$C' = \frac{(x - \alpha)\,(x - \alpha')\,(x - \alpha''')\ldots}{(\alpha'' - \alpha)\,(\alpha'' - \alpha')\,(\alpha'' - \alpha''')\ldots}$$

where the number of factors in each numerator and denominator is one less than the number of given points. Hence the general expression for y is

$$y = \frac{(x - \alpha)\,(x - \alpha'')\,(x - \alpha''')\ldots}{(\alpha - \alpha')\,(\alpha - \alpha'')\,(\alpha - \alpha''')\ldots}\,\beta$$
$$+ \frac{(x - \alpha)\,(x - \alpha'')\,(x - \alpha''')\ldots}{(\alpha' - \alpha)\,(\alpha' - \alpha'')\,(\alpha' - \alpha''')\ldots}\,\beta'$$
$$+ \frac{(x - \alpha)\,(x - \alpha')\,(x - \alpha''')\ldots}{(\alpha'' - \alpha)\,(\alpha'' - \alpha')\,(\alpha'' - \alpha''')\ldots}\,\beta''$$

Newton's method gives for the general expression for y

$y = \beta + a(x - \alpha) + b(x - \alpha)(x - \alpha') + c(x - \alpha)(x - \alpha')(x - \alpha'')$.

These two expressions are different only in form, as may be ascertained by developing the values of a, b, c, &c. and arranging the terms according to the quantities, β, β', β'', &c.

Either of the preceding values of y may be considered as a solution to this problem, viz. To determine the general relation which exists between two variable quantities, x, y, from knowing the relation which exists in the particular cases $x = \alpha, y = \beta$; $x = \alpha', y = \beta'$; $x = \alpha''$, $y = \beta''$; &c.

This is an important problem, being the foundation of the *method of interpolation*, since it enables us, from having a certain number of terms of a series given, the law of which is not known, to arrive at an approximate expression for the general term of that series, and thence to interpolate between the given terms as many more as we please, all governed by the same law. Of the two general expressions for this purpose just given, the former is the more commodious in calculation, because the several terms may be computed by logarithms. Nevertheless, the latter expression leads to a very neat and commodious formula, when we suppose the quantities α, α', α'', &c. to be in arithmetical progression, as is generally the case in practice.

Let h be the common difference of the progression, then $\alpha' = \alpha + h, \alpha'' = \alpha + 2h, \alpha'' = \alpha + 3h$, &c. let also $x = \alpha + h'$, then $x - \alpha = h', x - \alpha' = h' - h, x - \alpha'' = h' - 2h$, &c. Now, putting, for brevity, $\triangle\beta$, $\triangle\beta'$, $\triangle\beta''$, &c. for the several differences, $\beta' - \beta, \beta'' - \beta', \beta''' - \beta''$, &c. we have (2);

$a = \dfrac{\triangle\beta}{h}, a' = \dfrac{\triangle\beta'}{h}, a'' = \dfrac{\triangle\beta''}{h}$, &c. putting, in like manner, $\triangle^2\beta$, $\triangle^2\beta'$, &c. for the second differences, $\triangle\beta' - \triangle\beta$, $\triangle\beta'' - \triangle\beta'$, &c. have $b = \dfrac{\triangle^2\beta}{1 \cdot 2h^2}, b' = \dfrac{\triangle^2\beta'}{1 \cdot 2h^2}$, &c. substituting also $\triangle^3\beta$, &c. for the third differences, $\triangle^2\beta' - \triangle^2\beta$, &c. their results $c = \dfrac{\triangle^3\beta}{1 \cdot 2 \cdot 3h^3}$, &c. therefore the formula becomes

$$y = \beta + \frac{h'}{h}\triangle\beta + \frac{h'(h'-h)}{h \cdot 2h}\triangle^2\beta + \frac{h'(h'-h)(h'-2h)}{h \cdot 2h \cdot 3h}\triangle^3\beta + \ldots.$$

As an application of this formula, the following example is frequently given.

EXAMPLE.

To compute the logarithm of the number 3·1415926536 by means of a table of logarithms from 1 to 1000, calculated to ten places of decimals.

Regarding the logarithms in the table as the particular values of y, the corresponding numbers being the values of x, we shall have, by taking the successive differences, the following values

$\beta = \log. 3 \cdot 14 = \cdot 4969296481$	$\Delta\beta = \cdot 13809057$
$\beta' = \log. 3 \cdot 15 = \cdot 4983105538$	$\Delta\beta' = \cdot 13765288$
$\beta'' = \log. 3 \cdot 16 = \cdot 4996870826$	$\Delta\beta'' = \cdot 13721796$
$\beta''' = \log. 3 \cdot 17 = \cdot 5010592622$	$\Delta\beta''' = \cdot 13678578$
$\beta'''' = \log. 3 \cdot 18 = \cdot 5024271200$	

$\Delta^2\beta = -43769$	$\Delta^3\beta = 277$	$\Delta^4\beta = -3$;
$\Delta^2\beta' = -43492$	$\Delta^3\beta' = 274$	
$\Delta^2\beta'' = -43218$		

consequently $\Delta\beta = \cdot 0013809057$, $\Delta^2\beta = -\cdot 0000043769$
$\Delta^3\beta = \cdot 0000000277$, $\Delta^4\beta = -\cdot 0000000003$.

Now the constant difference, h, is $\cdot 01$, and $3 \cdot 14$ being taken for a, and $3 \cdot 1415926536$ for x, we have $h' = x - a = \cdot 0015926536$; hence $\frac{h'}{h} = \cdot 15926536$, $\frac{h' - h'}{2h} = \frac{h}{2h} - \frac{1}{2} = -\cdot 42036732$,
$\frac{h' - 2h}{3h} = \frac{h'}{3h} - \frac{2}{3} = -\cdot 61357821$, $\frac{h' - 3h}{4h} = \frac{h'}{4h} - \frac{3}{4} = -.71018366$.

These values, substituted in the formula

$$y = \beta + \frac{h'}{h}\Delta\beta + \frac{h'(h'-h)}{h \cdot 2h}\Delta^2\beta +$$
$$\frac{h'(h'-h)(h'-2h)}{h \cdot 2h \cdot 3h}\Delta^3\beta + \frac{h'(h'-h)(h'-2h)(h'-3h)}{h \cdot 2h \cdot 3h \cdot 4h}\Delta^4\beta,$$

give $y = \log. 3 \cdot 1415926536 = \cdot 4971498726$.

GENERAL SCHOLIUM.

(168.) Having discussed pretty fully the various properties of curves of the second order, it remains to make a few general remarks upon the higher orders of curves, or those of which the equations extend beyond the second degree.

If we have an equation of the nth degree, containing two variables, and which is not compounded of equations of inferior degrees, the curve which it represents is said to be of the nth order. But, if the equation is compounded of others of inferior degrees, then also its geometrical representation comprehends all the curves represented by the component equations. Such an assemblage of lines is called a complex line. For instance, the locus of the cubic equation $y^3 - axy^2 + bxy - abx^2 - cy + acx = 0$, which arises from the multiplication of the two equations $y - ax = 0$ (1), and $y^2 + bx - c = 0$ (2), is not a *simple* line of the third order, but a *complex* line, consisting of the straight line represented by the equation (1), and the parabola represented by equation (2). For the coordinates, (x, y), of every point in the straight line rendering the factor $y - ax$ equal to 0, the same coordinates must render 0 the product $(y - ax)(y^2 + bx - c)$, that is, they must always satisfy the proposed equation; hence this straight line must belong to the locus, and in the same manner is it shown that the parabola must also belong to the locus.

It appears, therefore, that for an equation of the nth degree to represent a curve of the nth order, it must be such that, when all the terms are arranged on one side, it may not admit of being resolved into rational factors.

The most general form of an equation between two variables of any proposed degree is that which beside constant quantities, contains every possible combination of the variables, under the condition, that, wherever their product enters, the sum of their exponents shall not exceed the required degree. Thus the most general form of the equation of the third degree is

$$Ay^3 + By^2x + Cyx^2 + Dx^3 + Ey^2 + Fyx + Gx^2 + Hy + Kx + L = 0$$

that of the fourth degree.

$$\left.\begin{array}{l} Ay^4 + By^3x + Cy^2x^2 + Dyx^3 + Ex^4 \\ \quad + Fy^3 + Gy^2x + Hyx^2 + Kx^3 \\ \quad\quad + Ly^2 + Myx + Nx^2 \\ \quad\quad\quad + Py + Qx \\ \quad\quad\quad\quad + R \end{array}\right\} = 0;$$

&c. so that the number of terms in a general equation of the nth degree will be equal to those in

$$(y + x)^n + (y + x)^{n-1} + (y + x)^{n-2} + \ldots . (y + x)^0.$$

Now the expansion of any power of a binomial consists of as many terms as there are units in its exponent, and one more (*Alg. p.* 160); hence the sum of the terms in the above series of expansions is that of the arithmetical progression

$1 + 2 + 3 + 4 + \ldots . n + 1 = \frac{1}{2}(n + 1)(n + 2)$; we infer, therefore, that, in the general equation of the nth degree, there are $\frac{1}{2}(n + 1)(n + 2)$ terms, and consequently the same number of constant coefficients; we may, however, without diminishing the generality of an equation, divide all its terms by the coefficient of any one of them, and thus reduce the number of arbitrary coefficients to

$\frac{1}{2}(n + 1)(n + 2) - 1 = \frac{1}{2}n(n + 3)$.

It follows from this, that a curve of the nth order may be made to pass through $\frac{1}{2}n(n+3)$, points arbitrarily assumed, for the coordinates of each point being successively substituted for x, y, in the general equation, will give rise to $\frac{1}{2}n(n + 3)$, equations in which the general coefficients are the unknown quantities, and which these equations are sufficient to determine; the values of the coefficients being thus ascertained the locus of the equation will pass through the proposed points, but, if these are so assumed as to render it impossible for any simple curve of the proposed order to pass through them, then the locus determined as above will be a *complex line* of the proposed degree. If the points are all in the same straight line, the equation of the locus will be found to be reducible to the form $(y + ax + b)^n = 0$, which represents n coinciding straight lines.

Let us now inquire in how many points it is possible for a straight line to intersect a curve of the nth order. Taking the general equation of the nth order, and putting $y = 0$, we have the equation $A' x^n + B' x^{n-1} + C' x^{n-2} + \dots P' x + Q = 0$, the roots of which are the values of so many abscissas of the points where the axis of x cuts the curve. As the values of the coefficients A', B', &c. are quite arbitrary, they may obviously be such as to render these n roots all possible and different from each other; hence a curve of the nth order *may* be cut by a straight line in n points, but not in more; there are, however, not *necessarily* n points of intersection; the number may be less, but cannot be more. If the term $A' x^n$ be absent from the equation of a curve of the nth order, or can by any transformation be removed, then there can at most be but $n - 1$ points of intersection between the curve and axis of x; if also the term $B' x^{n-1}$ be absent, then the number of intersections will be but $n - 2$, and so on. When any particular curve of the nth order is proposed, then the coefficients, A', B', &c. become fixed, and the number of intersections will be n, or $n - 2$, or $n - 4$, &c. according as the equation has n, or $n - 2$, or $n - 4$, &c. possible roots.*

Between the curves of the second order and those of the higher orders there exists a very intimate analogy. We can adduce here only the two following instances: 1. If two straight lines parallel to the axis of y, drawn in a curve of the nth order, be cut by the axis of x, so that the sum of the ordinates on one side be in each case equal to the sum of the ordinates on the other side, then every other line parallel to these will be cut by the axis in the same manner.

Let the equation of the curve be $y^n + (ax + b)\, y^{n-1} + \dots = 0$, and those of the two parallels, $x = p$, and $x = q$, then we have

$$y^n + (ap + b)\, y^{n-1} + \&c. = 0$$
$$y^n + (aq + b)\, y^{n-1} + \&c. = 0,$$

and, since in each case the sum of the negative ordinates is equal to the sum of the positive, we have

* If any roots of this equation are equal, it will intimate that the straight line passes through a *singular* point of the curve. Thus, if two roots are equal, the corresponding point will be either a point of contact or a double point, that is, a point in which two branches of the curve intersect. If three roots be equal, the corresponding point will be either a point of *inflexion*, or a triple point, &c The investigation of the singular points of curves belongs more properly to the *Differential Calculus.*

$ap + b = 0$, $aq + b = 0$ $\therefore a(p - q) = 0$ $\therefore a = 0$; hence, $b = 0$; consequently, whatever be the value of x, we must always have $ax + b = 0$, which establishes the proposition.

The line thus dividing parallel ordinates is called a *diameter.*

2. If to any line of the nth order two secants, parallel to the sides of a given angle, be drawn, then the continued products of the parts intercepted between their point of intersection and the curve will have a constant ratio.

For, taking these two secants as axes, and putting successively $y = 0$, and $x = 0$, the equation gives

$$A'x^n + B'x^{n-1} + Cx^{n-2} + \dots P'x + Q = 0$$
$$A\,y^n + B\,y^{n-1} + C\,y^{n-2} + \dots P\,x + Q = 0.$$

The roots of these equations give the parts of the two secants intercepted between their intersection, that is, the origin, and the curve.

Hence, dividing the first equation by A', we have for the product of its roots, or of the parts of one secant, the expression $Q' \div A'$. In like manner, for the product of the parts of the other secant we have the expression $Q \div A$. Hence these products are to each other as $Q \div A' : Q \div A$, or as $A : A'$, that is, they have a constant ratio. (*See art.* 160.)

For further particulars respecting the higher curves the student is referred to the comprehensive summary given by *Dr. Gregory* in the third volume of *Hutton's Mathematics*, see also *Lardner's Algebraic Geometry*, section xxi.

We here terminate the second principal division, or FIRST PART of the present performance; having now completed our inquiry into the general theory and properties of lines of the second order, usually called *analytical geometry of two dimensions.*

ANALYTICAL GEOMETRY

OF THREE DIMENSIONS.

SECTION I.

(169.) The preceding part of the present treatise has been occupied in discussing the properties of the plane curves, that is to say, of lines of which all the points are situated in the same plane. In these inquiries we have found nothing more to be necessary than to assume, in the same plane with the curve proposed, two fixed lines, or axes, and then to investigate the analytical expression which must characterize the position of every point in the curve, relatively to the assumed axes. This analytical representation of the proposed curve contains implicitly all its properties. In this second part of our subject we propose to extend our inquiries to the consideration of lines and surfaces not entirely situated in one plane, and where it will be necessary to employ three axes of reference, instead of two. We shall begin by determining the equation of a point situated in space.

CHAPTER I.

ON THE POINT AND STRAIGHT LINE SITUATED IN SPACE.

Equation of a Point.

(170.) Let ZAX ZAY, YAX, be the three planes which, for simplicity, we shall suppose to intersect at right angles, so that the line ZA will be perpendicular, to both AX and AY. Let, also, P be any point in space, whose position it is required to determine relatively to the axes, AX, AY, AZ.

Upon each of the assumed planes let fall, from the point P, the perpendiculars Pp, Pp', Pp'', then, if the lengths of these perpendiculars be given, the position of the point P will be determined. For, conceive the planes PC, PB, PD, to be drawn, forming with the planes DC, DB, the rectangular parallelopiped AP, then $AD=Pp$, $AC=Pp'$, and $AB=Pp''$, consequently the points B, C, D, are at given distances from the point A; if, therefore, through these points three planes parallel to the assumed planes be drawn, the point P will be that in which they all intersect.

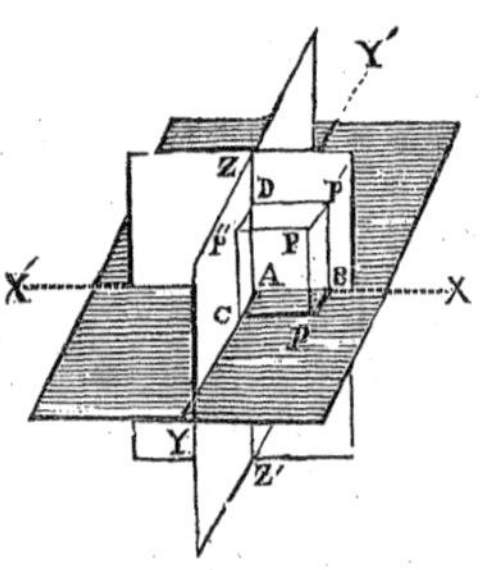

The three planes, ZAX, ZAY, YAX, in reference to which the position of the point has been determined, are called the *coordinate planes*, their intersections, AX, AY, AZ, are the axes of coordinates and the distances AB, AC, AD, of the three planes, parallel to the former, from the origin, A, are the three coordinates of the point P, where they intersect. The coordinates of any point are generally denoted by x, y, and z; if these are known, the point, as we have just seen, is determinable; hence the equations $x=a$, $y=b$, $z=c$, are the equations of a point.

If the three coordinate planes be produced beyond their intersections, there will obviously be formed about the point A eight triedral angles,* four above the horizontal plane YAX, and four below; hence to express analytically in which of these angles the proposed point is situated, we must prefix to its ordinates the signs which they must

---

* Angles formed by the meeting of three planes in a point.

take from considering the axes of ordinates as positive, in one direction, and as negative, in the opposite direction; thus, regarding the axes as positive in the directions AX, AY, AZ, they will be negative in the opposite directions, AX′, AY′, AZ′; hence we shall have the following variations of the signs of the coordinates for every possible position of the point P.

If $x = +a, y = +b, z = +c$, the point is in the angle AXYZ,
$x = -a, y = +b, z = +c$, AX′YZ,
$x = +a, y = -b, z = +c$, AXY′Z,
$x = +a, y = +b, z = -c$, AXYZ′,
$x = -a, y = -b, z = +c$, AX′Y′Z,
$x = -a, y = +b, z = -c$, AX′YZ′,
$x = +a, y = -b, z = -c$, AXY′Z′,
$x = -a, y = -b, z = -c$, AX′Y′Z′.

(171.) Of the three coordinate planes that which contains the axes AX, AY, and which is generally the horizontal plane, is called the *plane of xy*; that which contains the axes AX, AZ, is called the *plane of xz*; and the third, containing the axes AY, AZ, is called the *plane of yz*. If the proposed point be situated in the plane of xy, then its distance, z, from this plane being 0, its equation will be $x = a$, $y = b$, $z = 0$. If it be on the axis of x, that is, on the intersection of the planes of xy and xz, then its distance from each of these planes being 0, its position will be expressed by the equations $x = a$, $y = 0$, $z = 0$. But if it be at the origin, that is, at the common intersection of the three planes, then, its distance from each being 0, the equations of the point are $x = 0$, $y = 0$, $z = 0$. In like manner, if the point be situated in the plane of xz, its equations are $x = a$, $y = 0$, $z = c$; and if it be on the axis of x, or on the axis of z, we have, respectively $x = a$, $y = 0$, $z = 0$, $x = 0$, $y = 0$, $z = c$. Lastly, if the point be in the plane of yz, its equations are $x = 0$, $y = b$, $z = c$.

(172.) The points p, p', p'', where the perpendiculars from P meet the coordinate planes, are called the *projections* of P, on these planes. If the position of any two of these projections were given, it would be sufficient to determine the point P; for a perpendicular from either projection to the plane in which it is, necessarily passes through the point P, so that P will be at the intersection of two such perpendiculars; knowing, therefore, two projections, we can always, if required, determine the third.

Suppose, for instance, the projections p, p', on the planes of xy and xz be known, or, which is the same thing, that we have given the equations of these points, viz. $x = a$, $y = b$, and $x = a$, $z = c$; these two equations give for the third projection, p'', the equations $y = b$, $z = c$, and any two of these combined give the equations of P, viz. $x = a$, $y = b$, $z = c$.

If the coordinate planes had been oblique, instead of rectangular, the preceding equations would have been the same; but the coordinates a, b, c, would then have been oblique, and the projections of P would have been given by lines drawn from P, parallel to the coordinate planes.

On the Equation of the Straight line in Space.

(173.) If, through any given straight line situated in space, a plane, perpendicular to either of the coordinate planes, be drawn, the intersection of the two planes is called the *projection* of the proposed line. The plane thus drawn is called the *projecting plane*; there are, therefore, three projecting planes, each of which contains the proposed line, and one of its projections, consequently knowing two of the projections, we may draw two of the projecting planes, and, since the proposed line must be situated in each, their intersection will determine it; hence, in the straight line, as in the point, two projections are sufficient to determine it.

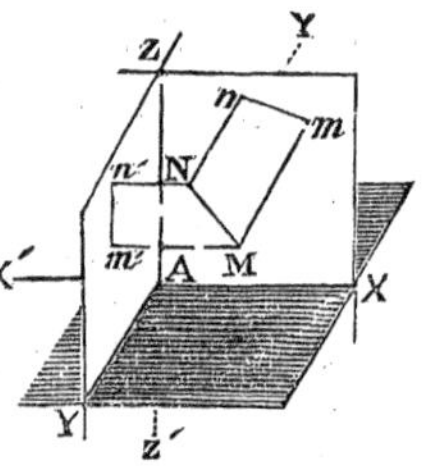

Let MN be a straight line in space, of which the projections on the planes of zx and of zy are mn and $m'n'$, and let the equations of these projections be $x = az + \alpha$ . . . . (1),
$y = bz + \beta$. . . . . (2).

Assume any point in the projecting plane, Nm, and through it draw in this plane a parallel to AY, then every point in this parallel being equally distant from the plane of zy, and also equidistant from the plane of xy, it follows that for every point in this line the coordinates x, z, are the same, and one of the points is in the line mn; but the coordinates x, z, for every point in mn are related, as in equation (1); hence also the coordinates x, z, of every point in the projecting plane, Nm, are related, as in equation (1). In a similar manner, the coordinates y, z, of any point in the line $m'\,n'$ are the same as those of any point in the projecting plane, Nm'. Hence, at the intersection, MN, of these planes, both the relations (1) and (2) must exist, so that these equations which, taken separately, characterize the two projections, represent, when taken together, the proposed line, therefore

$$\left.\begin{array}{l} x = az + \alpha \\ y = bz + \beta \end{array}\right\}$$

are the equations of the straight line in space.

Hence any assumed value being given to one of the coordinates, these equations will make known the other two, and thus the three coordinates of any point in the line may be obtained.

(174.) We have here supposed that the proposed line is projected on the two vertical planes ZX, ZY; if, however, one had been the horizontal plane, XY, then the equation of the projection on this plane would have exhibited the relation between the coordinates, x, y, of any point in the proposed line, and this, combined with either of the equations, (1), (2), would have equally characterized the proposed line. But the relation between x and y, or the equation of the projection on XY is readily obtained by eliminating z, in the equations (1), (2), this elimination gives the relation $y = \frac{b}{a}x - \frac{b\alpha - a\beta}{a}$ which is, therefore, the equation of the projection on the plane of xy, and in a similar manner may either projection be obtained from knowing the other two. But the projections usually employed are those on the vertical planes represented by equations (1), (2), in which the vertical axis, AZ, is considered as the axis of abscissas; the horizontal axis, AX, as the axis of ordinates for the projections on the plane of xz; and the horizontal axis, AY, as the axis of ordinates for the projections on the plane of yz. The constants, a and b, denote the tangents of the angles which the projections on the vertical planes make with the axis of z, and α, β, express the distances of the origin from the points where these projections intersect the axis of x and of y

PROBLEM I.

(175.) To determine the points where the coordinate planes are pierced by a given straight line,

Let the given straight line be represented by the equations

$$\left.\begin{aligned} x &= az + \alpha \\ y &= bz + \beta \end{aligned}\right\}$$

Then at the point where this line pierces the plane of xy, $z = 0$ substituting therefore this value of z, in the above equations, we obtain for the coordinates, x, y, of the same point, $x = \alpha$, $y = \beta$.

At the point where the line pierces the plane of xz, $y = 0$; putting, therefore, this value of y, in the second equation, and substituting the resulting expression for z in the first, we have for the coordinates, x, z, of this point $x = \frac{b\alpha - a\beta}{b}$, $z = -\frac{\beta}{b}$ and lastly at the point where the line pierces the plane of yz, $x = 0$, putting, therefore, this value x, in the first equation and substituting the resulting value of z, in the second, we obtain for the coordinates, y, z, of the same point

$$y = \frac{a\beta - b\alpha}{a},\ z = -\frac{\alpha}{a}.$$

PROBLEM II.

(176.) To find the equations of a straight line passing through a given point.

Let x', y', z' be the coordinates of the given point, and let the equations of the line sought be

$$\left.\begin{aligned} x &= az + \alpha \\ y &= bz + \beta \end{aligned}\right\} \dots (1),$$

then we must have the conditions

$$\begin{aligned} x' &= az' + \alpha \\ y' &= bz' + \beta, \end{aligned}$$

which give for α and β the values $\alpha = x' - az'$ and $\beta = y' - bz'$;

hence, by substitution, in equation (1), we have

$$\left.\begin{array}{l} x - x' = a(z - z') \\ y - y' = b(z - z') \end{array}\right\} \dots (2),$$

which are the equations sought, and characterize every straight line that can be drawn through the point (x', y', z'). If the given point be the origin, then $x' = 0$, $y' = 0$, $z' = 0$, and the equations of a line passing through it are therefore $x = az$, $y = bz$.

PROBLEM III.

(177.) To find the equations of a straight line which passes through two given points.

Let the two given points be (x', y', z'), and (x'', y'', z''), then the equations of the line passing through one of the points (x'', y'', z'') are

$$\left.\begin{array}{l} x - x'' = a(z - z'') \\ y - y'' = b(z - z'') \end{array}\right\} \dots (1),$$

and, in order that this line may pass also through the other point (x', y', z'), there must exist the conditions

$$x' - x'' = a(z' - z'') \text{ and } y' - y'' = b(z' - z''),$$

which determine for a and b the values $a = \dfrac{x' - x''}{z' - z''}$, $b = \dfrac{y' - y''}{z' - z''}$.

These values of a and b being substituted in equations (1), or in equations (2), last problem, either of which characterizes a line through one of the points, will furnish the equations sought, which are, therefore, indifferently either

$$\left.\begin{array}{l} x - x'' = \dfrac{x' - x''}{z' - z''}(z - z'') \\ y - y'' = \dfrac{y' - y''}{z' - z''}(z - z'') \end{array}\right\}$$

$$\left.\begin{array}{l} x - x' = \dfrac{x' - x''}{z' - z''}(z - z') \\ y - y' = \dfrac{y' - y''}{z' - z''}(z - z') \end{array}\right\}$$

If one of the points $(x'', y'', z'',)$ be the origin, then the first pair of equations become $x = \dfrac{x'}{z'}z$, $y = \dfrac{y'}{z'}z$, which remain the same, whether the other point be (x', y', z'), or $(-x', -y', -z',)$ so that if $(x', y', z',)$ be a point on a straight line passing through the origin, $(-x', -y', -z')$ will also be a point on the line.

PROBLEM IV.

(178.) To find the equations of the straight line which passes through a given point, and is parallel to a given line.

Let (x', y', z') be the given point, and let the equations of the given straight line be

$$\left.\begin{array}{l} x = a'z + \alpha' \\ y = b'z + \beta' \end{array}\right\}$$

Then the equations of any line passing through the given point are $\left.\begin{array}{l}x-x'=a(z-z')\\ y-y'=b(z-z')\end{array}\right\}$ and, in order that this line may be parallel to the former, its projections on the vertical planes must be parallel to the projections of the former line, in other words, they must cut the axis of z at the same angles, so that we must have $a=a'$, $b=b'$, therefore the equations required are $\left.\begin{array}{l}x-x'=a'(z-z')\\ y-y'=b'(z-z')\end{array}\right\}$

PROBLEM V.

(179.) To determine the conditions requisite for the intersection of two straight lines in space, and to find the coordinates of the point of intersection.

If two straight lines of which the equations are

$$\left.\begin{array}{l}x=az+\alpha\\ y=bz+\beta\end{array}\right\} \text{ and } \left.\begin{array}{l}x=a'z+\alpha'\\ y=b'z+\beta'\end{array}\right\}$$

intersect, the coordinates of the point of intersection will be the same for both lines; hence, in order to discover what relation must exist among the constants in these equations in this case, we must eliminate the variables, and we obtain, first by subtraction, $(a-a')z+\alpha-\alpha'=0$, $(b-b')z+\beta-\beta'=0$, and then by division, $z=\dfrac{\alpha'-\alpha}{a-a'}=\dfrac{\beta'-\beta}{b-b'}$, hence the relation among the constants, when the lines intersect, is fixed by the equation $(\alpha'-\alpha)(b-b')=(\beta'-\beta)(a-a')$. For the coordinates of the point of intersection we have, by substituting the value of z, just deduced, in the expressions for x and y, $x=\dfrac{a\alpha'-a'\alpha}{a-a'}$, $y=\dfrac{b\beta'-b'\beta}{b-b'}$, $z=\dfrac{\alpha'-\alpha}{a-a'}$.

If $a=a'$, and $b=b'$, these expressions for the coordinates of the point of intersection become infinite, therefore, this point being infinitely distant, the proposed lines must be parallel.

PROBLEM VI.

(180.) To find the analytical expression for the distance between two given points in space.

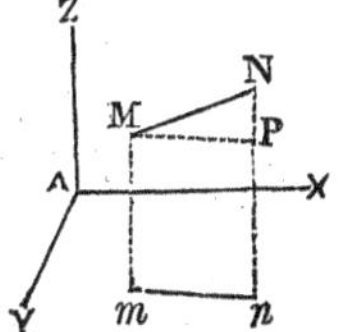

Let M and N be the given points, their coordinates being respectively $x', y'\ z'$, and x'', y'', z'', then, if the points M, N, be projected on the plane of xy, the coordinates x, y, of the projections m, n, will be the same as those of the proposed points; hence for the distance mn we have the expression (14) $mn^2=(x'-x'')^2+(y'-y'')^2$.

Now, if MP be drawn parallel to mn, NPM will be a right angle; hence $\text{MN}^2=mn^2+\text{PN}^2=mn^2+(\text{N}n-\text{M}n)^2$, that is, calling MN, D, we have $\text{D}=\sqrt{\{(x'-x'')^2+(y'-y'')^2+(z'-z'')^2\}}$.

If one of the points, as (x'', y'', z''), be the origin, then

$$D = \sqrt{\{(x'^2 + y'^2 + z'^2\}}.$$

This shows that, in a right angled parallelopiped, the square of the diagonal is equal to the sum of the squares of the three edges.

PROBLEM VII.

(181.) To find the relation which exists among the angles which any straight line makes with the axes of coordinates.

Parallel to any proposed line draw a line from the origin, and make its length, D, equal to the radius of the tables, or 1; then if from its extremity parallels be drawn to the three axes terminating in the planes, these parallels will obviously be the cosines of the respective angles which they form with D, or, which is the same thing, they will be the cosines of the angles that D forms with the axes. But the same parallels are the coordinates of the point from which they are drawn; hence we have, by substituting, in the expression for D, (last prob.) cos. α for x, cos. β for y, and cos. γ for z, the remarkable relation $\cos.^2 \alpha + \cos.^2 \beta + \cos.^2 \gamma = 1$ (1), in which α, β, γ denote the angles which any straight lines in space make the axes of x, y, z.

To determine the values of each cosine, let us suppose that a and b are the tangents of the angles which the projections of the proposed line makes with the axis of z, then the equations of the line D will be $x = az$, $y = bz$, therefore $\cos. \alpha = a \cos. \gamma$, $\cos. \beta = b \cos. \gamma$; substituting these values in 1, we obtain the expressions

$$\cos. \gamma = \frac{1}{\sqrt{\{a^2 + b^2 + 1\}}}, \cos. \beta = \frac{b}{\sqrt{\{a^2 + b^2 + 1\}}}$$

$$\cos. \alpha = \frac{a}{\sqrt{\{a^2 + b^2 + 1\}}}.$$

PROBLEM VIII.

(182.) To find the expression for the angle of intersection of two straight lines in space.

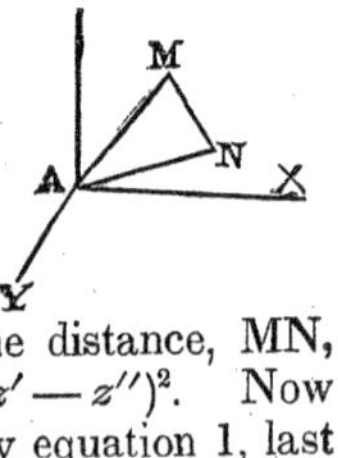

Let the two lines be represented by the equations $\left.\begin{array}{l} x = az + \alpha \\ y = bz + \beta \end{array}\right\}$ and $\left.\begin{array}{l} x = a'z + \alpha' \\ y = b'z + \beta' \end{array}\right\}$ and, parallel to them, draw, from the origin, the two lines, AM, AN, then we have to determine the angle MAN Make AM, AN, each equal to the radius, 1, of the tables, then, calling the coordinates of M, x', y', z', and those of N, x'', y'', z'', we have for the distance, MN, the expression $D^2 = (x' - x'')^2 + (y' - y'')^2 + (z' - z'')^2$. Now because AM, AN, are each equal to 1, we have, by equation 1, last prob. the conditions $x'^2 + y'^2 + z'^2 = 1$, and $x''^2 + y''^2 + z''^2 = 1$; therefore, substituting these values in the developement of the expression for D^2, it becomes $D^2 = 2 - 2(x'x'' + y'y'' + z'z'')$. By means of this expression, we arrive immediately at the expression for the cosine of the angle MAN, which we shall call V for, by trigonometry, $\cos. V = \frac{AM^2 + AN^2 - MN^2}{2AM \cdot AN} = \frac{2 - D^2}{2}$ which, by substituting the above value for D^2, becomes $\cos. V = x'x'' + y'y'' + z'z''$, and this, by last prop. is the same as $\cos. V = \cos. \alpha \cos. \alpha' + \cos. \beta \cos. \beta' + \cos. \gamma \cos. \gamma'$ (1), where α, β, γ, denote the angles which AM make with the axes of x, y, z, and α', β', γ', denote the angles which AN make with the same axes.

By substituting for the cosines of these angles their values in terms of a, b, and a', b', as given in last proposition, the expression (1) takes the form $\cos. V = \frac{aa' + bb' + 1}{\sqrt{(a^2 + b^2 + 1)(a'^2 + b'^2 + 1)}}$.

If the proposed lines are perpendicular to each other, that is, if $\cos. V = 0$, the numerator of this fraction must be 0, that is, we must have the condition $aa' + bb' + 1 = 0$.

It must be remarked that two straight lines in space may be inclined to each other without intersecting, although this is impossible when both are in the same plane; and the angle of inclination is always measured by that included by two parallels to them, drawn from one point; so that the foregoing expressions for V, the inclination of two straight lines in space, apply, whether they actually intersect or not.

It is also important to observe that the results in this and in the preceding problem do not preserve the same form, when the axes of coordinates are oblique, since the expression for D, which enters into these results, becomes obviously more complicated, when the planes are not rectangular. In all the other problems, the inclination of the axes will not affect the form of the results.

CHAPTER II.

ON THE PLANE.

(183.) If a straight line move in a direction parallel to itself along another straight line, given in position, the surface generated will be a *plane*.

The generating line is sometimes called the *generatrix*, and the line along which it moves the *directrix*.

The intersections of any plane with the coordinate planes are called its *traces*.

PROBLEM I.

(184.) To find the equation of the plane.

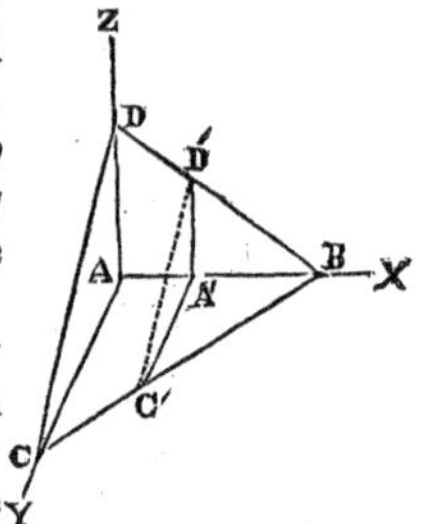

Let BC, BD, DC, be the traces of any proposed plane, which may, therefore, be supposed to be generated by the motion of DC along DB. Let the equation of the trace BD be $z = mx + p$ (1), and the equation of the trace DC, $z = ny + p$ (2), p being $=$ AD, the z of each trace at the origin, A.

Now, since the generating line is in every position D′C′ parallel to DC, the value of z in D′C′ will always be $z = ny + \beta$ (3).

At the point D′, where this line meets the trace BD, $y = 0$, because this point is in the plane of xz, so that the value of z at the same point is, from equation (3), $z = \beta$. But, by equation (1), the value of z at this point is $z = mx + p$; consequently $\beta = mx + p$ (4), x being the same for every point in D′C′ as for D′, because D′C′ is throughout equi-distant from the plane yz. Hence, substituting the expression for β, in equation (3), we shall have the following relation among the coordinates of any point in the proposed plane, viz. $z = mx + ny + p$ (5).

This, therefore, is the equation of the plane.

If the coordinate planes are rectangular, as, indeed, we shall always suppose them, m and n will denote the tangents of the angles which the traces BD, DC, make with the axes of x and y respectively. The symbol p denotes the value of z at the origin; if the proposed plane pass through the origin, then $p = 0$, and the equation is $z = mx + ny$.

The equation (5) is usually put under the form $Ax + By + Cz + D = 0$, being a complete equation of the first degree containing three variables, it comes from equation (5) by multiplying by the arbitrary quantity C, substituting A for mC, B for nC, D for pC, and then transposing all to one side.

PROBLEM II.

(185.) Having given the equation of a plane to determine the equations of its traces.

Let the equation of the plane be $Ax + By + Cz + D = 0$, then for every point in this plane, which is situated likewise in the plane of xy, that is, for every point in the trace on the plane of xy, we must have $z = 0$; hence the equation of this trace is $Ax + By + D = 0$ (1).

In like manner, for every point in the trace on the plane of xz we have $y = 0$; therefore the equation of this trace is $Ax + Cz + D = 0$ (2). And, similarly, the equation of the trace on the plane of yz is $By + Cz + D = 0$ (3).

If, in (1), we put $y=0$, the resulting value of x, viz. $x=-\frac{D}{A}$ will be the distance of the origin from the point where the axis of x pierces the proposed plane; or, putting $x=0$, we have $y=-\frac{D}{B}$ for the distance of the origin from the point where the axis of y pierces the plane. In like manner, for the point where the axis of z pierces the plane, we have $z=-\frac{D}{C}$; hence, when $D=0$ the plane must pass through the origin.

As to the angles which the traces make with the axes of x, y, we have for their trigonometrical tangents, as given by the three preceding equations, the expressions, $-\frac{A}{B}, -\frac{A}{C}, -\frac{B}{C}$.

PROBLEM III.

(186.) To find the equation of the plane which passes through three given points.

Let the three points be (x', y', z'), (x'', y'', z''), and (x''', y''', z'''), then, the form of the equation of the plane being $Ax+By+Cz+D=0$, or $\frac{A}{D}x+\frac{B}{D}y+\frac{C}{D}z=-1$ (1), we have to determine the values of A, B, C, D, so that the following conditions may be fulfilled, viz. $\frac{A}{D}x'+\frac{B}{D}y'+\frac{C}{D}z'=-1$, $\frac{A}{D}x''+\frac{B}{D}y''+\frac{C}{D}z''=-1$,

$$\frac{A}{D}x'''+\frac{B}{D}y'''+\frac{C}{D}z'''=-1.$$

By applying the common equations of algebra (see *Alg.* p. 69), we find for the values of the unknowns, $\frac{A}{D}, \frac{B}{D}, \frac{C}{D}$, the following expressions, viz

$$\frac{A}{D}=\frac{z'(y''-y''')-z''(y'-y''')+z'''(y'-y'')}{x'(y''z'''-y'''z'')-x''(y'z'''-y'''z')+x'''(y'z''-y''z')}$$

$$\frac{B}{D}=\frac{x'(z''-z''')-x''(z'-z''')+x'''(z'-z'')}{x'(y''z'''-y'''x'')-x''(y'z'''-y'''z')+x'''(y'z''-y''z')}$$

$$\frac{C}{D}=\frac{y'(x''-x''')-y''(x'-x''')+y'''(x'-x'')}{x'(y''z'''-y'''z'')-x''(y'z'''-y'''z')+x'''(y'z''-y''z')}$$

hence the equation of the plane, fulfilling the proposed condition, is determined by substituting these values in equation (1).

If the plane is required to pass through but one point (x', y', z'), then the equation of every such plane is $Ax+By+Cz+D=$

$Ax' + By' + Cz' + D$, or, rather, $A(x - x') + B(y - y') + C(z - z') = 0$.

PROBLEM IV.

(187.) To determine the conditions which must subsist in order that a straight line may be parallel to a plane.

Let the equation of the plane be $Ax + By + Cz + D = 0$, and the equations of the straight line $\left.\begin{array}{l} x = az + \alpha \\ y = bz + \beta \end{array}\right\}$

If these expressions for x and y be substituted in the equation of the plane, the resulting value of z will be that of a point common to both straight line and plane. This value is $z = -\dfrac{A\alpha + B\beta + D}{Aa + Bb + C}$ which, substituted for z, in the equations of the straight line, give the other two coordinates of the point where the straight line pierces the plane, on the supposition that they are *not* parallel. If the straight line have an indefinite number of points in common with the plane, that is, if it be wholly in the plane, then the foregoing expression for z, is susceptible of an indefinite number of values, that is, we must have $\dfrac{A\alpha + B\beta + D}{Aa + Bb + C} = \dfrac{0}{0}$ so that the *conditions of coincidence* are $A\alpha + B\beta + D = 0$, $Aa + Bb + C = 0 \ldots$ (1).

But, if the line is merely parallel to the plane, then, by drawing from the origin a line and plane, respectively parallel to the former, there will be coincidence, but then $\alpha = 0$, $\beta = 0$, and $D = 0$; hence the conditions (1) become simply $Aa + Bb + C = 0$ (2), which is the *condition of parallelism.*

Hence, if it be required to draw through a given point (x', y', z'), a straight line parallel to a plane, we have only to substitute, in equations (2), p. 226, any assumed value for one of the coefficients, a, b, and then to determine the other from the condition (2).

PROBLEM V.

(188.) To determine the conditions of parallelism of two planes.

Let the equations of two planes be $Ax + By + Cz + D = 0$, $A'x + B'y + C'z + D' = 0$.

If they intersect these equations, both exist from the line of intersection; hence, eliminating one of the variables, z for example, we have, for the other two coordinates of any point in this line, the relation $(AC' - A'C)\,x + (BC' - B'C)\,y + (DC' - D'C) = 0$ (1).

But (173) the relation between the coordinates x and y of any straight line in space, is the same as the relation between x and y in the projection of that line on the plane of xy, consequently equation (1) is that of the projection of the intersection of the two planes on the plane of xy, and similarly, by eliminating x or y from the proposed equations, we shall obtain the equations of the projection of the same intersection on the plane of yz, or of xz.

When, however, the proposed planes are parallel, the intersection, and, consequently, the projection of it is impossible, so that equation (1) cannot exist for any values of x and y. But so long as the coefficients of x and y, in that equation, exist, the equation itself may be satisfied, for by giving any arbitrary value to one of the variables, that of the other becomes determinable, so that the equation becomes impossible only when the coefficients of x and y become 0, that is, in order that the planes may be parallel, there must exist, the conditions $AC' - A'C = 0$, $BC' - B'C = 0$ (2).

The same conclusion is immediately deriveable from the expressions at (185), for the tangents of the angles which the traces of a plane make with the axes, for as the traces of two planes on either of the coordinate planes must be parallel, if the planes themselves are, it follows, from the expressions referred to. that we must have

$$\frac{A}{B} = \frac{A'}{B'}, \ \frac{B}{C} = \frac{B'}{C'}, \ \frac{A}{C} = \frac{A'}{C'}.$$

The first and second of these conditions, which, indeed, include the third, are the same as the conditions (2).

PROBLEM VI.

(189.) A point being given in space, to draw through it a plane parallel to a given plane.

Let the equation of the given plane be $Ax + By + Cz + D = 0$, then, representing the given point by $(x', y', z',)$, the equation of the required plane will take the form (186)

$$A'(x - x') + B'(y - y') + C'(z - z') = 0.$$

But, since the two planes are parallel, we have, by the conditions of parallelism, $A' = \frac{A}{C} C'$, $B' = \frac{B}{C} C'$; hence the equation becomes, by substitution, $A(x - x') + B(y - y') + C(z - z') = 0$, or $Ax + By + Cz - (Ax' + By' + Cz') = 0$, $= Ax + By + Cz + D' = 0$, where D' is put for $-(Ax' + By' + Cz')$; so that, if two planes are parallel, it is always possible to render the three first coefficients in their equations the same in each. If the point (x', y', z') is the origin, then $D' = 0$, and the equation is $Ax + By + Cz = 0$, which characterizes every plane passing through the origin.

PROBLEM VII.

(190.) To determine the conditions which must subsist, in order that a straight line may be perpendicular to a plane.

Let the equations of the projections of the straight line be $x = az + \alpha$ (1), and $y = bz + \beta$ (2), and the equation of the plane $Ax + By + Cz + D = 0$

Then, since the line is perpendicular to the plane, every plane passing through the line must be also perpendicular to the same plane; hence the planes which project the lines will each be perpendicular both to the proposed plane and to the coordinate plane on which the projection is made. But a plane which is perpendicular to two planes is perpendicular to their intersection; hence the projecting planes are perpendicular to the traces of the proposed plane; but, if a plane is perpendicular to a line, every line in that plane is perpendicular to the same line, and, as the projections of any line are in the projecting planes, it follows, therefore, that if these latter are perpendicular to any traces, so also are the projections. Now for the traces of the proposed plane we have the equations

$$\left.\begin{array}{l} Ax+Cz+D=0 \\ By+Cz+D=0 \end{array}\right\} \text{ or } \left\{\begin{array}{l} x=-(Cz\div A)-D \\ y=-(Cz\div B)-D, \end{array}\right.$$

and we have to express that the lines represented by these equations are respectively perpendicular to those represented by equations (1), (2). This is done by putting (11) $a=A\div C$ and $b=B\div C$ which are the conditions required.

PROBLEM VIII.

(191.) To draw a perpendicular from a given point to a plane, and to determine its length.

Let the plane be represented by the equation $Ax+By+Cz+D=0$ (1), then, if the given point be $(x', y', z';)$ the equations of the required line will take the form $x-x'=a(z-z')$, and $y-y'=b(z-z')$.

But, for this line to be perpendicular to the plane, we must have, by last problem, $a=A\div C$, $b=B\div C$, hence the equations of the proposed line are $x-x'=(A\div C)(z-z')$ and $y-y'=(B\div C)(z-z')$ (2).

If we knew the point where this perpendicular meets the plane, we could at once determine its length, from the expression at (180) for the distance between two given points. Now since at this unknown point the coordinates are the same both for the perpendicular and the plane, we shall determine it by finding what values of x, y, and z will satisfy the equations (1) and (2), when it will remain only to substitute these values in the expression

$$P=\sqrt{\{(x-x')^2+(y-y')^2+(z-z')^2\}} \quad \ldots \quad (3).$$

As, however, the expressions $x-x'$, $y-y'$, $z-z'$, occur both in the equations (2) and (3), it will be best to find the values of these which are common to those equations. For this purpose, assume $Ax'+By'+Cz'+D=D'$, and subtract it from (1), there results $A(x-x')+B(y-y')+C(z-z')+D'=0$. Substituting here the values of $x-x'$, $y-y'$, in (2), we get $z-z'=-\dfrac{CD'}{A^2+B^2+C^2}$

$$\therefore x-x'=-\frac{AD'}{A^2+B^2+C^2},\ y-y'=-\frac{BD'}{A^2+B^2+C^2},\ \therefore P=$$

$$\frac{D'\sqrt{\{A^2+B+C^2\}}}{A^2+B^2+C^2}=\frac{D'}{\sqrt{(A^2+B^2+C^2)}}=\frac{Ax'+By'+Cz'+D}{\sqrt{(A^2+B^2+C^2)}}.$$

If it were required to draw a plane through a given point (x', y', z'), that should be perpendicular to the line, through the same point then the equation to such plane would take the form (186) $A(x-x')+B(y-y')+C(z-z')=0$; and, therefore, from the relations $a=A \div C$, $b=B \div C$; the equation sought must be

$$a(x-x')+b(y-y')+(z-z')=0.$$

PROBLEM IX.

(192.) To determine the inclination of a given straight line to a given plane.

Let the projections of the given line be represented by the equations $x=az+\alpha$ and $y=bz+\beta$, the equation of the given plane being $Ax+By+Cz+D=0$, then, if from any point in the line a perpendicular be drawn to the plane, the angle included between this perpendicular and the other line will be the complement of the inclination of the latter to the plane, therefore the cosine of this angle will be the sine of the required inclination. Now, representing the perpendicular by the equations $x=a'z+\alpha'$, and $y=b'z+\beta'$, we have for the cosine of the included angle the expression (182)

$\cos. V=\dfrac{aa'+bb'+1}{\sqrt{(a^2+b^2+1)(a'^2+b'^2+1)}}$, therefore, substituting for a', b', the values a', $=A \div C$, $b'=B \div C$, and denoting the inclination of the proposed line and plane by I, we have

$\sin. I=\dfrac{Aa+Bb+C}{(a^2+b^2+c^2)(A^2+B^2+C^2)}$. If the line is parallel to the plane, $\sin. I=0$, therefore, as before determined, (187), the condition of parallelism is $Aa+Bb+C=0$.

PROBLEM X.

(193.) To determine the inclination of two given planes.

Let the equations of the given planes be

$Ax+By+Cz+D=0$ . . . . (1), $A'x+B'y+C'z+D'=0$ . . . .(2); then, if to each plane a perpendicular line be drawn, the inclination of these perpendiculars will be the inclination of the planes; hence, representing the perpendiculars by the equations

$\left.\begin{array}{l} x=az+\alpha \\ y=bz+\beta \end{array}\right\} \left.\begin{array}{l} x=a'z+\alpha' \\ y=b'z+\alpha' \end{array}\right\}$ we must have the relations $a=A \div C$, $b=B \div C$, $a'=A' \div C'$, $b'=B' \div C'$, and, therefore, for the angle, V, of inclination sought we have $\cos. V=\dfrac{aa'+bb'+1}{\sqrt{(a^2+b^2+1)(a'^2+b'^2+1)}}$,

$=\dfrac{AA'+BB'+CC'}{\sqrt{(A^2+B^2+C^2)(A'^2+B'^2+C'^2)}}$. If the two planes are parallel, then $\cos. V=1$, and we have the condition $(AA'+BB'+CC')^2 =(A^2+B^2+C^2)(A'^2+B'^2+C'^2)$, which reduces to $(AB'-BA')^2 +(AC'-CA')^2+(BC'-CB')^2=0$.

But the square of any quantity being always positive, the sum of any number of squares can never be 0, unless they themselves are 0; hence the final conditions are $AB' - BA' = 0$, $AC' - CA' = 0$, $BC' - CB' = 0$, as before determined (188).

If the planes are perpendicular to each other, then cos. $V = 0$, so that, in this case, we have the condition $AA' + BB' + CC' = 0$.

If one of the planes, the second, for instance, coincide with one of the coordinate planes, as the plane of xy, then, in equation (2), $z = 0$, and x and y may be any values whatever, consequently that equation cannot subsist, unless $A' = 0$, $B' = 0$, and $C' = 0$; hence, substituting these values in the expression for cos. V, we have

$\cos. V' = \dfrac{C}{\sqrt{\{A^2 + B^2 + C^2\}}}$ the inclination of the plane (1) to the plane of xy.

In like manner, $\cos. V'' = \dfrac{B}{\sqrt{\{A^2 + B^2 + C^2\}}}$ the inclination to the plane of xz and $\cos. V''' = \dfrac{A}{\sqrt{\{A^2 + B^2 + C^2\}}}$ the inclination to the plane of yz.

By adding the squares of these three last equations together, we obtain the relation $\cos.^2 V' + \cos.^2 V'' + \cos.^2 V''' = 1$, so that the relation (181) of the inclinations of a line to the coordinate axes is analogous to that of the inclinations of a plane to the coordinate planes.

The expressions for the inclinations of the second plane (2) to the coordinate planes will be obtained by accenting the letters A, B, C, in those just deduced, and if we represent them by cos. U', cos. U'', and cos. U''', the expression for cos. V will be the same as

$\cos. V = \cos. V' \cos. U' + \cos. V'' \cos. U'' + \cos. V''' \cos. U'''$

which result is also analogous to that (182).

When the angle V is right, we must have the condition

$$\cos. V' \cos. U' + \cos. V'' \cos. U'' + \cos. V''' \cos. U''' = 0.$$

SCHOLIUM.

(194.) The results of the last four problems become much more complicated, when oblique coordinates are employed, instead of rectangular; but of the other problems in this chapter the results preserve the same form, whether the coordinates are rectangular or not.

Before we conclude the present chapter, it will be necessary to prove the converse of the inference in prop. (1), viz. that every equation of the first degree containing three variables, such as $Ax + By + Cz + D = 0$ is the analytical representation of some plane.

For, putting this equation under the form $z = -\dfrac{A}{C} x - \dfrac{B}{C} y - \dfrac{D}{C}$ and then constructing, on two vertical planes, perpendicular to each other, and which may be designated as the planes of xz and of yz, two lines, of which the equations are

$z = -\frac{A}{C}x - \frac{D}{C}$ and $z = -\frac{B}{C}y - \frac{D}{C}$ the lines thus constructed may be regarded as the traces of some plane; hence, finding, by prob. 1, the plane of which these are the traces, we fall upon the equation

$$z = -\frac{A}{C}x - \frac{B}{C}y - \frac{D}{C}, \text{ or } Ax + By + Cz + D = 0.$$

SECTION II.

ON SURFACES OF THE SECOND ORDER.

(195.) A surface is said to be of the *second order*, when it may be analytically represented by an equation of the second degree, containing three variables.

CHAPTER I.

ON THE SPHERE, AND ON CYLINDRICAL AND CONICAL SURFACES.

PROBLEM I.

(196.) To determine the equation of the sphere

Let r represent the radius of a sphere, and α, β, γ, the coordinates of its centre; let, also, $(x, y, z,)$ denote any point on the surface of the sphere. Then, r being the distance between the points (α, β, γ) and (x, y, z), we have (180), $(x - \alpha)^2 + (y - \beta)^2 + (z - \gamma)^2 = r^2$, or, by developing $x^2 + y^2 + z^2 - 2\alpha x - 2\beta y - 2\gamma z + \alpha^2 + \beta^2 + \gamma^2 = r^2$, *the general equation of the sphere*, when related to rectangular axes.

If the origin is on the surface of the sphere, then $\alpha^2 + \beta^2 + \gamma^2 = 0$, and, therefore, the equation become $x^2 + y^2 + z^2 - 2\alpha x - 2\beta y + 2\gamma z = 0$.

If the origin is at the centre, then, the coordinates of the centre being each 0, the equation is $x^2 + y^2 + z^2 = r^2$. If one of the coordinate planes, as the plane of xy, passes through the centre, then $\gamma = 0$, and the equation is $(x - \alpha)^2 + (y - \beta)^2 + z^2 = r^2$; and, if one of the axes, as the axis of x, pass through the centre, then $\beta = 0$, and $\gamma = 0$, and the equation is $(x - \alpha)^2 + y^2 + z^2 = r^2$.

PROBLEM II.

(197.) To determine the intersection of a sphere with a plane.

Let p represent the distance of the intersecting plane from the centre of the sphere, and constitute three coordinate planes, originating at the centre, one of which, as the plane of xy, may be parallel to the cutting plane. Then every point in the intersecting plane will be given by the equation $z = p$, and, consequently, by the equation in last problem, all the points on the surface of the sphere, which are also common to the plane, must be given by the equation $x^2 + y^2 = r^2 - p^2$, which represents a circle; this is, therefore, the intersection.

PROBLEM III.

(198.) To determine the equation of the tangent plane passing through a given point on the surface of the sphere.

Let the given point be (x', y', z'), then the equation of a plane passing through it is $A(x - x') + B(y - y') + C(z - z') = 0 \ldots (1)$, and, since the same point is on the surface of a sphere, we must have $(x' - \alpha)^2 + (y' - \beta)^2 + (z' - \gamma)^2 = r^2 \ldots (2)$. Now, for the radius of this sphere, that is, for the line passing through the points $(\alpha, \beta, \gamma,)$ and $(x', y', z',)$ we have (177) the equations $x - x' = \frac{x' - \alpha}{z' - \beta}(z - z')$, and $y - y' = \frac{y' - \beta}{z' - \gamma}(z - z')$, and it remains to express that this line is perpendicular to the plane (1). By (190) the relations of perpendicularity are $A = aC$, $B = bC$, that is $A = \frac{x' - \alpha}{z' - \gamma}C$, $B = \frac{y' - \beta}{z' - \gamma}C$. Hence, substituting these values in (1), and dividing by C, we have for the tangent plane sought the equation

$(x' - \alpha)(x - x') + (y' - \beta)(y - y') + (z' - \gamma)(z - z') = 0 \ldots (3)$.

As this equation must exist in conjunction with (2), which is the same as $(x' - \alpha)(x' - \alpha) + (y' - \beta)(y' - \beta) + (z' - \gamma)(z' - \gamma) = r^2$, we having, by adding it to (3), the equation

$$(x' - \alpha)(x - \alpha) + (y' - \beta)(y - \beta) + (z' - \gamma)(z - \gamma) = r^2,$$

which alone characterizes a plane touching a sphere, of which the radius is r, and the centre $(\alpha, \beta, \gamma,)$ in the point (x', y', z').

If the origin is at the centre, then $\alpha = 0$, $\beta = 0$, and $\gamma = 0$, and the equation of the tangent plane is $x'x + y'y + z'z = r^2$.

Cylindrical Surfaces.

(199.) The name *cylindrical surface* is given to every surface which can be generated by a straight line moving parallel to itself, and, at the same time, describing with its extremity a curve line.

The curve described by the extremity of the generating line, is called the *directrix*, and, when it is a plane curve, is usually supposed, for simplicity, to be situated in one of the coordinate planes, the plane of xy.

PROBLEM IV.

(200.) To determine the general equation of a cylindrical surface.

Let the equations of the generating line in any position be

$$\left.\begin{array}{l} x = az + \alpha \\ y = bz + \beta \end{array}\right\} \therefore \left\{\begin{array}{l} \alpha = x - az \\ \beta = y - bz \end{array}\right.$$

then, since the line is always parallel to itself, the values of a and b will remain the same for any other position, but α, β, representing the x, y, of the point where the line meets the plane xy, necessarily vary with this point. Now this point is always in the directrix, and, as this is a given curve, the relation between x and y, that is, between α and β, is given. Hence, if in the equation expressing the relation between x and y, for every point in the directrix, that is, if in the equation of the directrix we substitute for x and y the foregoing values of α and β, the result will be the equation of the surface sought.

Thus if the equation of the directrix be represented by the function $F : (x, y) = 0$, that of the cylindrical surface will be $F : (x - az, y - bz) = 0$. For example, if it be required to find the equation of an oblique cylinder of circular base, then, supposing the base to be in the plane of xy, and the origin of the axes to be at the centre, the equation of the base, or of the directrix, will be $x^2 + y^2 = r^2$; therefore, substituting in this, $x - az$ for x, and $y - bz$ for y, we have for the equation sought $(x - az)^2 + (y - bz)^2 = r^2$. If the base had been an ellipse, characterized by the equation $A^2 y^2 + B^2 x^2 = A^2 B^2$, then the equation of the cylinder would have been $A^2 (y - bz)^2 + B^2 (x - az)^2 = A^2 B^2$.

If the cylinder is right instead of oblique, then $a = 0$, and $b = 0$, and the equation of the cylinder becomes then the same as the equation of the directrix, observing, however, that the equation of the directrix is always supposed to be accompanied by the condition $z = 0$, because the curve is considered as wholly in the plane of xy. But no such condition accompanies the equation of the right cylinder, on the contrary z may be taken of any value whatever, so that, while the equations $x^2 + y^2 = r^2$, $z = 0$, represent a circle on the plane of xy, the equations $x^2 + y^2 = r^2$, $z = \frac{0}{0}$ represent the right cylinder, having that circle for its base.

Conical Surfaces.

(201.) A *conical surface* is that generated by a straight line which constantly passes through the same point in space, and describes with its extremity a curve line.

The given point is called the *vertex*, or, sometimes, the *centre* of the conical surface, and the curve line, described by the generating line, is the directrix, which, when a plane curve, is usually supposed to be situated in the plane of xy.

From its mode of generation, it is obvious that a conical surface consists of two portions, united to each other by the vertex, which is the only point common to each. These two portions are called *sheets*, and thus the conical surface is said to be composed of two sheets.

PROBLEM V.

(202.) To determine the general equation of a conical surface.

Let $(x', y', z',)$ represent the vertex, or centre of the surface; then, since the generating line always passes through this point, its equations in any position will be (176)

$$\left.\begin{array}{l} x - x' = a(z - z') \\ y - y' = b(z - z') \end{array}\right\} \text{ or } \left\{\begin{array}{l} x = az + (x' - az') \\ y = bz + (y' - bz') \end{array}\right.$$

$\therefore a = \dfrac{x - x'}{z - z'}$, $b = \dfrac{y - y'}{z - z'}$. Now here, as in the preceding problem, $(x' - az')$, $(y' - bz')$, is the x, y, of the point were the generating line pierces the plane of xy, which point is, therefore, always in the directrix; but the xy of every such point is given by the equation of the directrix; hence, substituting, in the equation of the directrix, the values $(x - az')$, and $(y' - bz')$, for x and y, we shall obtain the equation of the surface. Thus, if the equation of the directrix be represented by the function $F : (x, y) = 0$, that of the conical surface will be $F : (x' - az', y' - bz') = 0$, where a and b involve the variable coordinates. By substituting for a and b their values above, the equation is $F : \left(\dfrac{x'z - xz'}{z - z'}, \dfrac{y'z - yz'}{z - z'}\right) = 0$.

Let it be required to find the equation of an oblique cone of circular base. Suppose the base to be situated in the plane of xy, and the axes to originate at its centre, then the equation of the base, or of the directrix, is $x^2 + y^2 = r^2$. Substituting, in this equation, $\dfrac{x'z - xz'}{z - z'}$ and $\dfrac{y'z - yz'}{z - z'}$ for x and y we have for the equation sought $(x'z - xz')^2 + (y'z - yz')^2 = r^2(z - z')^2$. If the cone is *right*, that is, if the axis of the cone coincides with the axis of z, then $x' = 0$, $y' = 0$, and this equation becomes $z'^2 x^2 + z'^2 y^2 = r^2(z - z')^2$.

(203.) Let it now be required to find the equation of a right cone, having an elliptical base.

Assuming, as before, the centre of the base for the origin, the equation of the directrix will be $A^2y^2 + B^2x^2 = A^2B^2$, and as the vertex is in the axis of z, we have $x' = 0$, and $y' = 0$; hence, instead of x and y, in the equation of the directrix, we must substitute $-\dfrac{xz'}{z - z'}$ and $-\dfrac{yz'}{z - z'}$ and we have for the equation of the surface $A^2y^2 + B^2x^2 = \left(\dfrac{z - z'}{z'}\right)^2 A^2B^2$. If we put Z for $z - z'$, m for $\dfrac{A}{z'}$ and n for $\dfrac{B}{z'}$, the equation of the elliptic right cone takes this simple form, $m^2y^2 + n^2x^2 = m^2n^2Z^2$.

(204.) Lastly, let it be required to find the equation of an oblique circular cone when the origin of the axes is not at the centre.

Here the equation of the directrix is $(x-\alpha)^2+(y-\beta)^2=r^2$, in which, if we substitute for x and y the values $\frac{x'z-xz'}{z-z'}$, $\frac{y'z-yz'}{z-z'}$ we have for the equation of the surface.

$$[x'z-xz'-\alpha(z-z')]^2+[y'z-yz'-\beta(z-z')]^2=r^2(z-z')^2.$$

If the origin of the axes be on the circumference of the base, and one of them, as the axis of x, pass through the centre, then $\alpha=r$, and $\beta=0$, and the equation of the directrix is $x^2+y^2=2rx$; therefore the equation of the conical surface is

$$(x'z-xz')^2+(y'z-yz')^2=2r(z-z')(x'z-xz').$$

CHAPTER II.

ON SURFACES OF REVOLUTION.

(205.) Every curve surface generated by the revolution of a curve round a fixed axis is called a *surface of revolution.*

Hence the characteristic of surfaces of revolution is this, viz. that every section made by a plane perpendicular to the fixed axis is a circle, whose centre is in that axis.

The equations to the several surfaces of revolution will take the most simple and commodious form, by supposing the fixed axis (sometimes called the axis of revolution) to coincide with one of the coordinate axes, as the axis of z. We shall here suppose this coincidence of the axis of revolution with the axis of z.

PROBLEM I.

(206.) To determine the general equation of a surface of revolution.

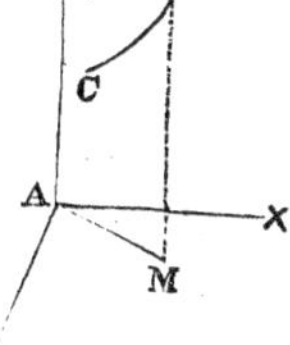

Let DC be any position of the generating curve, then, drawing DM, parallel to AZ, and joining AM, we shall have AM, MD, equal to the coordinates of the point D, as given by the equation of the plane curve, DC. Put AM $=r$, and MD $=z$, then, since in the equation of the curve one of the variables must be a function, F, of the other, we have $z=\mathrm{F}:r$. Now r is always equal to the radius of the circle, described by the point D round the axis AZ, and it is therefore related to the x and y of that point by the equation $r=\sqrt{\{x^2+y^2\}}$; hence the relation of the coordinates, x, y, z, of every point, D, in the surface of revolution, is given by the equation $z=\mathrm{F}:\sqrt{(x^2+y^2)}$; this, then, is the general equation of a surface of revolution.

(207.) As a first example, let it be required to find the equation of the surface generated by the revolution of any straight line about the axis of z. Here the equation of the generating line in any position, when referred to the axis of z as axis of ordinates, and a perpendicular to it from the origin, as axis of abscissas, is $z = \mathrm{F} : r = ar + b$, therefore, substituting $\sqrt{\{x^2 + y^2\}}$ for r, we have for the equation of the surface $(z - b)^2 = a^2 (x^2 + y^2)$. Now, b is the ordinate, z', of the generating line at the origin, and the tangent a is the same as $-z' \div r$; (r is obviously the abscissa of the point where the line cuts the axis of abscissas,) hence, by substitution, the equation becomes $(z - z')^2 r^2 = z'^2 x^2 + z'^2 y^2$, which agrees with the equation at prob. (5), as it ought, for the surface here considered is obviously that of a right cone.

(208.) Let it now be required to find the equation of the surface described by the revolution of an ellipse about one of its axes.

1. Let the minor axis be the fixed axis, coinciding with the axis of z, then the equation of the generating curve will be $A^2z^2 + B^2r^2 = A^2B^2$; hence, substituting $x^2 + y^2$ for r^2, we have $A^2z^2 + B^2 (x^2 + y^2) = A^2B^2$, the equation of the *ellipsoid of revolution.*

3. If the revolution be about the major axis, then the generating curve is represented by the equation $B^2z^2 + A^2r^2 = A^2B^2$, and the surface by the equation $B^2z^2 + A^2 (x^2 + y^2) = A^2B^2$. If $A = B$, we have $z^2 + x^2 + y^2 = A^2$ for the equation of a sphere, as before found.

The ellipsoid of revolution is generally called *spheroid*; a *prolate spheroid* when the revolution is about the major axis, and an *oblate spheroid* when the revolution is about the minor axis.

(209.) Let the surface be described by the revolution of an hyperbola about one of its axes.

Suppose, first, that the second or conjugate axis of the hyperbola is that which is fixed, then the equation of the generating curve is $A^2z^2 - B^2r^2 = - A^2B^2$, and, putting $x^2 + y^2$ for r^2, it becomes $B^2(x^2 + y^2) - A^2z^2 = A^2B^2$, the equation of *hyperboloid of revolution of a single sheet.*

Suppose, secondly, that the revolution is about the transverse axis, then it is obvious that the surface generated will consist of two sheets. The equation of the generating curve will be $B^2z^2 - A^2r^2 = A^2B^2$, and that of the surface, $B^2z^2 - A^2 (x^2 + y^2) = A^2B^2$, the equation of the *hyperboloid of revolution of two sheets.*

If the asymptotes revolve with the curve, then in each of the preceding cases there will be generated a conical surface of two sheets, asymptotic to the hyperboloid.

(210.) Let now the generating curve be a parabola revolving about its principal diameter, then the equation of the generating curve is $r^2 = pz$; and, consequently, for the surface we have $x^2 + y^2 = pz$, the equation of *the paraboloid of revolution*, (if the curve revolve about the other axis, the surface generated will be of the fourth order.)

CHAPTER III.

ON SURFACES OF THE SECOND ORDER IN GENERAL.

The equations which have just been shown to characterize surfaces of revolution are obviously only so many particular forms of the more general equations $Lx^2 + My^2 + Nz^2 = P$, (1), $Lx^2 + My^2 = Qz$, (2).

The first of these comprehends the ellipsoid and hyperboloid of revolution, and the second contains the paraboloid of revolution.

We here propose, by discussing these equations, to ascertain in general the nature of the characteristics of the two classes of surfaces to which the preceding belong, leaving it to be shown in the next chapter that these two classes, together with the cylinder, comprehend all the surfaces of the second order

(212.) Before we proceed to the discussion of the equations (1) and (2), we shall remark that the surfaces which they represent naturally divide themselves into two distinct classes; those which have a centre and those which have not. The former class are represented by equation (1) and the latter by equation (2). This may be readily shown; thus: Let (x', y', z'), be any point on a surface, represented by equation (1), and from this point let there be drawn a straight line through the origin, then we know (177) that $(-x', -y', -z')$, will be also a point on this line; but the same is likewise a point on the surface, for the equation (1) remains the same, whether the coordinates x, y, z, be positive or negative. Now these points are at the same distance from the origin, viz. $D = \sqrt{\{x'^2 + y'^2 + z'^2\}}$, therefore every straight line drawn through the origin, and terminating in the surface, is bisected at the origin, which point is, therefore, the centre of the surface.

Equation (2) cannot represent any surface which has a centre; for if it could, the origin of the rectangular axes might be removed to that centre. If the z of this proposed centre be c, then the z of any point in the surface would be $z + c$, so that the equation (2), when thus transformed, would still have a term containing only the first power of z; hence, if through this new origin a straight line from any point (x', y', z'), in the surface be drawn, the point $(-x', -y', -z')$, in the same line, equally distant from the origin, cannot belong to the surface, for we cannot change z into $-z$, in the equation of the surface, without producing a change in the sign of the term involving z.

(213.) If equation (1) be solved for x, we find two values numerically equal, but of contrary signs, so that the plane of yz divides into two equal parts every chord drawn parallel to the axis of x. What has been said of the plane of yz equally applies to the planes of xz, and of xy; viz. each of these bisects all the chords drawn parallel to the intersection of the other two. Any three planes, each possessing this property of bisecting every chord drawn parallel to the intersection of the other two, are called a system of *diametral planes.* Those which we have just noticed are no other than the rectangular coordinate planes; they are distinguished as the *principal* diametral planes.* The curves traced on these planes by their intersections with the surface, are called the *principal sections;* and the intersections of the same planes, that is, the axes of coordinates, are called the *principal axes of the surface.*

(214.) As to the surfaces represented by equation (2), we find, by proceeding as above, that they have but two principal diametral planes, viz. the planes of xz and of yz; nevertheless the traces of the surface on the three coördinate planes are called the principal sections, and the coordinate axes the principal axes of the surface.

On surfaces which have a Centre.

(215.) We shall now proceed to the discussion of equation (1), which it will be convenient to consider under each of the three following forms, viz.

$$Lx^2+My^2+Nz^2=P$$
$$Lx^2+My^2-Nz^2=P$$
$$Nz^2-My^2-Lx^2=P$$

which forms agree with those which we have already found to characterize the surfaces of revolution which have a centre.

The Ellipsoid.

Let us first take the form $Lx^2+My^2+Nz^2=P$, which characterizes a surface limited in every direction, for, if any straight line be drawn from the origin, its equations will be $x=az$, $y=bz$.

If these values of x and y be substituted in the equation of the surface, we shall have for the z of the point where the line pierces the surface the expression $z=\sqrt{\dfrac{P}{La^2+Mb^2+N}}$. Now, whatever values be given to a, b, the denominator of this expression can never become 0; hence the value of z, and consequently those of x and y, are real and finite, so that every diameter meets the surface. To determine the principal sections of the surface we must put successively $x=0$, $y=0$, $z=0$, in the proposed equation, and we have

$My^2+Nz^2=P$, the trace on the plane of yz,

$Lx^2+Nz^2=P$ $\qquad xz$,

$Lx^2+My^2=P$ $\qquad xy$

These equations characterize ellipses referred to their principal diameters, these diameters therefore coincide with the principal axes of the surface. If $P=0$, each ellipse will be reduced to a point, viz. the origin of the axes. By supposing P negative, the sections become imaginary, showing that in this case, no surface exists.

Let us now examine the sections parallel to these principal sections, and made by planes, whose respective distances from the principal planes may be represented by $x=\pm\alpha$, $y=\pm\beta$, $z=\pm\gamma$.

* It will be hereafter proved that there can be but one system of rectangular diametral planes.

The equations of these sections will be

$$L\alpha^2 + My^2 + Nz^2 = P, \text{ section parallel to } yz,$$
$$Lx^2 + M\beta^2 + Nz^2 = P \qquad xz,$$
$$Lx^2 + My^2 + N\gamma^2 = P \qquad xy.$$

These equations also represent ellipses referred to their principal diameters; hence their centres must be on the axes of coordinates.

Now, in order that these ellipses may exist, the quantities $P - L\alpha^2$, $P - M\beta^2$, $P - N\gamma^2$ must be positive, for, if such values, positive or negative, be given to α, β, and γ, as to render these expressions 0, then each ellipse is reduced to a point, and if greater values than these be given to α, β, and γ, the sections become impossible. Hence the surface is entirely comprised within six tangent planes, drawn parallel to the coordinate planes, and of which the distances from the origin or centre of the surface are $A = \sqrt{\frac{P}{L}}$, $B = \sqrt{\frac{P}{M}}$, $C = \sqrt{\frac{P}{N}}$, A, B, C, being put for the distances of the centre from the planes, which limit the surface in the directions of x, y, z, respectively; in other words, A, B, C, represent the principal semi-axes of the surface,* which, from the nature of the several sections, is called the *ellipsoid.*

From the foregoing expressions for the principal semi-axes, we get $L = \frac{P}{A^2}$, $M = \frac{P}{B^2}$, $N = \frac{P}{C^2}$, hence the equation of the surface becomes, by substitution, $A^2 B^2 z^2 + A^2 C^2 y^2 + B^2 C^2 x^2 = A^2 B^2 C^2$, or $\frac{x^2}{A^2} + \frac{y^2}{B^2} + \frac{z^2}{C^2} = 1$, *the equation of the ellipsoid*, related to its principal axes. If any two of the semi-axes, A, B, C, be equal, then also two of the coefficients, L, M, N, will be equal; and hence one system of parallel sections must be circles, and therefore the surface will be an ellipsoid of revolution. Thus, if $B = C$, then $M = N$, and we have $A^2 (z^2 + y^2) + B^2 x^2 = A^2 B^2$, or $\frac{x^2}{A^2} + \frac{y^2}{B^2} + \frac{z^2}{B^2} = 1$ for the equation of the *ellipsoid of revolution* about the axis of x. If $A = B = C$, the surface is spherical, having the equation $x^2 + y^2 + z^2 = A^2$.

Hence the varieties of the ellipsoid are the *ellipsoid of revolution, the sphere, a point,* and *an imaginary surface.*

The Hyperboloid of a single Sheet.

(216.) The second form of the general equation of central surfaces of second order is $Lx^2 + My^2 - Nz^2 = P$.

---

*The principal semi-diameters are also immediately obtained from the proposed equation of the surface; thus, at the point where the axis of x pierces the surface, y and z are 0; hence for this point the equation gives $x = A = \sqrt{(P \div L)}$. n like manner, $y = B = \sqrt{(P \div M)}$, and $z = C = \sqrt{(P \div N)}$.

The traces, or principal sections of the surface here represented, are

$My^2 - Nz^2 = P$, the trace on the plane of yz,
$Lx^2 - Nz^2 = P$ $\quad xz$,
$Lx^2 + My^2 = P$ $\quad xy$.

Of these sections the first two are hyperbolas, and the third an ellipse; and, as each curve is referred by its equation to its principal diameters, these diameters coincide with the principal axes of the surface. If $P = 0$, the hyperbolas each degenerate into a system of two intersecting straight lines, and the ellipse reduces to a point. The surface in this case will be an elliptic cone, as will be seen presently.

For the sections parallel to the principal sections, made by planes whose distances from the origin are $x = \pm \alpha$, $y = \pm \beta$, $z = \pm \gamma$, we have the equations $My^2 - Nz^2 = P - L\alpha^2$, section parallel to yz,

$Lx^2 - Nz^2 = P - M\beta^2$ $\quad xz$,
$Lx^2 + My^2 = P + N\gamma^2$ $\quad xy$.

Here, as before, the two former sections are hyperbolas, and the last an ellipse; and these sections are obviously always possible, however distant the intersecting planes may be from the origin; so that this surface is unlimited in every direction; it does not, however, meet the axis of z, for, putting both x and $y = 0$, in its equation, the resulting value of z is imaginary, viz. $z = \sqrt{-\frac{P}{N}}$. But, for the other principal semi-axes of the surface, the same equation gives $z = \sqrt{\frac{P}{L}}$, $y = \sqrt{\frac{P}{M}}$ calling these latter A, B, and the former $C\sqrt{-1}$, and introducing these terms into the equation of the surface, which, from the nature of its sections, and the continuity of surface, is called the hyperboloid of a single sheet, we have

$$A^2B^2z^2 - A^2C^2y^2 - B^2C^2x^2 = -A^2B^2C^2.$$

or $\frac{x^2}{A^2} + \frac{y^2}{B^2} - \frac{z^2}{C^2} = 1$, *the equation of the hyperboloid of one sheet* related to its principal axes.

It has already been seen, that when $P = 0$, in the proposed equation, the trace of the surface on the plane of xy is merely a point, but every section parallel to this plane on either side of it is an ellipse given by equation $Lx^2 + My^2 = N\gamma^2$; by making successively $x = 0$, and $y = 0$, we find for the semi-axes of the ellipse the values $y = \gamma\sqrt{\frac{N}{M}}$, $x = \gamma\sqrt{\frac{N}{L}}$ these increase with γ, that is, the elliptic sections increase as the intersecting plane recedes from the plane of xy; hence the surface can be no other than the elliptic right cone, having its vertex at the origin.

This conical surface is asymptotic to the hyperboloid, as may be thus proved. Let the hyperboloid and cone be both cut, be a plane, parallel to the plane of xy, at the distance, γ, from the origin, then the equations of the two elliptic sections will be, respectively, $Lx^2 + My^2 = P + N\gamma^2$ and $Lx^2 + My^2 = N\gamma^2$. Now, if these sections have any point in common, the coordinates, x, y, of that point will be the same in each; hence we must have

$P + N\gamma^2 = N\gamma^2 \therefore \frac{P}{N\gamma^2} + 1 = 1$, which is impossible, unless γ be infinite.

If $A = B$, in the equation of the hyperboloid, then $L = M$, and the sections parallel to the plane of xy become circles; hence the equation $C^2(x^2 + y^2) - A^2z^2 = A^2C^2$, or $\frac{x^2}{A^2} + \frac{y^2}{A^2} - \frac{z^2}{C^2} = 1$; characterizes *the hyperboloid of revolution of a single sheet* about the axis of z.

Hence the varieties of the hyperboloid of one sheet are the *hyperboloid of revolution*, and the *conical surface*. It appears from the above that the equation of the elliptic right cone, having its vertex at the origin, is $Lx^2 + My^2 = Nz^2$, or, substituting for the coefficients of this equation their values $L = \frac{P}{A^2}$, $M = \frac{P}{B^2}$, $N = \frac{P}{C^2}$, and dividing by P, it becomes $\frac{x^2}{A^2} + \frac{y^2}{B^2} = \frac{z^2}{C^2}$ or $A^2C^2y^2 + B^2C^2x^2 = A^2B^2z^2$, in which equation A, B, represent the semi-axes of the elliptic section which is at the distance, C, from the vertex. If we put m for $\frac{A}{C}$ and n for $\frac{B}{C}$ the equation becomes $m^2y^2 + n^2x^2 = m^2n^2z^2$, agreeing with the form at (203).

Or, if we put p for $\frac{C}{A}$, and q for $\frac{C}{B}$, the form is $qy^2 + px^2 = z^2$; hence, if it were required to express the equation of an elliptic right cone, of which the section, two feet from the vertex, has for principal semi-diameters the lengths 5 feet and 7 feet; then, since $A = 7$, $B = 5$, and $C = 2$, the equation is $\frac{2}{5}y^2 + \frac{2}{7}x^2 = z^2$.

The Hyperboloid of two Sheets.

(217.) The third species of central surfaces is represented by the equation $Nz^2 - My^2 - Lx^2 = P$.

For the principal sections we have the equations

$Nz^2 - My^2 = P$, the trace on the plane yz,
$Nz^2 - Lx^2 = P$ $\qquad xz$,
$My^2 + Lx^2 = -P$ $\qquad xy$.

Of these sections the first two are hyperbolas, referred to their principal diameters, which, therefore, coincide with the principal axes of the surface. The second section being imaginary, shows that the surface does not meet the plane of xy; hence of the former hyperbolic sections one branch of each hyperbola is situated above, and the other below, the plane of xy; so that the surface consists of *two sheets*. If $P = 0$, the hyperbolas degenerate into a system of straight lines, intersecting at the origin, and the trace on xy is a point. The sections parallel to the traces, and of which the distances from the origin are respectively $x = \pm \alpha$, $y = \pm \beta$, $z = \pm \gamma$, are given by the equations $Nz^2 - My^2 = P + L\alpha^2$, section parallel to yz,

$$Nz^2 - Lx^2 = P + M\beta^2 \qquad xz,$$
$$My^2 + Lx^2 = N\gamma^2 - P \qquad xy.$$

Hence the sections parallel to the planes of yz, and xz, are all hyperbolas, having their centres on the axis of x and y. The sections parallel to the plane of xy are ellipses, provided the distance $\pm \gamma$ of the cutting plane from the origin be not so small as to render $N\gamma^2 - P$ negative, for, if it be, the plane will not meet the surface. If

$$N\gamma^2 - P = 0; \text{ then } +\gamma = \sqrt{\frac{P}{N}}, \text{ and } -\gamma = -\sqrt{\frac{P}{N}}$$

therefore, at these distances, each section reduces to a point, and no part of the surface can be between these tangent planes; so that this value of γ is the vertical semi-axis of the surface; the two horizontal axes are imaginary. The expressions for them given by putting in the proposed equation, first $z = 0$, $y=0$, and then $z = 0$, $x = 0$, are $x = \sqrt{-\frac{P}{L}}$, $y = \sqrt{-\frac{P}{M}}$. Putting $A\sqrt{-1}$ and $B\sqrt{-1}$ for these semi-axes, and C for the former one, and then, as before, introducing these expressions in the original equation, we have

$$A^2B^2z^2 - A^2C^2y^2 - B^2C^2x^2 = A^2B^2C^2 \text{ or } \frac{x^2}{A^2} + \frac{y^2}{B^2} - \frac{z^2}{C^2} = -1$$

the equation of the hyperboloid of two sheets related to its principal axes.

When $P = 0$, in the equation proposed, it may be proved, as in art. (216), that the equation then represents an elliptical conic surface asymptotic to the hyperboloid. If $A = B$, then $L = M$, and the sections parallel to the plane of xy, become circles; hence the equation

$$A^2z^2 - C^2(x^2 + y^2) = A^2C^2, \text{ or } \frac{x^2}{A^2} + \frac{y^2}{A^2} - \frac{z^2}{C^2} = -1$$

represents the *hyperboloid of revolution of two sheets*. Hence this and the *elliptic right cone* are the varieties of the surface.

On Surfaces which have not a Centre.

(218.) We now proceed to examine the second class of surfaces and which are represented by the equation $Lx^2 + My^2 = Qz = 0$.

This equation involves in it the forms $Lx^2 + My^2 \quad Qz$, and $Lx^2 - My^2 = Qz$, or $My^2 - Lx^2 = Qz$. The last two equations represent the same surfaces, the axes of x and y being merely interchanged, so that we need discuss only the first two equations.

The Elliptic Paraboloid.

By putting, in the equation $Lx^2+My^2=Qz$, the successive values $x=0$, $y=0$, $z=0$, we have for the principal sections of the surface the equations $My^2=Qz$, the trace on the plane of yz,

$$Lx^2=Qz \qquad\qquad xz,$$
$$Lx^2+My^2=0 \qquad\qquad xy.$$

Hence the traces on the planes of yz and xz are parabolas, referred to their principal axes, which, therefore, coincide with the axes of the surface. Of these traces the principal diameter of the first coincides with the axis of z, and the second diameter with the axis of y; the principal diameter of the second trace coincides with the axis of z, and the second diameter with the axis of x. The third trace, or that on the plane of xy, is merely a point, viz. the origin of the axes.

For the sections parallel to the traces, and whose distances from the origin are respectively, $x=\pm\alpha$, $y=\pm\beta$, $z=\gamma$, we have the equations $My^2=Qz-L\alpha^2$, section parallel to yz,

$$Lx^2=Qz-M\beta^2 \qquad\qquad xz,$$
$$Lx^2+My^2=N\gamma \qquad\qquad xy.$$

The first two sections are, like the parallel traces, parabolas, whose vertices are not on the axes of the surface. The third section is an ellipse, whose centre is on the axis of z, and principal diameters parallel to the axis of x and y, and it will always be possible, however great the distance, γ, *above* the plane of xy may be; so that the surface has no limit above this plane, below it the sections are impossible. From the nature of its sections, this surface is called the *elliptic paraboloid.* If $L=M$, the elliptic sections become circular, and then the surface is one of revolution about the axis of z; hence the equation $x^2+y^2=pz$, where $p=\frac{Q}{L}$, represents the *elliptic paraboloid of revolution*, which is the only variety of the elliptic paraboloid.

The Hyperbolic Paraboloid.

(219.) The surface represented by the equation $Lx^2-My^2=Qx$ has for its traces the equations

$$My^2=-Qz, \text{ the trace on the plane of } yz,$$
$$Lx^2=Qz, \qquad\qquad xz,$$
$$Lx^2-My^2=0, \qquad\qquad xy.$$

The first two traces are parabolas, and the principal diameter of each coincides with the axis of z, the origin is the common vertex of both parabolas; but that in the plane of zy is *below*, and that in the plane of xz is *above*, the plane of xy. The trace on the plane of xy is merely a system of two straight lines intersecting at the origin, their equations being $x = \pm y\sqrt{\frac{M}{L}}$. For the sections made by the planes $x = \pm\alpha$, $y = \pm\beta$, $z = \pm\gamma$, we have the equations

$$My^2 = L\alpha^2 - Qz, \text{ section parallel to } yz,$$
$$Lx^2 = M\beta^2 + Qx, \qquad xz,$$
$$Lx^2 - My^2 = \pm Q\gamma \qquad xy.$$

The first two sections are parabolas, whose vertices are *not* on the axes, but the principal diameter of each is parallel to the axis of z; that of the first parabola is, however, in the direction of z negative; and that of the second in the direction of z positive. But, let us examine these sections more narrowly, and, first, it may be remarked that the coordinates y, z, of every point in the section parallel to yz are measured from the point $x = \pm\alpha$ of the axis of x; in other words, this point is the origin of the coordinates, y, z, of the section. Let us then remove this origin to the vertex of the parabola, which is done by substituting $z + \frac{L\alpha^2}{Q}$ for z, in the equation, which then becomes $My^2 = Qz$; this equation being the same as that of the trace on the plane of yz, it follows that all the sections parallel to this plane are equal parabolas. Similar remarks apply to the sections parallel to the plane of xz; but in these the vertices of the parabolas are all below the plane of xy, while, in the former, the vertices are above that plane, as the transformation shows. Moreover, the vertices of the former series are all in the plane of xz, and those of the latter all in the plane of yz. The third section, or that parallel to the plane of xy, is an hyperbola related to its principal diameters; its centre is, therefore, on the axis of z. The form of its equation, however, shows that when the section is above the plane of xy, the transverse diameter is parallel to the axis of x, and the conjugate parallel to the axis of y; but, when the section is below the plane of xy, then, on the contrary, the transverse axis is parallel to the axis of y, and the conjugate to the axis of x, and at equal distances, γ and $-\gamma$, above and below the plane of xy, the transverse axis of the one section is the same absolute length as the conjugate axis of the other, and vice versa.

Of all the hyperbolic sections above the plane of xy the vertices are situated on the parabolic trace, on the plane of xz, for, putting $y = 0$, in the equation $Lx^2 - My^2 = Q\gamma$, we have for the corresponding value of x, the semi-transverse axis of the hyperbola, or, which is the same thing, the distance of the vertex from the axis of z, the expression $x^2 = \frac{Q}{M}\gamma$. But, at the same distance, $z = \gamma$, from the plane of xy, there is a point in the parabolic trace, of which the distance from the axis of z is given by the same expression, viz. $z^2 = \frac{Q}{M}\gamma$; hence these two points coincide.

In a similar manner, it may be shown that of all the hyperbolic sections below the plane of xy, the vertices are situated on the parabolic trace on the plane of yz. Having thus seen that all the sections parallel to the plane of yz are parabolas, that these parabolas are all equal, and that their vertices are all on the parabolic trace on the plane of xz, it follows that if the parabola in the plane of yz be moved parallel to itself, its vertex always being in contact with the parabola in the plane of xz, the surface we are now considering will be generated. It will be also generated by keeping fixed the parabola which we have here supposed to move, and moving the other under like restrictions. From the nature of its sections, this surface is called the *hyperbolic paraboloid*. The equation of the asymptotes of any hyperbolic section $Lx^2 - My^2 = \pm Q\gamma$, is known to be (71) $Lx^2 - My^2 = 0$, or $x = \pm y\sqrt{(M \div L)}$. Hence, if these two lines are constructed on the plane of xy, the perpendicular planes, passing through them will be those which contain the asymptotes of all the hyperbolic sections. Now these lines to be constructed on the plane of xy are the very lines into which the hyperbolic section degenerates, when the cutting plane coincides with the plane of xy, as we have already seen by the equations of the traces; therefore planes drawn through these, perpendicular to the plane of xy, continually approach, but never meet the surface, except at their intersections with the plane of xy.

From articles (218) and (219) it appears that the traces on the planes of yz and xz of surfaces which have not a centre are parabolas, whose parameters are

$p = \pm \frac{Q}{M}$, and $p' = \frac{Q}{L}$, $\therefore M = \pm \frac{Q}{p}$, and $L = \frac{Q}{p'}$. Hence, substituting these values in the general equation $Lx^2 + My^2 = Qz$, of these surfaces, and dividing by Q, we have $\frac{x^2}{p'} \pm \frac{y^2}{p} = z$, or $px^2 \pm p'y^2 = pp'z$, for the equation of the paraboloid, which is elliptic, or hyperbolic, according as the upper or lower sign of p' has place.

Tangent Planes to Surfaces of the Second Order.

(220.) If a straight line meet a surface in but one point, it is said to be a *linear* tangent to the surface at that point, and, since an indefinite number of straight lines may be drawn through a given point, there is obviously no limit to the number of linear tangents at that point. The surface which is the locus of these tangents is called the tangent plane at the proposed point, and that it is a plane we shall presently see.

PROBLEM

To find the equation of a tangent plane, drawn through any point on a central surface of the second order.

Let x', y', z', be the coordinates of the proposed point on any surface, included in the general equation, $Lx^2+My^2+Nz^2=P$ . . . (1). Then any linear secant passing through the same point will be represented by the equations $\left.\begin{matrix} x-x'=a\,(z-z') \\ y-y'=b\,(z-z') \end{matrix}\right\}$ (2), in which a and b are quite arbitrary, because the secant may take any direction whatever. From equation (1) subtract $Lx'^2+My'^2+Nz'^2=P$ and there results $L\,(x^2-x'^2)+M\,(y^2-y'^2)+N\,(z^2-z'^2)=0$, or

$$L\frac{x-x'}{z-z'}(x+x')+M\frac{y-y'}{z-z'}(y+y')+N\,(z+z')=0.$$

When each secant becomes a tangent, then, at the point $(x', y', z',)$ of contact, $x=x'$, $y=y'$, and $z=z'$, and the various values of a and b, or, which is the same thing, of $\frac{x-x'}{z-z'}$, and $\frac{y-y'}{z-z'}$, no longer remain arbitrary in equation (2), or in the equation just deduced, but become subject to the relation $L\frac{x-x'}{z-z'}x'+M\frac{y-y'}{z-z'}y'+Nz'=0$ (3).

Now this equation remains the same for every point (x, y, z), in each tangent, and not exclusively for the point of contact with the surface, as is manifest from equation (2), which shows that the values $\frac{x-x'}{z-z'}, \frac{y-y'}{z-z'}$ are constant for every point (x, y, z), on the line represented by that equation. Hence equation (3) represents the surface in which all the lineal tangents through (x', y', z'), are situated; the surface is therefore a plane. By developing this equation, it takes the form $Lx'x+My'y+Nz'z=Lx'^2+My'^2+Nz'^2=P$, or (215) $\frac{x'x}{A^2}+\frac{y'y}{B^2}+\frac{z'z}{C^2}=1$ which is therefore, the *equation of the tangent plane*. It appears, from equation (3), that, if a straight line (2) touch the surface, the relation between the constants, a, b, must be such as to satisfy the equation $Lax'+Mby'+Nz'=0$.

The equations to the *normal*, or straight line drawn from the point of contact perpendicular to the tangent plane, are (191) $x-x'=\frac{Lx'}{Nz'}(z-z')$, $y-y'=\frac{My'}{Nz'}(z-z')$. If $M=N$, then the surface is of revolution about the axis of x, and the last of these equations, which represents the projection of the normal on the plane of zy, reduces to $z'y=y'z$, or $y=\frac{y'}{z'}z$; hence this projection passes through the origin, and, consequently, the normal must cut the axis of x.

In like manner it may be shown that if $L=N$, that is, if the surface revolve about the axis of y, the normal must cut this axis, and the projection on the plane of xy will in a similar way show that, when $L=M$, the normal will cut the axis of z, about which the surface revolves.

PROBLEM.

To find the equation of the tangent plane when the surface has not a centre.

By art. (219) the surfaces which have not a centre may be represented by the equation $px^2 \pm p'y^2 = pp'z$ . . . . (1).

Any linear secant drawn through a point (x', y', z') on this surface will be represented by the equations $\left\{ \begin{array}{l} x - x' = a(z - z') \\ y - y' = b(z - z') \end{array} \right\}$ . . (2). in which a, and b, varying with the direction of the secant, may take any values whatever. From equation (1) take $px'^2 \pm p'y'^2 = pp'z'$, and we have $p(x^2 - x'^2) \pm p'(y^2 - y'^2) = pp'(z - z')$,

or $p\dfrac{x - x'}{z - z'}(x + x') \pm p'\dfrac{y - y'}{z - z'}(y + y') = pp'$. Now, when each secant becomes a tangent then at the point (x', y', z'), of contact, we have $x = x'$, $y = y'$, and $z = z'$, and, therefore, the coefficients a, b, in equation (2), become subject to the condition $2pax' \pm 2p'by' = pp'$ . . . (3), for the values $\dfrac{x - x'}{z - z'}, \dfrac{y - y'}{z - z'}$, remain the same for every point in the same secant, or in the same tangent; hence, substituting these values for a, b, in (3), we have for the equation of the surface, in which all the linear tangents are situated

$2p\dfrac{x - x'}{z - z'}x' \pm 2p'\dfrac{y - y'}{z - z'}y' = pp'$; this represents a plane, which, when developed, becomes $2pxx' \pm 2pyy' = 2pp'z' + pp'(z - z') = pp'(z + z')$, the *equation of the tangent plane.*

Equation (3) expresses the conditions which the constants a, b, must have, in order that the straight line (2) may be a linear tangent to the surface.

On Conjugate Diametral Planes.

(221). We have already defined (213) a system of conjugate diametral planes to be such that each bisects all the chords drawn parallel to the intersection of the other two, and we have shown that such a system exists in every surface included in the general equation $Lx^2 + My^2 + Nz^2 = P$ (1), provided the coordinate axes are *rectangular.* It will be hereafter shown that there are an infinite number of *oblique* axes, to which every such surface may be referred, without altering the form of its equation; and hence we may infer, by imitating the reasoning at (212), that there are an infinite number of systems of oblique diametral planes in central surfaces of the the second order, these diametral planes being no other than the oblique coordinate planes. If then we suppose the equation $L'x^2 + M'y^2 + N'z^2 = P$ (2) to represent the same surface as equation (1), when the coordinates are transformed from rectangular to oblique, we may apply to this equation the same reasoning that we have employed in the discussion of equation (1), and the only difference in the results will be that what before were *rectangular* conjugate diameters, will here be *oblique* conjugates, both as regards the surface itself, and the several intersections. From the form of equation (1) it was shown (215, &c.) that a plane drawn through the extremity of a principal diameter (not imaginary), and parallel to the plane of the other two, was necessarily a tangent plane to the surface; hence, since (2) has the same form as (1), we infer that a plane through the extremity of one conjugate diameter, and parallel to the plane of the other two, touches the surface.

From these remarks and from what has been shown in (215), (216), &c. we may infer that, if A′, B′, C′, represent any system of semi-conjugates belonging to a central surface of the second order, the equation of that surface, referred to them as axes, will be $\frac{x^2}{A'^2} + \frac{y^2}{B'^2} + \frac{z^2}{C'^2} = 1$, and it will be an ellipsoid, if A'^2, B'^2, C'^2, are all positive, an hyperboloid of one sheet if only two of these are positive, and an hyperboloid of two sheets if but one is positive.

SECTION III.

CHAPTER I.

ON THE ORTHOGONAL PROJECTIONS OF PLANE SURFACES.

(222.) A plane figure is said to be orthogonally projected on a plane when each side of it is perpendicularly projected on the same plane.

THEOREM.

(223.) The projection of a plane surface on a plane is equal to the area of that surface multiplied by the cosine of its inclination to the plane of projection.

Since any plane figure may be divided into triangles, it will be sufficient to prove the truth of this theorem in the case of the triangle.

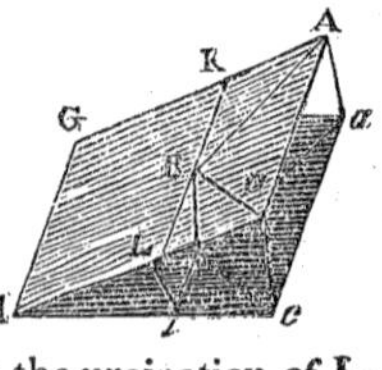

Let, then, ABC be any plane triangle, and let *abc* be the orthogonal projection of it on any plane, H*a*. Produce the plane of ABC to meet the plane of projection in GH and perpendicular to AC draw the two parallels, AG, CH; then, if through B the line KL be drawn parallel to AC, we shall have for the area of the triangle ABC the expression $\triangle ABC = \frac{1}{2}AC \cdot CL$; also, if l be the projection of L, we shall have $\triangle abc = \frac{1}{2}\text{AC} \cdot cl$. Now, if α represent the inclination of the two planes, that is, the angle CHc, we shall have

$$cl = \text{CL}\cos.\alpha, \therefore \triangle abc = \triangle\text{ABC} \cdot \cos.\alpha.$$

THEOREM II.

(224.) The square of the area of any plane figure is equal to the sum of the squares of its projections on three rectangular planes.*

Let S represent any plane surface, and S′, S″, S‴, its three projections on the planes of xy, xz, yz; then, putting α, α', α'', for the several inclinations of the plane of S to the coordinate planes, we have, by last theorem, $S'^2 = S^2\cos.^2\alpha$, $S''^2 = S^2\cos.^2\alpha'$, $S'''^2 = S^2\cos.^2\alpha''$. Now (193) $\cos.^2\alpha + \cos.^2\alpha' + \cos.^2\alpha'' = 1$; hence

$$S^2 = S'^2 + S''^2 + S'''^2.$$

Cor. If the projections of the same surface on 3 other rectangular planes be $S_1{}'$, $S_1{}''$, $S_1{}'''$, then, as before, $S^2 = S_1{}'^2 + S_1{}''^2 + S_1{}'''^2$; consequently $S'^2 + S''^2 + S'''^2 = S_1{}'^2 + S_1{}''^2 + S_1{}'''^2$, that is, *the sum of the squares of the projections is the same for every system of rectangular planes.*

THEOREM III.

(225.) If a surface be projected on three rectangular planes, and then these projections be orthogonally projected on a given plane, the sum of these last projections will be equal to the orthogonal projection of the surface on this given plane.

Let S represent the surface, and S′, S″, S‴, its projections on the rectangular planes. Let also s' be its projection on any other plane inclined at any angle, V, to the plane of S, then $s' = S\cos.V$, also $S' = S\cos.\alpha$, $S'' = S\cos.\alpha'$, $S''' = S\cos.\alpha''$. Now, if we represent the inclinations of the plane of s' to the rectangular planes, that is, to the planes of S′, S″, and S‴, by β, β', β'', we shall have for cos. V the expression (193),

$\cos.V = \cos.\alpha\cos.\beta + \cos.\alpha'\cos.\beta' + \cos.\alpha''\cos.\beta''$; multiplying this by S, we have $S\cos.V = s' = S'\cos.\beta + S''\cos.\beta' + S'''\cos.\beta''$, which expresses the property announced.

Cor. 1. Also, if S, T, U .... represent any number of surfaces situated in different planes, then we have, in a similar manner, the equations

$$s' = S'\cos.\beta + S''\cos.\beta' + S'''\cos.\beta'',$$
$$t' = T'\cos.\beta + T''\cos.\beta' + T'''\cos.\beta'',$$
$$u' = U'\cos.\beta + U''\cos.\beta' + U'''\cos.\beta'',$$

where T′, T″, T‴, and U′, U″, U‴, are the projections of T and U on the rectangular planes, while t', u', are the projections on the plane of s'. If we represent the sum of the projections on the plane of xy by M′, the sum of those on the plane of xz, by M″, and the sum of those on the plane yz by M‴, while the sum of the projections on the

* By the square of an area is meant the square of its numerical value.

fourth plane is represented by m', we have, by adding together, the foregoing equations $m' = M' \cos. \beta + M'' \cos. \beta' + M''' \cos. \beta''$ If we introduce a fifth plane, whose inclinations to the rectangular planes are γ, γ', and γ'', and of which the sum of the projections of S, T, U . . . . thereon is m'', we have $m'' = M' \cos. \gamma + M'' \cos. \gamma' + M''' \cos. \gamma''$. In the same way, for a sixth plane, $m''' = M' \cos. \delta + M'' \cos. \delta' + M''' \cos. \delta''$.

Thus *the sum of the projections of any series of areas on a plane is equal to the sum of the projections formed on the same plane, by first projecting all the figures on a system of rectangular planes, and then projecting these projections on the proposed plane.*

Cor. 2. If the three planes containing the projections m', m'', m''', are also perpendicular, we may consider these as the primitive planes, and we shall then have, conversely,

$$M' = m' \cos. \beta + m'' \cos. \gamma + m''' \cos. \delta$$
$$M'' = m' \cos. \beta' + m'' \cos. \gamma' + m''' \cos. \delta'$$
$$M''' = m' \cos. \beta'' + m'' \cos. \gamma'' + m''' \cos. \delta''.$$

THEOREM IV.

(226.) The same notation being employed, it is required to prove that $m'^2 + m''^2 + m'''^2 = M'^2 + M''^2 + M'''^2$, when both systems of planes are rectangular. By squaring each equation, in the above group, and adding together the results we have

$$\begin{aligned} M'^2+M''^2+M'''^2 = & \, m'^2 (\cos.^2 \beta + \cos.^2 \beta' + \cos.^2 \beta'') \\ & + m''^2 (\cos.^2 \gamma + \cos.^2 \gamma' + \cos.^2 \gamma'') \\ & + m'''^2 (\cos.^2 \delta + \cos.^2 \delta' + \cos.^2 \delta'') \\ & + 2m'm'' (\cos.\beta \cdot \cos.\gamma + \cos.\beta' \cdot \cos.\gamma' + \cos.\beta'' \cdot \cos.\gamma'') \\ & + 2m'm''' (\cos.\beta \cdot \cos.\delta + \cos.\beta' \cdot \cos.\delta' + \cos.\beta'' \cdot \cos.\delta'') \\ & + 2m''m''' (\cos.\gamma \cdot \cos.\delta + \cos.\gamma' \cdot \cos.\delta' + \cos.\gamma'' \cdot \cos.\delta'') \end{aligned}$$

Now, by art. (193), the factors of m'^2, m''^2, and m'''^2, are each equal to 1, and, by the same art. the factors of $2m'm''$, $2m'm'''$, and $2m''m'''$, are each equal to 0; hence this equation is the same as

$$M'^2 + M''^2 + M'''^2 = m'^2 + m''^2 + m'''^2.$$

This proves that, *if any number of plane surfaces, however situated in space, be projected on different systems of rectangular planes, and the projections on each plane be collected into one sum, then the squares of the three sums thus furnished by each system always amount to the same quantity.*

Cor. The expression for the sum of the projections on any one of the planes, as the plane of m', is $m' = \sqrt{\{M'^2 + M''^2 + M'''^2 - m''^2 - m'''^2\}}$, which sum will be the greatest possible, when $m'' = 0$, and $m''' = 0$, for then it becomes $m' = \sqrt{\{M'^2 + M''^2 + M'''^2\}}$.

Now there is nothing contradictory in supposing $m'' = 0$, and $m''' = 0$, for, since the projection of a surface is equal to that surface multiplied by the cosine of its inclination to the plane of projection, the projection must be considered as positive or negative, according as the cosine is positive or negative, or according as the inclination is acute or obtuse; hence the sum of the projections of any number of surfaces on one of the coordinate planes may become 0, on account of the negative projections equalling the positive, and when this is the case also with another coordinate plane, then, as we have just seen, the projections on the third plane amount to a greater sum than they would do under any other circumstances; this plane is, therefore, called *the plane of greatest projection.*

(227.) We may readily determine the direction of this plane, or its position in reference to the rectangular planes of M', M'', M''', by means of the conditions $m''=0$, $m'''=0$ which exist simultaneously with it, for, introducing these values in the group of equations originally employed, we have

$$M' = m' \cos. \beta, M'' = m' \cos. \beta', M''' = m' \cos. \beta'';$$

whence $\cos. \beta = \frac{M'}{m'} = \frac{M'}{\sqrt{\{M'^2 + M''^2 + M'''^2\}}}$; $\cos. \beta' = \frac{M''}{m'} = \frac{M''}{\sqrt{\{M'^2+M''^2+M'''^2\}}}$; $\cos. \beta'' = \frac{M'''}{m'} = \frac{M'''}{\sqrt{\{M'^2+M''^2+M'''^2\}}}$;
in which equations, β, β', β'', denote the inclinations of the plane of greatest projection to the arbitrary rectangular planes of M', M'', M''', and thus the position of this plane in reference to any system of rectangular planes is determinable, when the sums M', M'', M''', of the projections on this system of planes are known. The situation of the plane of greatest projections is not fixed in space, for the projections on any plane being the same as on any parallel plane, it follows that every plane having the requisite inclinations to the rectangular planes possesses the characteristic property of the *principal plane* or plane of greatest projection.

THEOREM V.

(228.) The sum of the projections on any plane equally inclined to the principal plane is constant.

Let T represent the sum of the projections of any number of surfaces inclined at an angle θ to the principal plane, and let ε, ε', ε'', represent its inclinations to the three primitive planes, then (227)
$T = M' \cos. \varepsilon + M'' \cos. \varepsilon' + M''' \cos. \varepsilon''$, but (228) $M' = m' \cos. \beta$, $M'' = m' \cos. \beta'$, $M''' = m' \cos. \beta''$; hence, by substitution $T = m'$ $(\cos. \beta \cos. \varepsilon + \cos. \beta' \cos. \varepsilon'' + \cos. \beta'' \cos. \varepsilon'')$.

Now the expression within the parentheses represents the cosine of the angle θ (206), therefore $T = m' \cos. \theta$, that is
$T = \sqrt{\{M'^2 + M''^2 + M'''^2\}} \cdot \cos. \theta$; so that T is constant, if θ is, and if $\cos. \theta$ increases or diminishes, T will increase or diminish proportionally.

Cor. On every plane perpendicular to the principal plane the sum of the projections is 0, for when $\theta = 90°$, then $T = 0$.

CHAPTER II.

ON THE TRANSFORMATION OF COORDINATES IN SPACE.

(229.) In order to transform the equation of a surface from one system of axes to another, we must first find expressions for the primitive coordinates in terms of the new, and these, substituted in the original equation, will lead to the transformation desired. The most simple of these transformations is that in which the new axes are parallel to the old, where the only change is in the position of the origin. In this case, if a, b, c, denote the coordinates of the new origin, and (x', y', z',) represent any point in the surface, in reference to the old axes, then the coordinates x, y, z, of the same point, in reference to the new, will obviously be $x = a + x'$, $y = b + y'$, $z = c + z'$, which are, therefore, the formulas to be employed, when the origin is merely altered.

If the direction of the axes be altered, then the formulas for substitution are not so readily obtained. We may affirm, however, that the values of the new coordinates must be linear functions of the old, that is, these values must be of the form $x = a + mx' + m'y' + m''z'$ $y = b + nx' + n'y' + n''z'$, and $z = c + px' + p'y' + p''z'$; for these expressions must be such that, when they are substituted in the equation of the plane, the result may not surpass the first degree, which it would, however, do, if either of the above equations surpassed the first degree. Having thus ascertained the form of the required expressions, it remains to determine the constant coefficients a, m, m', m'', &c. If we suppose $x' = 0$, $y' = 0$, $z' = 0$, we shall have for the coordinates of the new origin $x = a$, $y = b$, $z = c$; these, therefore, are easily determined. The remaining constants must depend on mutual inclinations of the two systems of axes; and, as these relations will not be disturbed by supposing the two origins to coincide, we shall, for greater simplicity, consider a, b, c, as absent from the foregoing expressions, then x, y, z, representing the primitive coordinates of any point in space, the new coordinates x', y', z', of the same point will be related to the former, as in the equations

$$x = mx' + m'y' + m''z'$$
$$y = nx' + n'y' + n''z'$$
$$z = px' + p'y' + p''z'.$$

Suppose the point to be situated on the axis of x', then $y' = 0$, $z' = 0$, and, consequently, $x = mx'$, $y = nx'$, $z = px'$,

$$\therefore m = \frac{x}{x'}, n = \frac{y}{x'}, p = \frac{z}{x'}.$$

(230.) 1. *Let the primitive axes be rectangular and the new ones oblique.*

Then for any point in the axis of x', x' will be the hypothenuse, and x the base, of a right-angled triangle, also x' and y will be the hypothenuse and base of a second right-angled triangle, and x', z, will, in like manner, be the hypothenuse and base of a third; hence, calling the angles which the axis of x' makes with the axes of x, of y, and of z, X, Y, and Z, respectively, we have

$$m = \frac{x}{x'} = \cos. X, \; n = \frac{y}{x'} = \cos. Y, \; p = \frac{z}{x'} = \cos. Z.$$

For a point situated on the axis of y', we have

$$x = m'y', \; y = n'y', \; z = p'y',$$

$$\therefore m' = \frac{x}{y'} = \cos. X', \; n' = \frac{y}{y'} = \cos. Y', \; p' = \frac{z}{y'} = \cos. Z',$$

where X′, Y′, Z′, denote the inclinations of the axis of y' to the axes of x, y, and z, respectively. In like manner, for any point on the axis of z' we have $x = m''z'$, $y = n''z'$, $z = p''z'$,

$$\therefore m'' = \frac{x}{z'} = \cos. X'', \; n'' = \frac{y}{z'} = \cos. Y'', \; p'' = \frac{z}{z'} = \cos. Z'',$$

X″, Y″, Z″, denoting the inclinations of the axis of z to the axes of x, y, z, respectively. Hence the expressions for the primitive coordinates in terms of the new are

$$\left.\begin{array}{l} x = x' \cos. X + y' \cos. X' + z' \cos. X'' \\ y = x' \cos. Y + y' \cos. Y' + z' \cos. Y'' \\ z = x' \cos. Z + y' \cos. Z' + z' \cos. Z'' \end{array}\right\} \ldots . (A);$$

the nine angles which enter into these expressions being subject to the conditions (181).

$$\left.\begin{array}{l} \cos.^2 X + \cos.^2 Y + \cos.^2 Z = 1 \\ \cos.^2 X' + \cos.^2 Y' + \cos.^2 Z' = 1 \\ \cos.^2 X'' + \cos.^2 Y'' + \cos.^2 Z'' = 1 \end{array}\right\} \ldots . (1),$$

the values in other respects being arbitrary. But if the angles which the new axes form among themselves are given or fixed, then six of the foregoing angles become dependent on the other three, which are arbitrary, for, let V = the angle $[x', y']$,
U = the angle $[y', z']$,
W = the angle $[z', x']$,
then (162) besides the preceding conditions we must also have

$$\left.\begin{array}{l} \cos. V = \cos. X \cos. X' + \cos. Y \cos. Y' + \cos. Z \cos. Z' \\ \cos. U = \cos. X \cos. X'' + \cos. Y \cos. Y'' + \cos. Z \cos. Z'' \\ \cos. W = \cos. X' \cos. X'' + \cos. Y' \cos. Y'' + \cos. Z' \cos. Z'' \end{array}\right\} (2);$$

we have thus altogether six equations, which are insufficient to determine the nine angles involved in them, but, by giving any arbitrary values not inconsistent with the conditions (1) to three of these, the remaining six are all deducible from the given equations. Hence, that we may be able to fix the positions of the new axes in space, we must know the angles which they make with each other, and three of the angles which they make with the primitive system. If we know only the angles which the new system of axes make with each other, then this system may take any position whatever, in reference to the primitive system.

(231.) 2. *Let both systems be rectangular.*

In this case cos. V = 0, cos. U = 0, cos. W = 0; hence the equations (2) become

$$\left.\begin{aligned}\cos. X\ \cos. X' + \cos. Y\ \cos. Y' + \cos. Z\ \cos. Z' &= 0\\ \cos. X\ \cos. X'' + \cos. Y\ \cos. Y'' + \cos. Z\ \cos. Z'' &= 0\\ \cos. X'\cos. X'' + \cos. Y'\cos. Y'' + \cos. Z'\cos. Z'' &= 0\end{aligned}\right\}\ldots (3),$$

so that three of the angles formed by the new system with the primitive being determined, the remaining six are given by the equations (1) and (3), and thus the constants in (A), the formula of transformation become known. If one of the new axes, as the axis of z', coincide with the primitive axis of z, in case 1, and V be the angle formed by the other two, both of which we shall here suppose situated in the plane of xy, the formulas of transformation become very simple, for since, in this case, cos. U = 0, cos. W = 0, cos. X″ = 0, cos. Y″ = 0, and cos. Z″ = 1, equations (2) give cos. Z = 0, cos. Z′ = 0; hence equations (1) reduce to

$$\left.\begin{aligned}\cos.^2 X + \cos.^2 Y &= 1\\ \cos.^2 X' + \cos.^2 Y' &= 1\end{aligned}\right\} \therefore \left\{\begin{aligned}\cos. Y &= \sin. X\\ \cos. Y' &= \sin. X'\end{aligned}\right.$$

and the formulas (A) are, in this case, $x = x' \cos. X + y' \cos. X'$, $y = x' \sin. X + y' \sin. X'$ (A′), which are the same as those already given at (40), to pass from a system of rectangular axes to any other system situated in the same plane.

(232.) *To pass from rectangular to polar coordinates.*

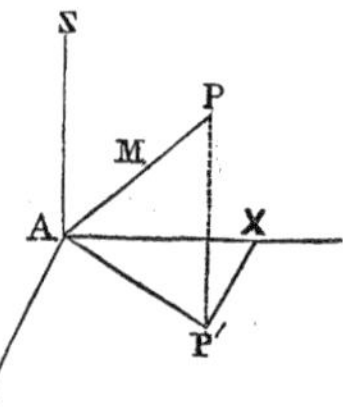

Let P be any point (x, y, z,) in space, and draw AP from the origin, then A may be considered as the pole, and AP = r the radius vector of the point P. Let AM be made equal to unity, or the radius of the tables; then, denoting the inclinations of AP to the axes of x, y, z, by α, β, γ, we shall have (181) for the coordinates of M, the values cos. α, cos. β, cos. γ, consequently, for the coordinates of P, we have the values

$x = r \cos. \alpha$, $y = r \cos. \beta$, $z = r \cos. \gamma$ (B), which must exist in conjunction with the condition $\cos.^2 \alpha + \cos.^2 \beta + \cos.^2 \gamma = 1$. Other formulas for transforming rectangular to polar coordinates may be obtained, in which only two angles enter, viz. the angle PAP′, formed by the radius vector and its projection on the plane of xy, and the angle P′AX, formed by this projection and the axis of x. Call the first of these angles θ, and the second φ, then, if r' represent the projection of r, the right-angled triangles PP′A, P′XA, give

$$r' = r \cos. \theta,\ x = r' \cos. \varphi,\ y = r' \sin. \varphi,\ z = r \sin. \theta;$$

hence $x = r \cos. \theta \cos. \varphi$, $y = r \cos. \theta \sin. \varphi$, $z = r \sin. \theta$ . . . . (C).

Note.—When the new origin does not coincide with the primitive, then the coordinates, a, b, c, of the new origin must be added to the expressions for x, y, z, in the preceding formulas.

With regard to the signs of the trigonometrical quantities, which enter the formulas (C), we must observe that, by supposing φ to vary from 0 to 360°, as in (111), and θ to vary from 0 to $\pm$ 90, while the sign of r always remains positive, the formulas will in all cases correctly mark out the position of the point whose coordinates they express. Hence the radius vector ought always to be considered positive. In like manner the position of the point (x, y, z,) is correctly determined by the signs of the cosines which enter the formulas (B), r being considered positive.

(233.) Of the preceding formulas of transformation, those which enable us to pass from one system of rectangular axes to another are the most important. To effect this transformation the knowledge of three angles only is requisite, as we have already seen (230); but, as nine angles enter the formulas (A), the remaining six must be determined from the equations of conditions (1) and (2). To remedy this inconvenience, new formulas were given by Euler, and afterwards employed by Lagrange and Laplace, which dispensed with the equations of condition, and gave at once the expressions for the new coordinates, in terms of the old, combined with the three given angles, viz. the angle formed by the axis of x' and the trace of the plane of $x'y'$ on the plane of xy, the angle formed by the same trace and the axis of x, and the angle at which the plane of $x'y'$ is inclined to the plane of xy. By means of these three angles, the values of all the nine angles which enter the formulas (A) may be expressed, and thus the equations (1) and (2) be dispensed with. We shall proceed to investigate these formulas.

Let AR be the trace of the plane of $x'y'$ on the plane of xy, and conceive a sphere, of which the centre is A, and radius AT $= 1$, to be pierced by the lines AX, AX′, AR; then the great circles which pass through the points of intersection, Q, T, R, will form a spherical triangle TQR, in which we observe the following circumstances, viz.

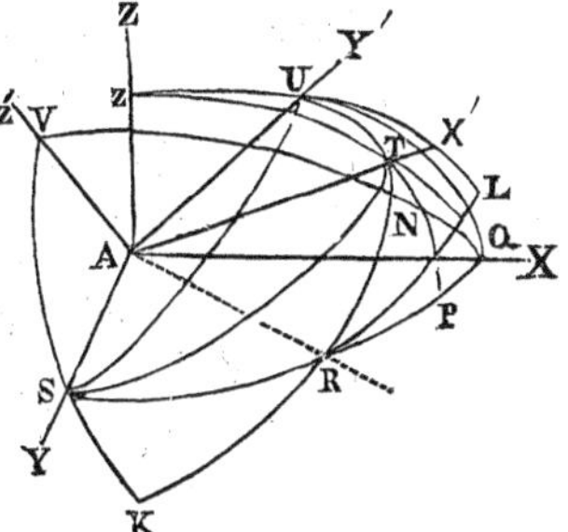

TR = angle formed by the axis of x' and its trace, AR, on xy,
TQ = angle formed by the axis of x' and axis of x,
RQ = angle formed by the trace and axis of x,
$\angle$ R = inclination of the plane of $x'y'$ to the plane of xy.

Hence, putting TR $= \psi$, TQ $=$ X, RQ $= \varphi$, and $\angle$R $= \theta$, we have, by the principles of spherical trigonometry, (*Gregory's Trig. p.* 84.) cos. TQ $=$ cos. X $=$ cos. ψ cos. $\varphi +$ sin. ψ sin. φ cos. θ.

Let now the great circle TS be drawn, then, in the spherical triangle TSR, we have ST $=$ angle formed by the axis of x' and axis of y, TR $= \psi$, SR $= 90° - \varphi$, and $\angle$ R $= 180° - \theta$,

$\therefore$ cos. ST $=$ cos. Y $=$ sin. φ cos. $\psi -$ cos. φ sin. ψ cos. θ.

From the point where AZ pierces the sphere draw a great circle through T, then ZTP $= 90°$, and $\angle$ P $= 90°$; hence the spherical triangle TPR gives TP $=$ complement of the angle formed by the axes of x' and z, $\therefore$ sin. TP $=$ cos. Z $=$ sin. θ sin. ψ; we have thus obtained expressions for three of the cosines which enter the formulas (A).

Again, let U be the point where the axis of y' pierces the sphere, and complete the spherical triangles URQ,* URS, in the first of which we have UQ $=$ angle formed by the axis of y' and axis of x.

UR $= 90° + \psi$, RQ, $= \varphi$, and $\angle$ R $= \theta$,

$\therefore$ cos. UQ $=$ cos. X$'$ $=$ cos. θ cos. ψ sin. $\varphi -$ sin. ψ cos. φ, and in the second triangle, URS, we have US $=$ angle formed by the axis of y', and axis of y, RS $= 90° - \varphi$, UR$=90° + \psi$, and $<$R$=180° - \theta$.

$\therefore$ cos. US $=$ cos. Y$'$ $= -$ cos. ψ cos. φ cos. $\theta -$ sin. ψ sin. φ.

Drawing, now, the quadrantal arc ZUL, the triangle URL, right angled at L, gives UL $=$ complement of the angle formed by the axes of y' and z,

UR $= 90° + \psi$, and $\angle$ R $= \theta$, $\therefore$ sin. UL $=$ cos. Z$'$ $=$ cos. ψ sin. θ, we have thus expressions for three more of the angles which enter (A), and there still remain three to determine. Let V be the point where the axis of z' pierces the sphere, and draw the arcs VQ and VSK, the former meeting TR in N, and the latter meeting the production of TR in K; then AZ$'$ being perpendicular to the plane of the circle TNRK, N and K will be right angles. The triangle NQR gives

NQ $=$ VQ $-$ VN $= -$ complement of the inclination of the axes of z' and x, RQ $= \varphi$, $\angle$R $= \theta$, $\therefore$ sin. NQ $= -$cos. X$''$ $=$ sin. φ sin. θ. The triangle KSR gives SK $=$ VK $-$ VS $= -$ complement of the inclinations of the axes of z' and y, SR $=$ SQ $-$ QR $=$ complement of φ, $\angle$R $= \theta$, $\therefore$ sin. SK $=$ cos. Y$''$$=$ cos. φ sin. θ. Moreover, since AZ, AZ$'$, are respectively perpendicular to the planes of xy and of $x'y'$, the inclinations of these two lines measures that of the planes, that is

$$Z'' = \theta \;\therefore\; \text{cos. } Z'' = \text{cos. } \theta.$$

(234.) Having now determined the expressions for all the cosines which enter the formulas (A,) we have, by substituting them therein, the following new formulas, for, transforming the equation of any surface from one rectangular system of coordinates to another, viz.

$$x = x' (\text{cos. } \theta \text{ sin. } \varphi \text{ sin. } \psi + \text{cos. } \varphi \text{ cos. } \psi)$$
$$+ y' (\text{cos. } \theta \text{ sin. } \varphi \text{ cos. } \psi - \text{cos. } \varphi \text{ sin. } \psi) - z' \text{ sin. } \theta \text{ sin. } \varphi.$$

* RT produced will pass through U, because AR, AX$'$, AY$'$, are in one plane.

$$y = x'(\sin.\varphi \cos.\psi - \cos.\theta \cos.\varphi \sin.\psi)$$
$$- y'(\cos.\theta \cos.\varphi \cos.\psi + \sin.\varphi \sin.\psi)$$
$$+ z' \sin.\theta \cos.\varphi,$$
$$z = x' \sin.\theta \sin.\psi + y' \sin.\theta \cos.\psi + z' \cos.\theta.$$

(235.) These formulas become much more simple, where $\psi = 0$, that is, when the axis of x' coincides with the trace AR, for, since, in this case, $\sin.\psi = 0$, and $\cos.\psi = 1$, the formulas become

$$x = x' \cos.\varphi + y' \sin.\varphi \cos.\theta - z' \sin.\theta \sin.\varphi$$
$$y = x' \sin.\varphi - y' \cos.\varphi \cos.\theta + z \sin.\theta \cos.\varphi$$
$$z = y' \sin.\theta + z' \cos.\theta.$$

Of Intersecting Planes.

(246.) The equation of a surface being given, let it be required to determine the equation of the intersection made by a plane whose inclination to, and trace on the plane of xy is given.

Call the inclination θ, and the angle formed by the trace and axis of x, φ; then, calling the trace the axis of x', and a perpendicular to it from the origin, and in the cutting plane, the axis of y', any point in the curve of intersection referred to these axes will be represented by $(x', y', 0,)$ and the same point referred to the original axes of the surface is (x, y, z); hence, by putting $z' = 0$, in the formulas above, we have $x = x' \cos.\varphi + y' \sin.\varphi \cos.\theta$, $y = x' \sin.\varphi - y' \cos.\varphi \cos.\theta$, $z = y' \sin.\theta$; and these expressions substituted in the equation of the surface will give the equation of the curve of intersection, when related to the axes of x' and y', taken as directed in the cutting plane.

The values of the angles θ and φ are immediately determinable, when the equation of the plane is given. For, if this equation be $Ax + By + Cz + D = 0$, then (193) $\cos.\theta = \dfrac{C}{\sqrt{\{A^2 + B^2 + C^2\}}}$, and the equation of the trace on the plane of xy being (185) $Ax + By + D = 0$, we have $\tan.\varphi = -\dfrac{A}{B}$.

Note. In the preceding transformations, we have supposed the origin to remain fixed, if, however, this be not the case, then the coordinates a, b, c, of the new origin must be introduced into those expressions.

(237.) Since the foregoing expressions for x, y, z, are linear functions of x', y', it follows that, when they are substituted in the equation of any surface of the second order, the result will be an equation also of the second order, between x and y'; hence every section of a surface of the second order is always a curve of the second order, and indeed, whatever be the order of the surface, no higher can be the order of the curve of intersection; it follows, moreover, that a straight line cannot cut a surface of the nth order in more than n points.

Let us suppose a conical surface: If a plane cut one sheet through, the section will obviously be a curve returning into itself, and as we know it must be of the second order, we immediately conclude that the section must be an ellipse. If the cutting plane be parallel to the generating line, in any position, then the plane can obviously meet only one sheet of the surface, the section will therefore consist of but one branch; hence it canbe no other curve than the parabola. If the plane be parallel to the axis of the cone, then both sheets will be cut, and the section will consist of two branches, and these will become two intersecting straight lines, when the axis coincides with the cutting plane; hence the section must be either an hyperbola, or one of its varieties. On these accounts the lines of the second order are frequently called *Conic Sections*. But the cone is not the only surface whose different sections furnish all the curves of the second order, as we shall presently see.

PROBLEM I.

(238.) To determine the nature of the different sections of a central surface of the second order.

In the general equation $Lz^2 + My^2 + Nx^2 = P$ (1), which comprehends the ellipsoid and hyperboloid, substitute for x, y, z, the values $x = x \cos. \varphi + y \cos. \theta \sin. \varphi + a$, $y = x \sin. \varphi - y \cos. \theta \cos. \varphi + b$, and $z = y \sin. \theta + c$, and it becomes of the form

$$Ay^2 + Bxy + Cx^2 + Dy + Ex + F = 0 \ldots. (2),$$

the coefficients of this equation having the following values:

$$A = L \sin.^2 \theta + M \cos.^2 \theta \cos.^2 \varphi + N \cos.^2 \theta \sin.^2 \varphi,$$
$$B = 2(N - M) \cos. \theta \sin. \varphi \cos. \varphi,$$
$$C = M \sin.^2 \varphi + N \cos.^2 \varphi,$$
$$D = 2(Lc \sin. \theta - Mb \cos. \theta \cos. \varphi + Na \cos. \theta \sin. \varphi),$$
$$E = 2(Mb \sin. \varphi + Na \cos. \varphi),$$
$$F = Lc^2 + Mb^2 + Na^2 - P.$$

Now we know (124) that the curve represented by the equation (2) will be an ellipse, an hyperbola, or a parabola, according as $B^2 - 4AC$ is negative, positive, or 0. The value of this expression is

$$B^2 - 4AC = -4MN \cos.^2 \theta - 4LM \sin.^2 \theta \sin.^2 \varphi - 4LN \sin.^2 \theta \cos.^2 \varphi \quad (3).$$

Hence, if the surface is an ellipsoid, that is, if L, M, N, are all positive, $B^2 - 4AC$ must be negative, consequently every section of an ellipsoid made by a plane is an ellipse, or else one of its varieties.

If the surface is an hyperboloid of one sheet, then of the coefficients L, M, N, two are positive, and the third negative; but, if the hyperboloid have two sheets, then two of the coefficients are negative; and one positive, so that the expression for $B^2 - 4AC$ must consist either of two positive terms and one negative, or else of two negative terms and one positive. In either case, the aggregate of the terms may be either positive, negative, or 0. For, dividing each term by $\sin.^2 \theta$, and abstracting from the signs, they may be represented by $Q \cot.^2 \theta$,

$R\sin.^2\varphi$, $S\cos.^2\varphi$, and it is plain that an infinite variety of values may be given to φ, that will render possible either of the conditions
$Q\cot.^2\theta > (R\sin.^2\varphi + S\cos.^2\varphi)$, $Q\cot.^2\theta < (R\sin.^2\varphi \backsim S\cos.^2\varphi)$ (4),
or
$Q\cot.^2\theta = (R\sin.^2\varphi + S\cos.^2\varphi)$, $Q\cot.^2\theta = R\sin.^2\varphi \backsim S\cos.^2\varphi)$ (5).

By the conditions (4) it appears that any term may be made to exceed numerically the sum of the other two, and, consequently, the aggregate of the three terms may take the sign of any of them, that is, it may be either positive or negative. The conditions (5) show that either term may become equal to the sum of the other two, so that, whichever two have the same sign, their aggregate may become equal to the third, when the aggregate of the whole will be 0.

Hence *the section of an hyperboloid by a plane may*, like the cone which is a variety of it, *be either an ellipse, an hyperbola, or a parabola.*

Cor. As none of the constants a, b, c, enter into the expression (3,) the curve of intersection must continue of the same kind, however these constants may be altered, provided only, that the angles θ, φ, remain the same, that is, *parallel sections always give the same kind of curve.*

PROBLEM II.

(239.) To determine the nature of the sections in surfaces which have not a centre.

Substituting the formulas employed in last problem, in the general equation $Lz^2 + My^2 = Qx$ (1), we obtain a result of the form $Ay^2 + Bxy + Cx^2 + \&c. = 0$, in which $A = L\sin.^2\theta + M\cos.^2\theta\cos.^2\varphi$, $B = -2M\cos.\theta\sin.\varphi\cos.\varphi$, $C = M\sin.^2\varphi$, $\therefore$ $B^2 - 4AC = -4LM\sin.^2\theta\sin.^2\varphi$ .... (2).

If the surface is the elliptic paraboloid, then the coefficients L, M, being of the same sign, the expression for $B^2 - 4AC$ must be negative, unless $\theta = 0$, or $\varphi = 0$, when the expression becomes $= 0$. Hence the intersections of an elliptic paraboloid must be either ellipses or parabolas, or else varieties of these curves.

If the paraboloid is hyperbolic, then L, M, having contrary signs, the expression (2) can never be negative, so that the intersections of the parabolic hyperboloid must be either hyperbolas or parabolas, or else varieties of these curves.

Cor. It follows here, as in last problem, that because the expression (2) is independent of the values of the constants, a, b, c, *parallel sections always give the same kind of curve.*

PROBLEM III.

(240.) To determine the locus of the centres of any parallel sections.

Let the surface be central, then, when the sections are central curves, let us suppose the coordinates in each to originate at the centre, and let us represent $(x', y', z',)$ the centre of any parallel section when referred to the axes to which the surface is referred, then x', y', z', mean the same thing here as a, b, c, in the preceding problems. Now, since the axes of the section are here supposed to originate at the centre, the terms Dy, Ex, in equation (2), prob. 1, will be absent (126) showing that, in this case, we must have

$D = 2\,(Lz' \sin.\theta - My' \cos.\theta \cos.\varphi + Nx' \cos.\theta \sin.\varphi) = 0,$

$E = 2\,(My' \sin.\varphi + Nx' \cos.\varphi) = 0.$ These equations being linear, each separately represents a plane; but, as they exist together, they denote the straight line, which is their intersection; hence the locus of (x', y', z',) is a straight line. If the surface have no centre, we should, in the same manner, find that, when the parallel sections have centres, these are all situated in the same straight line.

Neither of the preceding equations having a constant term, it follows (185) that the planes which they represent both pass through the origin, this point then belongs to their intersection. Hence *the centres of parallel sections are all on the same diameter of the surface.*

CHAPTER III.

DISCUSSION OF THE GENERAL EQUATION.

(241.) We shall now proceed to examine the equation of the second degree of three variables, in its most general form, and show that it can never represent any surface not among those which we have already examined. This general equation is

$Az^2 + By^2 + Cx^2 + Dzy + Ezx + Fxy + Gz + Hy + Kx + L = 0$ (1),

and we shall for simplicity suppose it to refer the surface which it represents to rectangular axes. This supposition will not in the least diminish the generality of our reasoning, since, if the axes were originally oblique, they might be transformed to rectangular, by the substitution of certain *linear* functions of the new coordinates in place of the old, so that the *degree* of the equation would remain the same and its *generality* could not exceed that of equation (1).

Let us now transform these rectangular axes to another system, also rectangular, by substituting, in equation (1), the values (275)

$$\left.\begin{aligned} x &= x' \cos. X + y' \cos. X' + z' \cos. X'' \\ y &= x' \cos. Y + y' \cos. Y' + z' \cos. Y'' \\ z &= x' \cos. Z + y' \cos. Z' + z' \cos. Z'' \end{aligned}\right\} \text{(A)},$$

then the resulting equation must be of the form $A'z'^2 + B'y'^2 + C'x'^2 + D'z'y' + E'z'x' + F'x'y' + G'z' + H'y' + K'x' + L = 0$ (2), in which the coefficients are functions of the nine angles which enter the formulas (A).

Now it has been seen (230) that these nine angles are subject to only six conditions, and that, therefore, in order to fix their values, three more conditions must be introduced among them, and the only limit to the choice of these conditions is that they must not be inconsistent with the other six.

Let us here suppose the three conditions $D' = 0$, $E' = 0$, $F' = 0$ then, if it can be shown that these may exist conjointly with the conditions (1) and (2), art. (230), we may immediately infer that the general equation (1) may always be reduced to the more simple form

$$A'z'^2 + B'y'^2 + C'x'^2 + G'z' + H'y' + K'x' + L = 0,$$

by merely altering the directions of the rectangular axes.

The expressions for the coefficients D', E', F', may be obtained without substituting the expressions (A) in every term of the equation (1), the last four terms obviously have no influence on these coefficients; and, instead of actually squaring the expressions (A) for the first three terms, we need only attend to the products two and two of the three terms, in each, these products being the only parts of the squares concerned in the formation of the three coefficients under consideration. Also, in the three following terms of equation (1), which contain the products of the expressions (A), two and two, the partial products arising from multiplying any term by that in the same vertical row, are not concerned in these coefficients, and are, therefore not to be attended to. Availing ourselves of these considerations, we find for the terms $F'x'y'$, $E'x'z'$, $D'x'y'$, the expressions

$$\left.\begin{array}{l} 2A\cos.Z\cos.Z' \\ 2B\cos.Y\cos.Y' \\ 2C\cos.X\cos.X' \\ D\cos.Z\cos.Y' \\ D\cos.Y\cos.Z' \\ E\cos.Z\cos.X' \\ E\cos.X\cos.Z' \\ F\cos.Y\cos.X' \\ F\cos.X\cos.Y' \end{array}\right| x'y' + \left.\begin{array}{l} 2A\cos.Z\cos.Z'' \\ +2B\cos.Y\cos.Y'' \\ +2C\cos.X\cos.X'' \\ +D\cos.Z\cos.Y'' \\ +D\cos.Y\cos.Z'' \\ +E\cos.Z\cos.X'' \\ +E\cos.X\cos.Z'' \\ +F\cos.Y\cos.X'' \\ +F\cos.X\cos.Y'' \end{array}\right| x'z' + \left.\begin{array}{l} 2A\cos.Z'\cos.Z'' \\ +2B\cos.Y'\cos.Y'' \\ +2C\cos.X'\cos.X'' \\ +D\cos.Z'\cos.Y'' \\ +D\cos.Y'\cos.Z'' \\ +E\cos.Z'\cos.X'' \\ +E\cos.X'\cos.Z'' \\ +F\cos.Y'\cos.X'' \\ +F\cos.X'\cos.Y'' \end{array}\right| z'y'$$

Hence the three equations of condition are

$$\left.\begin{array}{l} 2A\cos.Z \\ +D\cos.Y \\ +E\cos.X \end{array}\right| \cos.Z' + \left.\begin{array}{l} 2B\cos.Y \\ +D\cos.Z \\ +F\cos.X \end{array}\right| \cos.Y' + \left.\begin{array}{l} 2C\cos.X \\ +E\cos.Z \\ +F\cos.Y \end{array}\right| \cos.X' = 0$$

$$\left.\begin{array}{l} 2A\cos.Z \\ +D\cos.Y \\ +E\cos.X \end{array}\right| \cos.Z'' + \left.\begin{array}{l} 2B\cos.Y \\ +D\cos.Z \\ +F\cos.X \end{array}\right| \cos.Y'' + \left.\begin{array}{l} 2C\cos.X \\ +E\cos.Z \\ +F\cos.Y \end{array}\right| \cos.X'' = 0$$

and

$$\left.\begin{array}{l} 2A\cos.Z' \\ +D\cos.Y' \\ +E\cos.X' \end{array}\right| \cos.Z'' + \left.\begin{array}{l} 2B\cos.Y' \\ +D\cos.Z' \\ +F\cos.Z' \end{array}\right| \cos.Y'' + \left.\begin{array}{l} 2C\cos.X' \\ +E\cos.Z' \\ +F\cos.Y' \end{array}\right| \cos.X'' = 0$$

or more briefly

$$\left.\begin{array}{l} M\cos.Z' + N\cos.Y' + P\cos.X' = 0 \\ M\cos.Z'' + N\cos.Y'' + P\cos.X'' = 0 \\ M'\cos.Z'' + N'\cos.Y'' + P'\cos.X'' = 0 \end{array}\right\} \text{(B)}.$$

These, then, are the equations which must exist in conjunction with the following, if the transformation in view is possible,

$$\left.\begin{array}{l}\cos. X \cos. X' + \cos. Y \cos. Y' + \cos. Z \cos. Z' = 0 \\ \cos. X \cos. X'' + \cos. Y \cos. Y'' + \cos. Z \cos. Z'' = 0 \\ \cos. X' \cos. X'' + \cos. Y' \cos. Y'' + \cos. Z' \cos. Z'' = 0\end{array}\right\} \ldots (C)$$

$$\left.\begin{array}{l}\cos.^2 X + \cos.^2 Y + \cos.^2 Z = 1 \\ \cos.^2 X' + \cos.^2 Y' + \cos.^2 Z' = 1 \\ \cos. X'' + \cos. Y'' + \cos. Z'' = 1\end{array}\right\} \ldots (D)$$

Eliminating N from the first two of equations (B), by multiplying the first, by $\cos. Y''$, and the second by $\cos. Y'$, and then, subtracting the first result from the second, we get

$$\left.\begin{array}{l}M(\cos. Y' \cos. Z'' - \cos. Z' \cos. Y'') \\ + P(\cos. Y' \cos. X'' - \cos. X' \cos. Y'')\end{array}\right\} = 0 \ldots (E);$$

eliminating P from the same equations, we have

$$\left.\begin{array}{l}M(\cos. X' \cos. Z'' - \cos. Z' \cos. X'') \\ + N(\cos. X' \cos. Y'' - \cos. Y' \cos. X'')\end{array}\right\} = 0 \ldots (F).$$

In like manner, by eliminating first cos. Y, and then cos. X, from the first two of equations (C) we have

$$\left.\begin{array}{l}\cos. X(\cos. Y' \cos. X'' - \cos. X' \cos. Y'') \\ + \cos. Z(\cos. Y' \cos. Z'' - \cos. Z' \cos. Y'')\end{array}\right\} = 0 \ldots (G),$$

and
$$\left.\begin{array}{l}\cos. Y(\cos. X' \cos. Y'' - \cos. Y' \cos. X'') \\ + \cos. Z(\cos. X' \cos Z'' - \cos. Z' \cos. X'')\end{array}\right\} = 0 \ldots (H).$$

Putting, for simplicity
$$\left\{\begin{array}{l}\cos. Y' \cos. Z'' - \cos. Z' \cos. Y'' = Q, \\ \cos. Y' \cos. X'' - \cos. X' \cos. Y'' = R, \\ \cos. X' \cos. Z'' - \cos. Z' \cos. X'' = S,\end{array}\right.$$

the four preceding equations become $MQ + PR = 0$, $MS - NR = 0$, $Q \cos. Z + R \cos. X = 0$, $S \cos. Z - R \cos. Y = 0$, in which the quantities Q, R, S, are the only ones containing the accented cosines. These three quantities may be eliminated from the four equations thus. The first and second give $Q = -\frac{PR}{M}$, $S = \frac{NR}{M}$, which values, substituted in the remaining two, give

$$P \cos. Z - M \cos. X = 0, \quad N \cos. Z - M \cos. Y = 0,$$

or, replacing, P, M, and N, by their values, these equations are

$$\left.\begin{array}{l}(2C \cos. X + E \cos. Z + F \cos. Y) \cos. Z \\ -(2A \cos. Z + D \cos. Y + E \cos. X) \cos. X\end{array}\right\} = 0 \ldots (I.)$$

$$\left.\begin{array}{l}(2B \cos. Y + D \cos. Z + F \cos. X) \cos. Z \\ -(2A \cos. Z + D \cos. Y + E \cos. X) \cos. Y\end{array}\right\} = 0 \ldots (K)$$

Now these two equations, together with $\cos.^2 X + \cos.^2 Y + \cos.^2 Z = 1$ are sufficient to determine the three angles, cos. X, cos. Y, cos. Z.

For, put $m = \frac{\cos. X}{\cos. Z}$ and $n = \frac{\cos. Y}{\cos. Z}$ $\therefore \cos. Z = \frac{1}{\sqrt{\{1 + m^2 + n^2\}}}$, then, dividing the first and second equations by $\cos.^2 Z$, they become

$$2(C - A) m + E(1 - m^2) - Dmn + Fn = 0,$$
$$2(B - A) n + D(1 - n^2) - Emn + Fm = 0.$$

From this last we get $\frac{2(B - A) n + D(1 - n^2)}{En - F} = m$, which value of m, substituted in the preceding equation, gives, after reduction, a cubic equation; for, although the second term of the equation will furnish a term containing the fourth power of n, viz. $D^2 En^4$, yet, when the whole result is multiplied by $(En - F)^2$, to clear it of fractions, this term will obviously be also given with contrary sign in the value of Dmn, and is thus destroyed. This being a cubic equation, there necessarily exists at least one real value for n, and, consequently, the value of m is real; and hence also the values of cos. Z, and of cos. X $= m$ cos. Z, and cos. Y $= n$ cos. Z. We have thus proved the reality of the three cosines cos. X, cos. Y, cos. Z.

If now we go back to the equations (B,) and (C), and proceed with the first and third of each group, exactly as we have done with the first and second, taking care, however, to put the first of (B) under the form M' cos. Z $+ N'$ cos. Y $+ P'$ cos. X $= 0$, to which it is obviously identical, we shall in the same way, establish the reality of cos. X$'$, cos. Y$'$, cos. Z$'$; and lastly, employing the second and third equations of each group, we demonstrate the reality of cos. X$''$, cos. Y$''$, cos. Z$''$. Hence we may infer that it is always possible to reduce the general equation (1) to the form

$$A'z^2 + B'y^2 + C'x^2 + G'z + H'y + K'x + L = 0 \quad . \; . \; . \; . \; . \; . \quad (1'),$$

by altering the position of the rectangular axes to which the surface represented by it is referred.

(242.) We shall now show that the equation in this form may be finally reduced to the more simple form $A'z^2 + B'y^2 + C'x^2 + P = 0$ (A$'$). For, let the origin be removed, by substituting in (1$'$) the values

$$x = x' + a, \; y = y' + b, \; z = z' + c,$$

then, putting P for the last term, it becomes

$$A'z'^2 + B'y'^2 + C'x'^2 + \left.\begin{matrix}2A'c \\ + G'\end{matrix}\right| z' + \left.\begin{matrix}2B'b \\ + H'\end{matrix}\right| y' + \left.\begin{matrix}2C'a \\ + K'\end{matrix}\right| x' + P = 0,$$

in which the coefficients of z', of y', and of x', vanish when we have the conditions $c = -\dfrac{G'}{2A'}$, $b = -\dfrac{H'}{2B'}$, $a = -\dfrac{K'}{2C'}$, which are always possible when the coefficients A$'$, B$'$, C$'$, are neither of them 0. But, if one of these coefficients be 0, then it will not be possible to remove the variable, whose square is absent; thus, if $A' = 0$, then, that the coefficient of z' may be 0, there must be $c = -G' \div 0$, that is, the new origin must be infinitely distant from the old, in the direction of the axis of z; this origin, therefore, is not determinable. Nothing, however, hinders us from removing the other two variables, and thus reducing this equation to the form $B'y^2 + C'x^2 + P = 0 \; . \; . \; . \; . \; (B')$, and, as the quantity c enters P, and is still arbitrary, we may determine it from the condition $P = 0$, which will finally reduce the equation to $B'y^2 + C'x^2 + G'z = 0 \; . \; . \; . \; . \; (C')$.

But, if not only $A' = 0$, but also $G' = 0$, in equation (1$'$), that is, if one of the variables z be entirely absent from the equation, then

(C′) is simply $B'y^2 + C'x^2 + P = 0$, in which case the surface is obviously (200) a cylinder, whose base or directrix on the plane of xy is either an ellipse or hyperbola, according as the signs of B′ and C′ are like or unlike. If two of the coefficients A′, B′, C′ are 0, as $A' = 0$, $B' = 0$, then the removal of the terms containing the first powers of the variables, whose squares are absent, is impossible, but the conditions $a = -\frac{K'}{2C'}$, and $P = 0$, may still exist, and will reduce the equation to $C'x^2 + G'z + K'y = 0$. When all three of the squares are absent from (1), the equation represents a plane.

(243.) From the preceding discussion it follows that any surface of the second order may be represented by one or other of the following equations, viz. $Lx^2 + My^2 + Nz^2 + P = 0$, $Lx^2 + My^2 + Qz = 0$, $Lx^2 + My^2 + P = 0$, and $Lx^2 + Gz + Hy = 0$.

All the surfaces represented by the first two of these equations have been fully considered in Chapter iii. They were found to comprehend ellipsoids, hyperboloids of one and two sheets, paraboloids, elliptic and hyperbolic, and, as varieties of these, the sphere and the cone. With regard to the remaining two equations, the first we have seen characterizes cylindrical surfaces, whose directrices on the plane of xy are either ellipses or hyperbolas. The other equation is also that of a cylindrical surface, but of this the directrix is a parabola, for, suppose it to be cut by a series of planes parallel to the plane of zy, that is, by planes of which the equations are $x=k$, $x=k'$, $x=k''$, &c. then for the sections we have the equations $Gz + Hy = -Lk^2$, $Gz + Hy = -Lk'^2$, $Gz + Hy = -Lk''^2$, &c. and these all representing parallel straight lines, it follows that we may conceive the surface to be generated by a straight line moving parallel to itself. This surface must, therefore, be a cylinder. For the directrix or trace on the plane of xy put $z = 0$, in its equation, and we have $Lx^2 + Hy = 0$, which represents a parabola, or one of its varieties.

(244.) We may obviously infer from this discussion that the only conical surface of the second order is the elliptic cone, of which the circular is a variety, in other words, this conical surface is always such that a system of coordinate planes may be found that will render the trace on the plane of xy an ellipse or a circle. But we may assume any curve of the second order on the plane of xy, and thus, agreeably to art. (201), generate a conical surface which will also be of the second order. It follows, therefore, that, by giving different inclinations to the plane of xy, the elliptic cone will furnish for traces on that plane all the curves of the second order, as was also shown from other considerations at art. (237).

(245.) We shall now briefly discuss the general equation (1), in order to ascertain criteria by which we may know, without the trouble of transforming it, the nature of the surface, which it represents.

Solving the equation for z, we have, by putting Q for the quantity under the radical, $z = -\frac{Dy + Ex + G}{2A} \pm \frac{1}{2A}\sqrt{Q}$.

Also, solving for y, we have $y = -\frac{Dz + Fx + H}{2B} \pm \frac{1}{2B}\sqrt{Q'}$;

and, in like manner, for x, $x = -\frac{Ez + Fy + K}{2C} \pm \frac{1}{2C}\sqrt{Q''}$.

Representing the rational parts of these expressions by Z, Y, and X, respectively, we have the three equations,

$$Z = -\frac{Dy+Ex+G}{2A}, \ Y = -\frac{Dz+Fx+H}{2B}, \ \& \ X = -\frac{Ez+Fy+K}{2C},$$

which represent three planes which, from the foregoing general expressions for z, y, and x, are obviously such that they bisect the chords drawn parallel to the axes of z, y, x; each of these planes, therefore, passes through the centre, that is, it is diametral; so that the centre of the surface, supposing it to have one, must be at the intersection of these planes. Calling this intersection (x', y', z'), the preceding equations give

$$\left.\begin{array}{l} 2Az' + Dy' + Ex' + G = 0, \\ 2By' + Dz' + Fx' + H = 0, \\ 2Cx' + Ez' + Fy' + K = 0, \end{array}\right\} \ldots . \ (1),$$

from which we get, by elimination, the three expressions for x', y', and z'. The common denominator of these expressions we find to be

$$AF^2 + BE^2 + CD^2 - DEF - 4ABC;$$

hence, if this be infinite, the surface will have a centre, but if it be $=$ 0, then the surface will not have a centre, that is, if the surface have a centre, $AF^2 + BE^2 + CD^2 - DEF - 4ABC \gtrless 0$ (2). If the surface has not a centre, $AF^2 + BE^2 + CD^2 - DEF - 4ABC = 0$ (3).

If $G = 0$, $H = 0$, and $K = 0$, then (*Algebra p.* 62) the numerators of the expressions for x', y', z', each become 0; hence, if in this case the condition (3) do not exist, the coordinates of the centre each become 0, that is, the centre of the surface must coincide with the origin, as is also at once manifest from the equations (1) themselves, which represent planes passing through the origin when the constants G, H, K, are absent, therefore, whatever be the inclinations of the axes, provided they originate at the centre of the surface, the most general form of the equation is

$$Az^2 + By^2 + Cx^2 + Dzy + Ezx + Fxy + L = 0.$$

If, however, the condition (3) exists at the same time that we have $G = 0$, $H = 0$, $K = 0$, then the expressions for $(x', y', z',)$ become each $= \frac{0}{0}$, intimating that there are an indefinite number of centres, or points, in which the three planes (1) meet; hence they must intersect in one common straight line, passing through the origin, and which is the locus of all the centres of the surface. This surface having innumerable centres in the same straight line can be no other than an elliptic or hyperbolic cylinder.

(246.) Having given criteria (2) and (3) for discovering when the equation represents a central surface, and when it does not, let us now inquire by what means we may know when a surface is ascertained to be central, whether it is limited or unlimited, that is, whether it belongs to the class of ellipsoids or hyperboloids.

If the surface be limited in every direction, then to whatever point in the surface a line from the origin be drawn, this line will always have a finite length; but, should the surface be unlimited in any direction, then, as some of its points will be infinitely distant from the origin, lines drawn to them from the origin will not be finite, and it is hence obvious that while, in the former case, every section of the surface is a limited curve, every section, in the latter case, which passes through either of the infinite lines which we have supposed to be drawn, will necessarily be an unlimited curve. This being the case with every such section, we need consider only those made by planes perpendicular to one of the coordinate planes, when, it is found that no one of these can possibly give an unlimited curve, we may conclude that the surface is itself limited in every direction.

Now the equation of any plane drawn perpendicularly to the plane of xz, and passing through the origin, is $x = \alpha z$. Combining this with the equation of the surface (241), we have for the intersection the equation $(A + C\alpha^2 + E\alpha) z^2 + (D + F\alpha) zy + By^2 + (G + K\alpha)z + Hy + L = 0$, and, in order that this may represent always a limited curve, we must have (124) $(D + F\alpha)^2 - 4(A + C\alpha^2 + E\alpha) B \angle 0$ (4.) whatever be the value of α, that is,

$$(F^2 - 4BC)\alpha^2 + 2(DF - 2BE)\alpha + D^2 - 4AB \angle 0,$$

or, dividing by $F^2 - 4BC$, and then, dividing the expression into factors, $(\alpha - \beta)$, $(\alpha - \beta')$, (*see art.* 135), we have $(F^2 - 4BC)(\alpha - \beta)(\alpha - \beta') \angle 0$. Now this expression cannot preserve the same sign for every value of α, unless $(\alpha - \beta)(\alpha - \beta')$ does, and that the sign of this may always remain the same, that sign must be positive, and the values of β and β' imaginary (*Alg. p.* 180); hence, that the condition (4) may exist, we must have, in the first place, $F^2 - 4BC \angle 0$, and, in the second place, the roots β, β', of the equation $\alpha^2 + \frac{2(DF - 2BE)}{F^2 - 4BC}\alpha + \frac{D^2 - 4AB}{F^2 - 4BC} = 0$ must be imaginary, that is to say, the quantity under the radical which enters into the expressions for these roots, which quantity we find to be $(DF - 2BE)^2 - (F^2 - 4BC)(D^2 - 4AB)$, must be negative.

Hence we may conclude that, when the general equation represents a surface limited in every direction, there must exist the relations $F^2 - 4BC \angle 0$, $(DF - 2BE)^2 - (F^2 - 4BC)(D^2 - 4AB) \angle 0$.

CHAPTER IV.

MISCELLANEOUS PROPOSITIONS ON SURFACES OF THE SECOND ORDER.

PROPOSITION I.

(247.) To prove that in a central surface of the second order there is an infinite number of systems of conjugate diameters.

Taking the general equation of these surfaces when referred to their rectangular conjugates, viz. $Ax^2 + By^2 + Cz^2 = D \ldots (1)$, and substituting therein for x, y, z, the values in the formulas (A), at p. 259, we have for the same surface, when related to oblique axes, the equation

$$A'x^2 + B'y^2 + C'z^2 + 2\,(A \cos. X \cos. X' + B \cos. Y \cos. Y' + C \cos. Z \cos. Z')\,xy$$
$$+ 2\,(A \cos. X \cos. X'' + B \cos. Y \cos. Y'' + C \cos. Z \cos. Z'')\,xz$$
$$+ 2\,(A \cos. X' \cos. X'' + B \cos. Y' \cos. Y'' + C \cos. Z' \cos. Z'')\,yz = D;$$

in which equation A', B', C', are put, for brevity, to represent certain functions of the cosines. Now the nine cosines which enter into this equation are subject to the three conditions (1), p. 259, and, in order that the three last terms of the equation may vanish, they must fulfil the additional conditions

$$A \cos. X \cos. X' + B \cos. Y \cos. Y' + C \cos. Z \cos. Z' = 0,$$
$$A \cos. X \cos. X'' + B \cos. Y \cos. Y'' + C \cos. Z \cos. Z'' = 0,$$
$$A \cos. X' \cos. X'' + B \cos. Y' \cos. Y'' + C \cos. Z' \cos. Z'' = 0.$$

Hence, if these six conditions have place, the cosines are such as to render the transformed equation of the form $A'x^2 + B'y^2 + C'z^2 = D$ (2). But to fix the values of these cosines three more conditions are requisite, they being nine in number, which conditions, being arbitrary, may be infinitely varied; hence the position of the axes of reference may be infinitely varied, without altering the form of the equation (1), so that (221) the systems of conjugate diameters are infinite in number

PROPOSITION II.

(248.) In any central surface of the second order the longest of the principal diameters exceeds any other, and the shortest principal diameter is shorter than any other diameter.

Let A, B, C, be the principal semi-diameters of the surface, and of which A is either the longest or the shortest, then the equation of the surface is (221) $\frac{x^2}{A^2} + \frac{y^2}{B^2} + \frac{z^2}{C^2} = 1$, and the equation of a concentric sphere passing through the extremity of the semi-diameter, A, is $x^2 + y^2 + z^2 = A^2 \;\therefore\; x^2 = A^2 - y^2 - z^2$. If this sphere have any other points in common with the surface, the y, z, of those points will be given by substituting this value of x^2 in the above equation. This substitution gives $\left(\frac{1}{B^2}-\frac{1}{A^2}\right)y^2+\left(\frac{1}{C^2}-\frac{1}{A^2}\right)z^2=0$, which equation is impossible, except in the single case $y=0$, $z=0$, since by hypothesis, the two terms in the first member of this equation are either both negative or both positive, and therefore can never destroy each other.

PROPOSITION III.

(249.) To express the area of a triangle in space, by means of the coordinates of its vertices.

Since (224) the square of the area of any plane figure is equal to the sum of the squares of its projections on three rectangular planes, we shall be able to express the area of the triangle, provided we can express the areas of its projections in terms of the given coordinates.

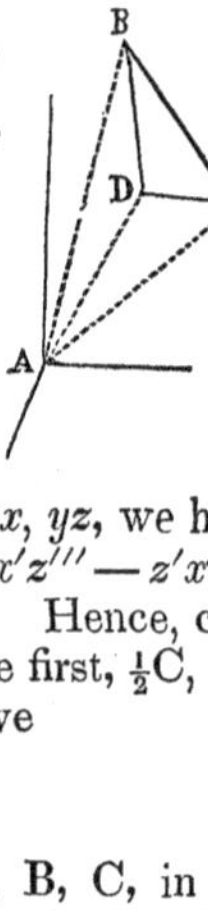

Let BCD be the given triangle in space, and draw lines from the origin, A, to its vertices B, C, D, of which the coordinates are x', y', z'; x'', y'', z''; x''', y''', z'''. Now the projection of the triangle BCD on either of the coordinate planes is equal to the projection of ABC minus the projections of ABD, ACD, that is, for the projection of BCD, on the plane of xy, we have (21)

$$\frac{x'y''-y'x''}{2}\quad\frac{x''y'''-y''x'''}{2}\quad\frac{x'y'''-y'x'''}{2},$$

or $\frac{1}{2}(x'y''-y'x''+y''x'''-x''y'''+y'x'''-x'y''')$.

In like manner, for the projections on the planes of zx, yz, we have the expressions $\frac{1}{2}(z'x''-x'z''+x''z'''-z''x'''+x'z'''-z'x''')$, and $\frac{1}{2}(y'z''-z'y''+z''y'''-y''z'''+z'y'''-y'z''')$. Hence, calling this last expression $\frac{1}{2}$A, the preceding, $\frac{1}{2}$B, and the first, $\frac{1}{2}$C, and putting a for the area of the proposed triangle, we have

$$a=\frac{\sqrt{\{A^2+B^2+C^2\}}}{2}$$ for the expression sought.

(250.) *Corollary.* If we determine the values A, B, C, in the equation $Ax+By+Cz=D$ of the plane, passing through the three points (x', y', z'), (x'', y'', z''), and (x''', y''', z'''), by means of the three equations of condition $Ax'+By'+Cz'=D$, $Ax''+By''+Cz''=D$, and $Ax'''+By'''+Cz'''=D$, we find for them precisely the expressions above, and for D we have

$$D=(x'y''-y'x'')z'''+(y'z''-z'y'')x'''+(z'x''-x'z'')y'''$$

It appears, therefore, that the coefficients, A, B, C, of the variables, in the equation of a plane passing through three points, denote the doubles of the projections of the triangle, whose vertices are these points, upon the planes of yz, zx, and xy.

PROPOSITION IV.

(251.) To express the equation of a plane by means of the perpendicular, let fall upon it from the origin and the inclinations of this perpendicular to the axes.

Representing the plane by the equation $Ax + By + Cz = D$, we have for the portions of the axes of x, y, z, intercepted between it and the origin, the respective values $\frac{D}{A}, \frac{D}{B}, \frac{D}{C}$. Now, if ρ be the perpendicular from the origin upon the plane, and α, β, γ, be the respective angles it makes with the axes of x, y, z, then

$$\rho = \frac{D}{A}\cos.\alpha = \frac{D}{B}\cos.\beta = \frac{D}{C}\cos.\gamma,$$

$$\therefore A = D\frac{\cos.\alpha}{\rho},\ B = D\frac{\cos.\beta}{\rho},\ C = D\frac{\cos.\gamma}{\rho}.$$

Hence, substituting these values, in the equation of the plane, and multiplying by ρ, we have $x\cos.\alpha + y\cos.\beta + z\cos.\gamma = \rho$ (1) for the equation sought. When the plane passes through the origin, $\rho=0$, and the equation is $x\cos.\alpha + y\cos.\beta + z\cos.\gamma = 0$ . . . . (2).

PROPOSITION V.

(252.) If the vertex of a pyramid be at the origin of three rectangular planes, and its base be projected upon them, then if any point be assumed in the plane of the base, the three pyramids whose bases are the projections, and vertices the assumed point, will together be equivalent to the original pyramid.

Let a represent the base of the proposed pyramid and ρ a perpendicular upon it from the origin, then, if α, β, γ, be the inclinations of this perpendicular to the axes of x, y, z, we shall have

$a\cos.\alpha$ = the projection of a on the plane of yz,
$a\cos.\beta$ .... xz,
$a\cos.\gamma$ .... xy.

Now, by multiplying the equation of the plane (1), last proposition, by a, there results $xa\cos.\alpha + ya\cos.\beta + za\cos.\gamma = a\rho$ . . . . . (1),

$\therefore \frac{1}{3}xa\cos.\alpha + \frac{1}{3}ya\cos.\beta + \frac{1}{3}za\cos.\gamma = \frac{1}{3}a\rho$ . . . . (2).

The first member of this equation denotes three pyramids, whose bases are the projections above, and whose common vertex is any point (x, y, z,) in the plane (1), last proposition, and whose perpendicular altitudes are respectively x, y, and z; the second member represents the proposed pyramid; hence the truth of the theorem.

(253.) *Corollary* 1. Comparing equation (1), above with $Ax+By+Cz = D$, since (prop. 3 cor.) when a is a triangle, we have $A = 2a\cos.\alpha$, $B=2a\cos.\beta$, $C=2a\cos.\gamma$, $\therefore D=2a\rho$, that is, the expression $(x'y''-y'x'')z''' + (y'z''-z'y'')x''' + (z'x''-x'z'')y'''$ . . (3) represents six times the volume of the triangular pyramid, whose vertex is at the origin, and of which the corners of the base are the points (x', y', z'), (x'', y'', z''), and (x''', y''', z'''), or, which is the same thing, the expression represents a parallelopiped of which three contiguous edges meet at the origin and terminate in the points (x', y', z'), (x'', y'', z''), (x''', y''', z''').

(254.) *Cor.* 2. But those parts of the foregoing expression which are within the parentheses are obviously the projections of one of the faces of this parallelopiped, viz. that face whose contiguous sides terminate in the points (x', y', z'), (x'', y'', z''), upon the three coordinate planes; and these projections are severally multiplied by the perpendiculars z''', x''', y''', let fall upon them from the point (x''', y''', z'''). Hence *the expression* (3) *represents the sum of the volumes of three parallelopipeds, having these projections for bases, and* x''', y''', z''', *for altitudes.*

(255.) *Cor.* 3. It follows, therefore, that, *if the vertex of a triangular pyramid be at the origin of three rectangular planes, and either of its faces be projected on them, then the three pyramids constituted on these bases, and having a common vertex in that corner of the original pyramid's base, which is opposite to the projected face, shall together be equal to the original pyramid.*

PROPOSITION VI.

(256.) To determine the position of a plane, so that, if a given triangle be projected orthogonally upon it, the projection may be similar to a given triangle.

Let the plane of projection pass through a vertex of the given triangle, and let the perpendiculars, dropped from the other two vertices upon that plane, be z', z''; let also the sides of the given triangle be A, B, C, and those of the triangle to which the projection is to be similar a, b, c; then, on account of this similarity, the sides of the projected triangle will be $\sqrt{\{A^2 - z'^2\}}$, $\frac{b}{a}\sqrt{\{A^2 - z'^2\}}$, $\frac{c}{a}\sqrt{\{A^2 - z'^2\}}$. Both the two latter projections are also $\sqrt{\{B^2 - z''^2\}}$, $\sqrt{\{C^2 - (z' - z'')^2\}}$, therefore $\frac{b^2}{a^2}(A^2 - z'^2) = B^2 - z''^2$, and $\frac{c^2}{a^2}(A^2 - z'^2) = C^2 - (z' - z'')^2$.

Hence we have these two equations to find z' and z''.

Substituting, in the second, the value of z''^2, furnished by the first, we have, after transposing,

$$\frac{A^2c^2 - A^2b^2}{a^2} + B^2 - C^2 + \frac{a^2 + b^2 - c^2}{a^2} z'^2 = 2z'\sqrt{\left(B^2 - \frac{A^2b^2}{a^2} + \frac{b^2}{a^2}z'^2.\right)}$$

Squaring each side, and putting in the result single letters for the known coefficients, we have $p^2 + pqz'^2 + q^2z'^4 = rz'^2 + sz'^4$,

$\therefore z'^4 + \frac{pq - r}{q^2 - s} z^2 = -p^2$. This quadratic determines z', and thence we get z'', as also the three sides of the projected triangle, and thus the position of the required plane becomes known.

PROPOSITION VII.

(257.) If the equations

$$\left.\begin{array}{l|l} a'^2 + b'^2 + c'^2 = 1 & a'a'' + b'b'' + c'c'' = 0 \\ a''^2 + b''^2 + c''^2 = 1 & a'a''' + b'b''' + c'c''' = 0 \\ a'''^2 + b'''^2 + c'''^2 = 1 & a''a''' + b''b''' + c''c''' = 0 \end{array}\right\} \ldots\ldots (1)$$

exist, so also do the equations

$$\left.\begin{array}{l|l} a'^2 + a''^2 + a'''^2 = 1 & a'b' + a''b'' + a'''b''' = 0 \\ b'^2 + b''^2 + b'''^2 = 1 & a'c' + a''c'' + a'''c''' = 0 \\ c'^2 + c''^2 + c'''^2 = 1 & b'c' + b''c'' + b'''c''' = 0 \end{array}\right\} \ldots\ldots (2).$$

For, assume $\left\{\begin{array}{l} x = a't + a''u + a'''v \\ y = b't + b''u + b'''v \\ z = c't + c''u + c'''v \end{array}\right\} \ldots (3).$

Then squaring each equation, and adding the results, we have in virtue of the proposed conditions $x^2 + y^2 + z^2 = t^2 + u^2 + v^2 \ldots (4)$.

Let us now determine from (3) the values of t, u, v in terms of x, y, z. In order to this, multiply the equations (3) respectively by a', b', c', add the results, and we shall obtain t.

Similarly, multiply by a'', b'', c'', and we shall get u, &c. thus

$$\left.\begin{array}{l} t = a'\,x + b'\,y + c'\,z, \\ u = a''\,x + b''\,y + c''\,z, \\ v = a'''x + b'''y + c'''z, \end{array}\right\} \ldots\ldots (5).$$

Substitute these values in equation (4), and we shall have, by comparing the coefficients of the like terms, the equations announced.

PROPOSITION VIII.

(258.) If the conditions (1), in last proposition, exist, then also the following equations have place, viz.

$$(a'b'' - b'a'')^2 + (a''b''' - b''a''')^2 + (a'''b' - b'''a')^2 = 1,$$
$$(b'c'' - c'b'')^2 + (b''c''' - c''b''')^2 + (b'''c' - c'''b')^2 = 1,$$
$$(c'a'' - a'c'')^2 + (c''a''' - a''c''')^2 + (c'''a' - a'''c')^2 = 1,$$
$$(a'b'' - b'a'')\,c''' + (b'c'' - c'b'')\,a''' + (c'a'' - a'c'')\,b''' = 1.$$

Put, for brevity, the first member of this last equation, $= l$, then if we determine the values of t, u, v, in equation (3), last proposition, not by the process there directed, but by the usual algebraical method (see Algebra p. 61, where the operation is given at length,) we shall obtain these results, viz.

$$t = \frac{(b''c''' - c''b''')\,x + (b'''c' - c'''b')\,y + (b'c'' - c'b'')z}{l},$$

$$u = \frac{(c''a''' - a''c''')\,x + (c'''a' - a'''c')\,y + (c'a'' - a'c'')\,z}{l},$$

$$v = \frac{(a''b''' - b''a''')\,x + (a'''b' - b'''a')\,y + (a'b'' - b'a'')\,z}{l},$$

Now these values must be identical with those marked (5), in the preceding investigation, provided the conditions there announced have place here.

Hence, comparing the coefficients of the like terms, we have

$$b''c''' - c''b''' = la',\ b'''c' - c'''b' = la'',\ b'c'' - c'b'' = la''',$$
$$c''a''' - a''c''' = lb',\ c'''a' - a'''c' = lb'',\ c'a'' - a'c'' = lb''',$$
$$a''b''' - b''a''' = lc',\ a'''b' - b'''a' = lc'',\ a'b'' - b'a'' = lc'''.$$

Adding together the squares of the three equations in each horizontal row, in the last for example, we have, in virtue of the given conditions, $(a''b''' - b''a''')^2 + (a'''b' - b'''a')^2 + (a'b'' - b'a'')^2 = l^2$. It is easy to see that this equation may be put under the form

$$(a'^2 + a''^2 + a'''^2)\ (b'^2 + b''^2 + b'''^2) - (a'b' + a''b'' + a'''b''')^2 = l^2.$$

But, by the conditions (2), the first member of this equation is 1, therefore $l = 1$. Hence the truth of the first and fourth of the equations announced; and, by proceeding in like manner with the other two horizontal rows of equations above, we establish the truth of the two remaining equations announced.

PROPOSITION IX.

(259.) In a central surface of the second order the sum of the squares of any system of conjugate diameters is equivalent to the sum of the squares of the principal diameters.

Let A, B, C, represent the principal semi-diameters, and (x', y', z'), (x'', y'', z''), (x''', y''', z'''), the extremities of any system of semi-con-jugates A′, B′, C′. Then the equation of a tangent plane through the extremity of A′, is (p. 252) $\frac{x'}{A^2}x + \frac{y'}{B^2}y + \frac{z'}{C^2}z = 1$, and this plane is parallel to both B′ and C′ (p. 220-1). Now the equations of B′ are $x = \frac{x''}{z''}z$, $y = \frac{y''}{z''}z$, and that this line may be parallel to the plane, we must have the condition (187) $\frac{x'x''}{A^2} + \frac{y'y''}{B^2} + \frac{z'z''}{C^2} = 0$.

Hence we have

From the equation of the surface.	*From the equation of the tangent planes.*
$\frac{x'^2}{A^2} + \frac{y'^2}{B^2} + \frac{z'^2}{C^2} = 1$	$\frac{x'x''}{A^2} + \frac{y'y''}{B^2} + \frac{z'z''}{C^2} = 0$
$\frac{x''^2}{A^2} + \frac{y''^2}{B^2} + \frac{z''^2}{C^2} = 1$	$\frac{x'''x'}{A^2} + \frac{y'''y'}{B^2} + \frac{z'''z'}{C^2} = 0$
$\frac{x'''^2}{A^2} + \frac{y'''^2}{B^2} + \frac{z'''^2}{C^2} = 1$	$\frac{x''x'''}{A^2} + \frac{y''y'''}{B^2} + \frac{z''z'''}{C^2} = 0$

Consequently, (prop. 7), $x'^2 + x''^2 + x'''^2 = A^2$
$y'^2 + y''^2 + y'''^2 = B^2$
$z'^2 + z''^2 + z'''^2 = C^2$

Adding these equations, $A'^2 + B'^2 + C'^2 = A^2 + B^2 + C^2.$

PROPOSITION X.

(260.) In a central surface of the second order the sum of the squares of the faces of the parallelopiped whose edges are any system of semi-conjugate diameters, is equal to the sum of the squares of the faces of the rectangular parallelopiped whose edges are the semi-principal diameters; also the volume of the former is equal to the volume of the latter.

Since the conditions furnished by the equations of the surface and by the equations of the tangent planes at the extremities of the conjugate diameters, as exhibited in last proposition, agree with the conditions in prop. 7, a', b', c', being here replaced by $\frac{x'}{A}, \frac{y'}{B}, \frac{z'}{C}$, &c. we may derive from these conditions the equations announced in prop. 8, which, in the present case, are

$$(x'y''-y'x'')^2+(x''y'''-y''x''')^2+(x'''y'-y'''x')^2=A^2B^2$$
$$(y'z''-z'y'')^2+(y''z'''-z''y''')^2+(y'''z'-z'''y')^2=B^2C^2$$
$$(z'x''-x'z'')^2+(z''x'''-x''z''')^2+(z'''x'-x'''z')^2=A^2C^2$$
$$(x'y''-y'x'')z'''+(y'z''-z'y'')x'''+(z'x''-x'z'')y'''=ABC.$$

Now the first vertical row of terms in the three first equations exhibits the sum of the squares of the projections of the parallelogram whose contiguous sides are A′, B′; the second vertical row is the sum of the squares of the projections of the parallelogram whose sides are B′, C′, and the third row is the sum of the squares when the sides are A′, C′. Hence, by adding the three equations together, it follows (224) that the sum of the squares of the sides of the parallelopiped whose edges are A′, B′, C′, is equal to the sum of the squares of the sides of the parallelopiped whose edges are A, B, C. Again, the first member of the fourth equation expresses the volume of the parallelopiped whose edges are A′, B′, C′, (prop. 5, cor. 1); hence this parallelopiped is equal to that whose edges are A, B, C.

Cor. From this theorem we may immediately infer that there can be but one system of rectangular conjugates except the surface be of revolution. For, if A′, B′, C′, could be mutually at right angles, as well as A, B, C, then from these theorems, and the composition of equations, the equations $(x+A^2)(x+B^2)(x+C^2)=0$ and $(x+A'^2)(x+B'^2)(x+C'^2)=0$ are identical; hence their roots are the same; therefore A, B, C, are respectively equal to A′, B′, C′, which (248) is impossible, when the surface is not of revolution.

PROPOSITION XI.

(261.) In a central surface of the second order the squares of the reciprocals of any system of rectangular diameters are together equal to the squares of the reciprocals of the principal diameters.

Let A, B, C, be the principal semi-diameters of the surface, and $A_{\prime}$, $B_{\prime}$, $C_{\prime}$, any other semi-diameters mutually at right angles. Then the equation of the surface is $\frac{x^2}{A^2}+\frac{y^2}{B^2}+\frac{z^2}{C^2}=1$. To transform this equation, so as to refer the surface to another system of axes, we must substitute for x, y, z, the values (A), p. (259), which gives

$$\left(\frac{\cos.^2 X}{A^2}+\frac{\cos.^2 Y}{B^2}+\frac{\cos.^2 Z}{C^2}\right)x^2+\left(\frac{\cos.^2 X'}{A^2}+\frac{\cos.^2 Y'}{B^2}+\frac{\cos.^2 Z'}{C^2}\right)$$

$$y^2+\left(\frac{\cos.^2 X''}{A^2}+\frac{\cos.^2 Y''}{B^2}+\frac{\cos.^2 Z''}{C^2}\right)z^2+Pxy+Qxz+Ryz=1,$$

where P, Q, R, are put to represent certain functions of the inclinations of the new axes to the old.

Now, to determine the lengths of the semi-diameters which coincide with these new axes, put successively, in this equation, $z=0$, $y=0$; $z=0$, $x=0$; $x=0$, $y=0$; and there results

$$\frac{1}{x^2}=\frac{1}{A_{,}^2}=\frac{\cos.^2 X}{A^2}+\frac{\cos.^2 Y}{B^2}+\frac{\cos.^2 Z}{C^2}$$

$$\frac{1}{y^2}=\frac{1}{B_{,}^2}=\frac{\cos.^2 X'}{A^2}+\frac{\cos.^2 Y'}{B^2}+\frac{\cos.^2 Z'}{C^2}$$

$$\frac{1}{z^2}=\frac{1}{C_{,}^2}=\frac{\cos.^2 X''}{A^2}+\frac{\cos.^2 Y''}{B^2}+\frac{\cos.^2 Z''}{C^2}$$

$$\text{Adding, } \frac{1}{A_{,}^2}+\frac{1}{B_{,}^2}+\frac{1}{C_{,}^2}=\frac{1}{A^2}+\frac{1}{B^2}+\frac{1}{C^2}$$

because, by hypothesis, the two systems of axes are rectangular, and (181) the sum of the squares of the cosines of the inclinations of any straight lines to three rectangular axes is always unity.

PROPOSITION XII.

(262.) Three rectangular planes constantly touch a central surface of the second order; required the locus of their point of intersection.

Let the equations of the surface and of a tangent plane at the point (x', y', z'), be $\frac{x^2}{A^2}+\frac{y^2}{B^2}+\frac{z^2}{C^2}=1$, $\frac{x'x}{A^2}+\frac{y'y}{B^2}+\frac{z'z}{C^2}=1$; then, putting successively $y=0, z=0$; $x=0, z=0$; $x=0, y=0$; we have for the parts of the axes included between this plane and the origin, the values $\frac{A^2}{x'}$, $\frac{B^2}{y'}$, $\frac{C^2}{z'}$. Hence, if ρ be the perpendicular from the origin upon the plane, and α, β, γ, denote the angles it makes with these three lines, we have $\rho=\frac{\cos.\alpha\, A^2}{x'}=\frac{\cos.\beta\, B^2}{y'}=\frac{\cos.\gamma\, C^2}{z'}$,

$$\therefore \cos.\alpha=\frac{\rho x'}{A^2},\ \cos.\beta=\frac{\rho y'}{B^2},\ \cos.\gamma=\frac{\rho z'}{C^2}$$

$$\therefore A^2\cos.^2\alpha+B^2\cos.^2\beta+C^2\cos.^2\gamma=\rho^2\left(\frac{x'^2}{A^2}+\frac{y'^2}{B^2}+\frac{z'^2}{C^2}\right)=\rho^2.$$

In a similar manner, if ρ', ρ'', be perpendiculars from the origin upon two tangent planes at the points (x'', y'', z''), and (x''', y''', z'''), and if the first make angles α', β', γ', with the axes and second α'', β'', γ'', we have $A^2 \cos.^2 \alpha' + B^2 \cos.^2 \beta' + C^2 \cos.^2 \gamma' = \rho'^2$; and $A^2 \cos.^2 \alpha'' + B^2 \cos.^2 \beta'' + C^2 \cos.^2 \gamma'' = \rho''^2$. Now if the three tangent planes may be mutually rectangular, ρ, ρ', ρ'', will be mutually rectangular; so that the sum of the squares of the cosines which they make with the axis of x, with the axis of y, and with the axis of z, is in either case equal to unity (181). Hence, by addition, $A^2 + B^2 + C^2 = \rho^2 + \rho'^2 + \rho''^2$. If R represent the distance of the intersection $(x, y, z,)$ of the tangent planes from the origin, then $R^2 = \rho^2 + \rho'^2 + \rho''^2$; but $R^2 = x^2 + y^2 + z^2$; hence $x^2 + y^2 + z^2 = A^2 + B^2 + C^2$ is the equation of the locus of $(x, y, z,)$ which is a sphere concentric with the proposed surface, and of which the radius is

$$R = \sqrt{\{A^2 + B^2 + C^2\}}.$$

PROPOSITION XIII.

(263.) Chords are drawn to a surface of the second order so as all to pass through a fixed point; what is the locus of their middle points?

Assume the fixed point for the origin, let the axis of x pass through the centre of the surface, and let the other two be parallel to the two diameters conjugate to this, then the equation of the surface will be $Az^2 + By^2 + Cx^2 + Fx = G$, and the equations of any chord through the origin are $x = mz$, $y = nz$ substituting these values in the equation of the surface, we have at the points common to both $(A + Bn^2 + Cm^2) z^2 + Fmz = G$, and half the sum of the two values of z given by this equation must be the z of the middle of the chord, that is, by the theory of equations, this z is $z = -\dfrac{\frac{1}{2}Fm}{A + Bn^2 + Cm^2}$; or, substituting for m and n the values $x \div z$, $y \div z$, as given by the equations of the chord, we have, after reduction, $Az^2 + By^2 + Cx^2 + \frac{1}{2}Fx = 0$ for the equation of the locus, which is, therefore, a surface of the second order, similar to the proposed. If, in this equation, we make any two of the variables 0, we have 0 for one value of the third; thus showing that the surface, if completed, would pass through the origin. The coordinates of the centre of the proposed surface are (242) $x' = -\dfrac{F}{2C}$, $y' = 0$, $z' = 0$, and these same coordinates satisfy the equation of the locus; hence the locus passes through the centre of the proposed surface, if it have a centre. If we subtract the equation of the locus from that of the proposed surface, there results $\frac{1}{2}Fx = G$ $\therefore x = \dfrac{2G}{F}$; this therefore is the value of the abscissa x belonging to every point common to both surfaces; consequently the two surfaces intersect in a plane parallel to the plane of yz, and at the distance

$\frac{2G}{F}$ from the fixed point. If this point be upon the proposed surface, then $G = 0$; hence, in that case, the two surfaces merely touch at that point.

PROPOSITION XIV.

(264.) Planes passing through a fixed point cut a surface of the second order; what is the locus of the centres of all the sections?

It is not difficult to perceive that the locus will be in this case the same as in last proposition, but we shall give an independent investigation. Assuming the same axes as before, we have, for the equations of the surface, and of any plane through the origin,
$Az^2 + By^2 + Cx^2 + Fx = G$ (1), $z = mx + ny$ (2). Substituting this value of z, in the first equation, we have for the x, y, of the intersection the equation $(Am^2 + C)x^2 + 2Amnxy + By^2 + Fx = G$ . . (3).

Let the x, y, of the centre of this section be x', y', then, if the origin be removed to this centre, x, y, must be changed into $x + x'$, $y + y'$, which changes (3) into $(Am^2 + C)(x + x') + 2Amn(x + x')(y + y') + B(y + y')^2 + F(x + x') = G$. Now the origin being by hypothesis at the centre, the coefficients of x and y must vanish from this equation. These coefficients, without developing all the terms, are readily seen to be $2Am(mx' + ny') + 2Cx' + F = 0$; $2An(mx' + ny') + 2By' = 0$, that is, from equation (2), $\left.\begin{array}{r}2Amz' + 2Cx' + F = 0\\ 2Anz' + 2By' = 0\end{array}\right\}$

$\therefore \left\{ m = -\frac{Cx' + \frac{1}{2}F}{Az'};\ n = -\frac{By'}{Az'}\right.$. These values of m and n, substituted in the value of z', (2), give $Az'^2 + By'^2 + Cx'^2 + \frac{1}{2}Fx' = 0$ for the equation of the locus, which is the same as that deduced in last prop., and to which, therefore, the same remarks apply.*

PROPOSITION XV.

(265.) Three straight lines mutually at right angles meet in a point and constantly touch a surface of the second order; what is the locus of the point?

Let the equation of the surface be $Ax^2 + By + Cz^2 = D$, and let x', y', z', be the coordinates of a point in the locus, then, if the origin be removed to this point, and the touching lines be taken for axes, the equation of the surface will (230) be transformed into

$$\left.\begin{array}{r}A\cos.^2 X\\ B\cos.^2 Y\\ C\cos.^2 Z\end{array}\right| x^2 + \left.\begin{array}{r}A\cos.^2 X'\\ B\cos.^2 Y'\\ C\cos.^2 Z'\end{array}\right| y^2 + \left.\begin{array}{r}A\cos.^2 X''\\ B\cos.^2 Y''\\ C\cos.^2 Z''\end{array}\right|$$

*A very simple *geometrical* solution of this problem is given in the Gentleman's Diary for 1830, by *Mr. T. S. Davies*, of Bath, the able and ingenious proposer of the problem. We take this opportunity of recommending to the student's attention the very instructive researches on *Geometry of three Dimensions*, which this gentleman has published in Leybourn's Repository, Nos. 20 and 21.

$$\begin{array}{l|l|l} z^2 + \text{A cos. X}x' & x^2 + \text{A cos. X}'x' & 2y \\ \quad\text{B cos. Y}y' & \quad\text{B cos. Y}'y' & \\ \quad\text{C cos. Z}z' & \quad\text{C cos. Z}'z' & \end{array}$$

$$\begin{array}{l|l} +\text{A cos. X}''x' & 2z + Pxy + Qxz + Ryz = D - Ax'^2 - By'^2 - Cz'^2 \\ \text{B cos. Y}''y' & \\ \text{C cos. Z}''z' & \end{array}$$

where P, Q, R, are put for brevity to represent certain functions of the cosines. Now, in order to determine the parts of the axes intercepted between the origin and the surface, we must put successively in this equation $y=0, z=0$; $x=0, z=0$; $x=0, y=0$; and there results

$$(\text{A cos.}^2 \text{X} + \text{B cos.}^2 \text{Y} + \text{C cos.}^2 \text{Z})\, x^2 + 2\,(\text{A cos. X}x' + \text{B cos. Y}y' + \text{C cos. Z}z')\, x + D' = 0,$$

$$(\text{A cos.}^2 \text{X}' + \text{B cos.}^2 \text{Y}' + \text{C cos.}^2 \text{Z}')\, y^2 + 2\,(\text{A cos. X}'x' + \text{B cos. Y}'y' + \text{C cos. Z}'z')\, y + D' = 0,$$

$$(\text{A cos.}^2 \text{X}'' + \text{B cos.}^2 \text{Y}'' + \text{C cos.}^2 \text{Z}'')\, z^2 + 2\,(\text{A cos. X}''x' + \text{B cos. Y}''y' + \text{C cos. Z}''z')\, z + D' = 0,$$

where D' is put for $Ax'^2 + By'^2 + Cz'^2 - D$. These equations furnish two values for each of the quantities x, y, z, corresponding to the *two* points in which each axis *cuts* the surface; but, if we introduce the conditions that each axis merely *touches* the surface, the two points coincide; and, therefore, in this case, the two roots of each equation become equal. Hence, by the theory of equations,

$$D'(\text{A cos.}^2 \text{X} + \text{B cos.}^2 \text{Y} + \text{C cos.}^2 \text{Z}) = (\text{A cos. X}\, x' + \text{B cos. Y}\, y' + \text{C cos. Z}\, z')^2$$

$$D'(\text{A cos.}^2 \text{X}' + \text{B cos.}^2 \text{Y}') + \text{C cos.}^2 \text{Z}') = (\text{A cos. X}'x' + \text{B cos. Y}'y' + \text{C cos. Z}'z')^2$$

$$D'(\text{A cos.}^2 \text{X}'' + \text{B cos.}^2 \text{Y}'' + \text{C cos.}^2 \text{Z}'') = (\text{A cos. X}''x' + \text{B cos. Y}''y' + \text{C cos. Z}''z')^2$$

Adding these equations together, we have, in virtue of the conditions (1) and (3), p. 259, 60, $D'(A + B + C) = A^2x'^2 + B^2y'^2 + C^2z'^2$, that is, $(Ax'^2 + By'^2 + Cz'^2 - D)\ (A+B+C) = A^2x'^2 + B^2y'^2 + C^2z'^2$ whence $A\,(B+C)\,x'^2 + B\,(A+C)\,y'^2 + C\,(A+B)\,z'^2 = D(A+B+C)$ the equation of the locus, which is a surface of the second order concentric with the proposed.

PROPOSITION XVI.

(267.) If N represent the normal at any point of the earth's surface, supposed to be an oblate spheroid, and if λ denote the latitude or angle under the normal, and equatorial diameter 2A, prove that

$$N = \frac{A\,(1 - e^2)}{\sqrt{\{1 - e^2 \sin.^2 \lambda\}}}.$$

PROPOSITION XVII.

If N be produced to meet the polar diameter, show that the whole length, R, is

$$R = \frac{A}{\sqrt{\{1 - e^2 \sin.^2 \lambda\}}}.$$

PROPOSITION XVIII

If P represent the perpendicular from the centre of an ellipsoid on a tangent plane, prove that $\frac{1}{P^2}=\frac{x^2}{A^4}+\frac{y^2}{B^4}+\frac{z^2}{C^4}$.

PROPOSITION XIX.

If a conical surface envelope a surface of the second order, prove that the curve of contact is of the second order.

PROPOSITION XX.

Parallel planes cut a surface of the second order; required the locus of the foci of the sections.

PROPOSITION XXI.

Given the position of two lights of known intensities (m, n,) to determine the surface of which every point shall be equally illuminated by both lights, the law of intensity varying inversely as the square of the distance.

Before closing the present volume, we shall very briefly advert to a class of curves denominated *curves of double curvature*, a name by which all lines are designated which cannot be traced upon a plane but only upon a curve surface. The simplest analytical representation of such a line is analogous to that already employed for the straight line in space, viz. by the combination of the two equations denoting the two surfaces which project the proposed line on two coordinate planes. In the case of the straight line, these projecting surfaces are planes; in lines of double curvature they are obviously cylindrical surfaces. The equations to a curve of double curvature are, therefore, those two combined which, taken separately, represent the two traces of the projecting cylinders. A line of double curvature generally presents itself, in mathematical inquiries, as the intersection of two curve surfaces, and for these surfaces we are, for simplicity, to substitute the two intersecting cylinders of which we have just spoken. This is effected thus: Let one of the variables, as z, be eliminated from the equations of the two intersecting surfaces, and there will result a function of x and y as $F:(x, y)=0$, and this represents the cylinder projecting the intersection on the plane of xy. In like manner, eliminate another variable as y, and we have $f:(x, z)=0$ for the representative of the cylinder, projecting the same curve on the plane of xz; so that the equations sought are $F:(x, y)=0, f:(x, z)=0$.

Suppose, for example, it were required to express the equations of the curve of double curvature formed by the intersection of a sphere, which is pierced by a right cylinder, supposing the radius of the base to be equal to that of the sphere, and that their surfaces are in contact.

Taking the centre of the sphere for the origin, the line coinciding with the cylindrical surface for the axis of z, and the perpendicular to this touching the same surface, for the axis of y, we have, since the given cylinder projects the curve of intersection on the plane of xy into a semicircle, this equation to the projection, viz. $y^2 = rx - x^2$ . . . (1), which also expresses the relation between the x, y, of every point in the cylinder, and consequently between the x, y, of every point on the sphere belonging to the intersection. The equation of this sphere is $z^2 + y^2 + x^2 = r^2$; hence the relation between the x, z of the intersection is (1), $z^2 = r^2 - rx$ (2), which therefore, is the equation of the projection on the plane of xz, consequently the proposed curve of double curvature is expressed analytically by the equations (1) and (2) combined, viz.

$$\left.\begin{array}{l} y^2 = rx - x^2 \\ z^2 = r^2 - rx \end{array}\right\}$$

If we had to determine the tangent line at any point in a curve of double curvature, we should first determine the two tangents through the projections of that point, then perpendicular planes through these tangents would obviously be in contact with the two projecting cylinders throughout the lengths of these cylinders, and would each pass through the proposed point in their intersection, and through no other; hence the intersection of these planes would be the linear tangent sought, and would, therefore, be analytically represented by the two equations, which, separately, represent the tangents to the projections. We cannot enter into further particulars respecting these curves here, since their full discussion requires the aid of the transcendental analysis; but the preceding remarks may serve to convey a notion of this class of lines and of the manner of representing them by equations.

The calculation by trigonometry is hence very easy, for AC : sin. AHO :: CH : sin. CAH=sin. $\frac{1}{8}\text{P}\times\frac{\text{CH}}{\text{AC}}$ ∴ the angle CAH is found and CHO=$\frac{1}{8}$P+CAH, is known, ∴ CAB=2 CAO is found; we have rad. : sin. CAB :: AC : BC ∴ BC is found; ∴ and hence AB is easily known also.

Note A. page 64.

Remark. If $cd=ab$ the value of x would be infinite, which cannot be if a, b, c, d, are finite ∴ if $cd=ab$ we must also have $d^2+c^2=a^2+b^2$ which gives $x=\frac{0}{0}$ which shows that x is indeterminate, as is evident from the equations $cd=ab$, $d^2+c^2=a^2+b^2$, which give either $d=a$, $c=b$, that is AD=AB, DC=CB and the triangle ADC, ABC are identical; so that DE is indeterminate unless something further is given; or the above equations are satisfied by making $d=b$, $c=a$, or AD=BC, DC=AB, in which case the quadrilateral is a parallelogram and the triangles ADC, ABC are indeterminate although identical, for there are not sufficient data to find DE=x. Again, if cd and ab are unequal but $d^2+c^2=a^2+b^2$, or $\text{AD}^2+\text{DC}=\text{AB}^2+\text{BC}^2$ then $x=0$ and the angles at D and B are right, whence the problem is easily solved, for the area $\text{A}=\frac{1}{2}(cd+ab)$; it may also be observed, that the above value of x indicates an easy geometrical construction of the quadrilateral when x is determinate. Finally, should x come out negative it shows that the angles at D and B are obtuse.

Note B. page 70.

Otherwise let the angle CAB=2φ, then EAB=φ, also ACB=$90°-2\varphi$, DCB=$45°-\varphi$, put AE=a, CD=b, then (supposing radius=unity,) we have a cos. φ=AB=AC cos. 2φ, also b cos. $(45°-\varphi)$=BC=AC sin. 2φ; hence we have by division $\frac{b\cos.(45°-\varphi)}{a\cos.\varphi}=\frac{\sin.2\varphi}{\cos.2\varphi}$, or we have $\frac{b}{a}(1+\tan.\varphi)=\sqrt{2}\times\tan.2\varphi=\frac{2\sqrt{2}\times\tan.\varphi}{1-\tan.^2\varphi}$; hence $\tan.^3\varphi+\tan.^2\varphi+\left\{\frac{(2\sqrt{2})a}{b}-1\right\}\tan.\varphi=1$; the same as before cited upon principles more simple.

Prob. 19, p. 207. Let straight lines be drawn from the centres of the given circles to any point in the locus, then it is evident that their difference equals the difference of the radii of the given circles; hence, by p. 141, &c., the locus is an hyperbola whose foci are at the centres of the given circles, and whose transverse axis $=$ the difference of their radii.

Prob. 21, p. 207. For solution, see Prob. 8, p. 112.

Prob. 24. p. 18. For solution, see Prob. 4, p. 197.

Problem 25. Page 18. Let $y'=ax+b$ (1), $y''=a'x+b'$ (2), be the equations of the given straight lines, and let $y=$ the geometric mean between y', y'', then $y^2=y'y''=(ax+b).(a'x+b')$, or by reduction $y^2=aa'x^2+(ab'+ba')x+bb'$, put $aa'=A$, $ab'+ba'=B$, $bb'=c$, and we have $y^2-Ax^2-Bx=c$ which is the equation of a line of the second order: if A is positive it is evident by chapter 2, that the curve is an hyperbola, but if A is negative, that it is an ellipse, and if $A=o$ it is a parabola.

Proposition 18. Page 284. Let $\frac{x^2}{A^2}+\frac{y^2}{B^2}+\frac{z^2}{C^2}=1$ (1), denote the equation of the surface of an ellipsoid, the origin of the coordinates being at the centre and the axes of the ellipsoid being the axes of x, y, z; let $p=$ the perpendicular from the centre to a plane which touches the ellipsoid at any point (x, y, z), it is required to show that $\frac{1}{P^2}=\frac{x^2}{A^4}+\frac{y^2}{B^4}+\frac{z^2}{C^4}$? By p. 252, $x'\frac{x}{A^2}+y'\frac{y}{B^2}+z'\frac{z}{C^2}-1=0$, (2), or the equation of the tangent plane at the point (x, y, z), and by p. 234, for the perpendicular from the point whose coordinates are x', y', z', to the plane $Ax+By+Cz+D=0$, the expression $P=\frac{Ax'+By'+Cz'+D}{\sqrt{A^2+B^2+C^2}}$; which is adapted to the present question by making x', y', z', each $=o$, (since the perpendicular is drawn from the centre of the ellipsoid whose coordinates are each $=0$;) $\therefore P=\frac{D}{\sqrt{A^2+B^2+C^2}}$, and by comparing the equation $Ax+By+$ &c. with (2) we have $A=\frac{x}{A^2}$, $B=\frac{y}{B^2}$, $C=\frac{z}{C^2}$, $D=-1$, hence $P=\frac{-1}{\sqrt{\frac{x^2}{A^4}+\frac{y^2}{B^4}+\frac{z^2}{C^4}}}$ $\therefore \frac{1}{P^2}=\frac{x^2}{A^4}+\frac{y^2}{B^4}+\frac{z^2}{C^4}$ as required.

Prop. 19. By p. 252, the general equation of the tangent plane to a central surface when referred to its axes as coordinates, their origin being at the centre, is $x\frac{x'}{A^2}+y\frac{y'}{B^2}+z\frac{z'}{C^2}=1$, (1), x', y', z' being the coordinates of any point in the plane which is not at the point of contact with the surface, and A, B, C are the semi-axes of the surface; its equation being $\frac{x^2}{A^2}+\frac{y^2}{B^2}+\frac{z^2}{C^2}=1$, (2); whence it may be observed that if each of the terms in the first member of (2) is positive, the surface is that of an ellipsoid, but if one or two of the terms in the first member is negative, the surface is that of an hyperboloid. Let then the point (x', y', z') be given, and put $\frac{x'}{A^2}=a$, $\frac{y'}{B^2}=b$, $\frac{z'}{C^2}=c$, and (1) becomes $ax+by+cz=1$, (3), which is the equation of a plane, in which evidently is situated the curve of contact of all the tangent planes to the given surface drawn from the point $(x'$ $y', z')$, and it is evident that the successive intersections of the tangent planes form a conical surface which envelopes the given surface. Now, since the curve of contact is the line of common section of the given surface, and the plane denoted bv (3), it is evident by pp. 264, 265, that the curve of contact is a line of the second order. Again: If the given surface is not central, its general form is $px^2 \pm p'y^2 = pp'z$, (4), (see p. 253.), and the equation of any tangent plane through the point x', y', z' is $zpxx' \pm zp'yy' = pp'(z+z')$; hence, it may be shown, as before, that the curve of contact is a line of the second order; (see art. 237, p. 263.).

Prop. 20. It is evident by pp. 264, 265, that the sections are similar figures, and when they have centres, by p. 266, that those centres are all on the same diameter of the surface. It is also easy to see that the principal axes of the sections lie in the same plane whose section with the given surface is a line of the second order (p. 263, art. 237.), having the line of the centres for one of its diameters, to which the semi-axes of the sections are ordinate. Hence, since the sections are similar, their semi-axes will have a constant ratio to their focal distances, $\therefore$ the locus of the foci will be a line of the second order, of the same kind as the section made by the plane of the axes, having the line of the centre for a diameter, and lying in the plane of the axes. It is hence evident that if the sections have no centres, the locus of their foci will be a curve equal to that cut out of the surface by the plane of the axes of the sections, provided the surface has no centre; but if the surface has a centre, the locus will be a line of the same kind as that cut out of the surface by the plane of the axes of the sections.

Prop. 21. Given the position of two lights of known intensities (m, n,) to determine the surface of which every point shall be equally illuminated by both lights, the law of intensity varying inversely as the square of the distance.

By the law of intensity varying inversely as the square of the distance, I suppose is meant that the quantity of light received varies as the intensity directly and the square of the distance inversely. On this supposition I shall consider the question. Let A denote the position of the light whose intensity is m, and B, that of the other, and let r, r' respectively denote the distances of any point of the surface from A and B, then by the question $\frac{m}{r^2} = \frac{n}{r'^2} \therefore \frac{r}{r'} = \sqrt{\frac{m}{n}}$, or $r : r' :: \sqrt{m} : \sqrt{n}$, or the ratio of r to r' is given. Let then x, y, z denote the rectangular coordinates of the point, their origin being at A, the axis of x being in the straight line AB, and the axis of y in any plane drawn at pleasure through AB, and the axis of perpendicular to the plane (x, y), put AB $= a$, then $r = \sqrt{x^2+y^2+z^2}$, $r' = \sqrt{(a-x)^2+y^2+z^2} \therefore x^2+y^2+z^2 : a^2-2ax+x^2+y^2+z^2 :: m : n$, or $n(x^2+y^2+z^2) = m(a^2-2ax+x^2+y^2+z^2)$, or supposing $n > m$ we have $(n-m) \times (x^2+y^2+z^2) + 2max = ma^2$, or $x^2+y^2+z^2+\frac{2ma}{n-m}x = \frac{ma^2}{n-m}$, or $x^2+\frac{2max}{n-m}+\frac{m^2a^2}{(n-m)^2}+y^2+z^2 = \frac{ma^2}{n-m}+\frac{m^2a^2}{(n-m)^2} = \frac{nma^2}{(n-m)^2}$ or $\left(x+\frac{ma}{n-m}\right)^2+y^2+z^2 = \frac{nma^2}{(n-m)^2}$, which is the equation of the surface sought, it being a spherical surface whose radius $= \frac{a\sqrt{(nm)}}{n-m}$, and the distance of its centre from the point A on the line AB $= -\frac{ma}{n-m} = \frac{ma}{m-n}$, which shows that the centre is on the negative side of the origin. If $m = n$ we have $a^2-2ax = 0$, or $x = \frac{a}{2} \therefore$ the locus is a plane at right angles to the middle of AB. *Note.*—This question might have been made much more difficult by considering the obliquity at which the light strikes the elements of the surface sought; but as this does not appear to have been the author's intention, we shall here leave the subject at present for those who may feel disposed to prosecute the inquiry. (See Williams' Key.)

Problems 18, 20, 22, and 23, in art. 159, are not solved, for want of room; and also in art. 267, Prop. 16 and 17, p. 283.—*Ed.*

www.ingramcontent.com/pod-product-compliance
Lightning Source LLC
LaVergne TN
LVHW010217110826
845151LV00004B/1109
* 9 7 8 1 4 2 5 5 2 7 3 3 4 *